Artificial Intelligence Advances for Medical Computer-Aided Diagnosis

Artificial Intelligence Advances for Medical Computer-Aided Diagnosis

Editor

Mugahed A. Al-antari

Basel • Beijing • Wuhan • Barcelona • Belgrade • Novi Sad • Cluj • Manchester

Mugahed A. Al-antari
Artificial Intelligence and
Data Science
Daeyang AI Center
Sejong University
Seoul
Korea, South

Editorial Office
MDPI AG
Grosspeteranlage 5
4052 Basel, Switzerland

This is a reprint of the Special Issue, published open access by the journal *Diagnostics* (ISSN 2075-4418), freely accessible at: www.mdpi.com/journal/diagnostics/special_issues/ NP8BWO8966.

For citation purposes, cite each article independently as indicated on the article page online and using the guide below:

Lastname, A.A.; Lastname, B.B. Article Title. *Journal Name* **Year**, *Volume Number*, Page Range.

ISBN 978-3-7258-1644-6 (Hbk)
ISBN 978-3-7258-1643-9 (PDF)
https://doi.org/10.3390/books978-3-7258-1643-9

Contents

About the Editor

Mugahed A. Al-antari

Dr. Mugahed A. Al-antari (IEEE EMBS) is an assistant professor in the Department of Artificial Intelligence and Data Science, Daeyang AI Center, College of AI Convergence, Sejong University, Seoul, Korea. In August 2019, Dr. Al-antari received his Ph.D. in Biomedical Engineering from the College of Electronics and Information at Kyung Hee University (KHU), Republic of Korea, where his thesis earned an excellent award for the topic of AI-based breast cancer detection, segmentation, and classification. Dr. Al-Antari has over seven years of teaching experience in the Departments of Biomedical Engineering, Artificial Intelligence, Software and Data Science. Dr. Al-antari currently leads domestic and international (Korea–Turkey) collaborative joint research projects funded by the government of the Republic of Korea. Dr. Al-antari has over five years of professional experience as a senior researcher and executive engineering director at YOZMA BMTech Group, Seoul, developing mini- and whole-body DXA machines using X-ray pencil and fan beams. During his PostDoc position with Prof. Sungyoung Lee, CEO of UCLab, the Minister Award was received twice for best research project (AI Doctor Platform) from the Ministry of Information and Communication Technology (ICT), Korea. Dr. Al-Antari's recent AI-based publications have garnered significant attention from international journal editorials and are recognized by distinguished editorial boards as foundational for AI-based modern medicine. Dr. Al-Antari's current research interests include explainable artificial intelligence (XAI), deep learning, machine learning, pattern recognition, medical signal and image processing, medical imaging of dual-energy X-ray absorptiometry (DXA), and AI-based NLP for healthcare applications.

Preface

In the current era of medical innovation, the confluence of artificial intelligence (AI) and machine learning (ML) with medical diagnostics is revolutionizing healthcare. These technologies are enabling unprecedented accuracy, speed, and predictive capabilities in disease detection, classification, and treatment planning. This Special Issue is a testament to the transformative potential of AI and ML in the diagnostic field.

This Special Issue presents a curated collection of cutting-edge research exploring the integration of AI and ML technologies across various diagnostic modalities. The contributions in this issue highlight innovative algorithms, models, and applications that enhance diagnostic capabilities in fields such as radiology, pathology, genomics, and personalized medicine. By showcasing both theoretical advancements and practical implementations, this Special Issue provides a comprehensive overview of the current trends and future directions in AI-driven diagnostics, fostering further research and collaboration in this dynamic area of healthcare.

This compilation of twelve research articles, gathered between March 2023 and December 2023, includes one editorial cover letter, nine regular research articles, one review article, and one categorized as "other". Each contribution offers unique insights into leveraging AI and ML to overcome current challenges in diagnostic medicine, enhance clinical decision-making, and ultimately improve patient outcomes.

The rapid advancements in AI and ML have demonstrated immense potential in various areas such as image analysis, pattern recognition, and the integration of vast datasets for individualized patient care. This Special Issue features contributions from leading researchers exploring diverse facets of AI and ML applications in diagnostics. The collected works encompass theoretical advancements, algorithm development, and practical implementations, demonstrating the broad impact and future potential of these technologies.

As we move forward, the intersection of AI and ML with medical diagnostics continues to evolve, heralding a new era of precision and efficiency in healthcare. This Special Issue aims to capture the latest advancements and applications of these technologies in the diagnostic process, providing a valuable resource for researchers, clinicians, and technologists. By advancing diagnostic accuracy and efficiency, AI and ML have the potential to significantly enhance patient care and health outcomes.

We hope that this collection of high-quality research will serve as a valuable resource for those in the diagnostics field, fostering further innovation and collaboration in the quest to advance medical diagnostics. For more detailed insights, we encourage readers to explore each article and join the ongoing dialogue on the transformative potential of AI and ML in healthcare.

Mugahed A. Al-antari
Editor

Editorial

Advancements in Artificial Intelligence for Medical Computer-Aided Diagnosis

Mugahed A. Al-antari 

AISSLab, Department of Artificial Intelligence and Data Science, Daeyang AI Center, College of Software & Convergence Technology, Sejong University, Seoul 05006, Republic of Korea; en.mualshz@sejong.ac.kr

Abstract: Rapid advancements in artificial intelligence (AI) and machine learning (ML) are currently transforming the field of diagnostics, enabling unprecedented accuracy and efficiency in disease detection, classification, and treatment planning. This Special Issue, entitled "Artificial Intelligence Advances for Medical Computer-Aided Diagnosis", presents a curated collection of cutting-edge research that explores the integration of AI and ML technologies into various diagnostic modalities. The contributions presented here highlight innovative algorithms, models, and applications that pave the way for improved diagnostic capabilities across a range of medical fields, including radiology, pathology, genomics, and personalized medicine. By showcasing both theoretical advancements and practical implementations, this Special Issue aims to provide a comprehensive overview of current trends and future directions in AI-driven diagnostics, fostering further research and collaboration in this dynamic and impactful area of healthcare. We have published a total of 12 research articles in this Special Issue, all collected between March 2023 and December 2023, comprising 1 Editorial cover letter, 9 regular research articles, 1 review article, and 1 article categorized as "other".

Citation: Al-antari, M.A. Advancements in Artificial Intelligence for Medical Computer-Aided Diagnosis. *Diagnostics* **2024**, *14*, 1265. https://doi.org/10.3390/diagnostics14121265

Received: 9 June 2024
Accepted: 14 June 2024
Published: 15 June 2024

1. Introduction

The intersection of artificial intelligence (AI) and machine learning (ML) with medical diagnostics is rapidly evolving, heralding a new era of precision and efficiency in healthcare. This Special Issue of *Diagnostics* aims to capture the latest advancements and applications of AI and ML in the diagnostic process, spanning a diverse array of medical fields. AI and ML technologies are reshaping the landscape of diagnostics by providing tools that enhance the accuracy, speed, and predictive power of medical evaluations. These technologies have shown immense potential in areas such as radiology, pathology, genomics, and personalized medicine, where they are utilized for tasks ranging from image analysis and pattern recognition to the integration of vast datasets for individualized patient care. This Special Issue features contributions from leading researchers who explore various facets of AI and ML applications in diagnostics. The collected works encompass theoretical advancements, algorithm development, and practical implementations, demonstrating the broad impact and future potential of these technologies. The papers included in this Special Issue were gathered over a period spanning from March 2023 to December 2023, resulting in a diverse collection of high-quality research. We present a total of twelve articles: one Editorial cover letter, nine regular research articles, one review article, and one article categorized as "other". Each contribution offers unique insights into how AI and ML can be leveraged to overcome current challenges in diagnostic medicine, enhance clinical decision-making, and ultimately improve patient outcomes. Through this Special Issue, we aim to provide a comprehensive overview of the state of the art in AI-driven diagnostics, highlighting both achievements and ongoing challenges in the field. We hope that this collection will serve as a valuable resource for researchers, clinicians, and technologists, fostering further innovation and collaboration in the quest to advance medical diagnostics. For more detailed insights, we encourage readers to explore each article and join the ongoing dialogue on the transformative potential of AI and ML in healthcare.

2. Overview of the Published Articles

This Special Issue on "Artificial Intelligence Advances for Medical Computer-Aided Diagnosis" presents a diverse collection of research that illustrates the significant impact of AI and ML on medical diagnostics. Below is a summary of each of the 12 articles published in this Special Issue. The Editorial cover letter, prepared by Mugahed A. Al-antar, who served as the Guest Editor (GE) of this Special Issue, provides an introduction to the Special Issue, outlining the scope of, the significance of, and the emerging trends in the integration of AI and ML in diagnostics. It sets the stage for the subsequent articles by emphasizing the transformative potential of these technologies in healthcare [1].

Following this, Gil-Rios et al. [2] introduce a new method for automatically classifying coronary stenosis using a feature selection technique. Their proposed method achieved a 99% discrimination rate using only four features, suggesting that this method could be useful in a clinical decision support system.

Ogunpola et al. [3] tackle the global health issue of cardiovascular diseases by focusing on improving early detection, particularly of myocardial infarction, using machine learning. Their study addresses the problem of imbalanced datasets, which can bias predictions, and employs seven classifiers: K-Nearest Neighbors, Support Vector Machine, Logistic Regression, Convolutional Neural Network, Gradient Boost, XGBoost, and Random Forest. The optimized XGBoost model achieved an outstanding performance with an accuracy of 98.50%, a precision of 99.14%, 98.29% recall, and an F1 score of 98.71%, significantly enhancing diagnostic accuracy for heart disease.

Lee et al. [4] evaluate the diagnostic accuracy of two AI techniques, namely KARA-CXR and ChatGPT, in chest X-ray reading. Using 2000 chest X-ray images, their study assessed accuracy, false findings, location inaccuracies, count inaccuracies, and hallucinations. They found that KARA-CXR significantly outperformed ChatGPT regarding diagnostic accuracy (70.50% vs. 40.50%, according to one observer), false findings, and non-hallucination rate (75% vs. 38%). Both systems showed moderate inter-observer agreement. Their study highlights the superior performance of KARA-CXR in medical imaging diagnostics compared to ChatGPT.

Marquez et al. [5] focused on using machine learning to differentiate between positive and negative influenza patients in Mexico, where influenza has been a persistent issue since 2009. They used a dataset of 15,480 records from 2010 to 2020, containing clinical and demographic data of patients tested with RT-qPCR. This study evaluated various classification methods and found that Random Forest and Bagging classifiers perform best in terms of accuracy, specificity, sensitivity, precision, F1 score, and the AUC. These methods show promise for aiding clinical diagnosis in settings where molecular tests are impractical.

Ali et al. [6] investigated detecting Parkinson's disease (PD) using voice attributes from both PD patients and healthy individuals. They employed filter feature selection to remove quasi-constant features and tested several classification models, including Decision Tree, Random Forest, and XGBoost models, on two datasets. Remarkable results were achieved on Dataset 1, with the Decision Tree and Random Forest methods achieving impressive accuracy. Ensemble learning methods (voting, stacking, and bagging) were applied to further enhance the results, and genetic selection was also tested for accuracy and precision. Their study found higher precision in predicting PD patients and better overall performance when using Dataset 1 compared to Dataset 2.

Zakareya et al. [7] propose a new deep learning model for breast cancer classification, incorporating features like granular computing and attention mechanisms to achieve enhanced accuracy and reduce the workload of doctors. Their model's effectiveness is demonstrated by its superior performance compared to other models on two case studies, achieving 93% and 95% accuracy on ultrasound and histopathology images, respectively.

Al-rimy et al. [8] examined the use of DenseNet169 for knee osteoarthritis detection using X-ray images. An adaptive early-stopping technique with gradual cross-entropy loss estimation is proposed to improve the model's performance. This approach prevents overfitting and optimizes the number of training epochs. The proposed model demonstrates superior accuracy, precision, recall, and loss compared to existing solutions, indicating the effectiveness of adaptive early stopping and GCE for accurately detecting knee OA.

Fatih Uysal [9] developed a hybrid artificial intelligence system to detect monkeypox in skin images. An open-source, multi-class image dataset was used, addressing data imbalance through augmentation and preprocessing. State-of-the-art deep learning models were employed for detection, and a unique hybrid model combining high-performing models with LSTM was created. The resulting system achieved 87% test accuracy and a 0.8222 Cohen's kappa score.

Al-Haidari et al. [10] propose a new deep learning framework for MR image reconstruction based on conditional Generative Adversarial Networks (CGANs) and U-Net. A hybrid spatial and k-space loss function is introduced to improve image quality by minimizing the L1 distance in both spatial and frequency domains. The proposed framework outperformed the traditional SENSE technique and individual U-Net/CGAN models in terms of the PSNR, while maintaining a comparable SSIM. Their CGAN-based framework also showed the best reconstruction performance, proving useful for practical cardiac image reconstruction by providing an enhanced image quality.

Bhakar et al. [11] reviewed recent computational intelligence-based approaches for identifying disease severity levels. Their study focuses on Parkinson's disease and diabetic retinopathy, but also briefly covers other diseases. This review examines the methodology, dataset used, and disease type of each approach, evaluating performance metrics such as accuracy and specificity. It also presents public repositories for further research in this field.

Alnashwan et al. [12] conducted a systematic review to examine the latest advancements in artificial intelligence (AI) and computational intelligence for the detection and treatment of stuttering. They analyzed 14 journal articles from 2019 onward to investigate how AI can accurately determine and classify manifestations of stuttering, as well as how computational intelligence can contribute to developing innovative assessment tools and intervention strategies. Their review highlights the potential of AI and computational intelligence to revolutionize the assessment and treatment of stuttering, enabling personalized and effective approaches for improving the lives of people who have a stutter.

3. Limitations

Although the articles published in this Special Issue present valuable contributions to the field of medical diagnostics leveraging AI and machine learning (ML) techniques, some limitations can be identified across these articles:

(1)   Many of these studies have specific focuses or use case scenarios, potentially limiting the generalizability of their findings to broader contexts or populations;

(2)   Some articles may have methodological limitations, such as biases in data collection, imbalanced datasets, or reliance on specific classifiers, which could affect the reliability and applicability of their results;

(3)   While many studies report impressive results in performance metrics such as accuracy, precision, recall, and F1 scores, the evaluation criteria and datasets used may not fully reflect real-world clinical scenarios, leading to potential overestimations of performance;

(4)   The interpretation of results may vary across the studies, being potentially influenced by the researchers' perspectives or biases, which could impact the conclusions drawn from the data.

4. Future Directions

Medical image diagnosis powered by artificial intelligence (AI) represents the cutting edge of improvements in healthcare. We should continually strive to enhance the health and wellbeing of our community by providing smart solutions that benefit both patients and physicians. AI technology has significantly advanced medical image diagnosis, improving accuracy, efficiency, and accessibility in healthcare scenarios. AI algorithms can analyze

X-rays, CT scans, MRI scans, and patient records, such as those of patients with coronary stenosis or Parkinson's disease, to detect abnormalities such as tumors, fractures, and infections. For example, AI can identify early-stage cancers that might be missed by the human eye. AI can interpret echocardiograms (ECGs) and cardiac MRI scans to detect heart conditions like arrhythmias, heart failure, and coronary artery disease. In light of this, this Special Issue highlights and disseminates the latest breakthroughs in AI-driven diagnostic technologies, with the aim of showcasing innovative research, fostering interdisciplinary collaboration, and exploring the transformative impact of AI on medical imaging and diagnosis. By bringing together contributions from leading experts, we hope to advance this field of research, improve diagnostic accuracy and efficiency, and ultimately enhance patient care and health outcomes.

The recent rise of large language models (LLMs) like Gemini and medGemini [13,14] or ChatGPT and BiomedGPT [15] presents a transformative opportunity for the healthcare industry, particularly in the realm of diagnosis. These powerful tools can significantly enhance existing diagnostic systems by providing more insightful explanations, ultimately leading to more accurate and efficient patient care. One key contribution of LLMs lies in their ability to generate detailed textual reports. By analyzing vast amounts of medical data, including patient history, test results, and relevant research, LLMs can produce comprehensive reports that not only outline a diagnosis but also explain the reasoning behind it. This detailed explanation assists healthcare professionals by offering a transparent and step-by-step breakdown of how the LLM arrived at its conclusions, fostering trust in the system and allowing doctors to leverage the LLM's insights while incorporating their own medical expertise. While LLMs cannot directly generate heat maps (i.e., saliency maps) for medical images, they can play a significant role in the process. When integrated with computer vision models, LLMs are able to analyze medical images (X-rays, CT scans, MRIs, etc.), and they can be trained to detect and extract relevant features, anomalies, and patterns from these images. Through this, LLMs have the ability to generate textual descriptions of these highlighted areas that explain why they are important for a certain diagnosis. This provides a form of visual explanation alongside the LLM's textual report. The combined power of detailed textual explanations and visual heat maps can empower healthcare professionals in several ways:

(1) It improves diagnostic accuracy by offering a comprehensive understanding of the LLM's reasoning;
(2) It facilitates communication with patients by providing clear and accessible explanations of the diagnostic process [16,17];
(3) It fosters collaboration between humans and machines, allowing medical professionals to leverage the strengths of LLMs while retaining their own irreplaceable role in healthcare decision making [15].

However, it is crucial to acknowledge that LLMs are still under development, and their integration into healthcare systems requires careful consideration. Ensuring data privacy and security is paramount, and ongoing human oversight remains essential. In conclusion, LLM-based tools like Gemini and GPTs (Generative Pre-Trained Transformers) hold immense potential to revolutionize healthcare diagnostics. As mentioned above, the detailed textual explanations and visual heat maps provided by these tools can empower healthcare professionals by allowing them gain a deeper understanding of the diagnostic process, leading to improved accuracy and efficiency in patient care. As LLM technology continues to evolve, the future of healthcare diagnosis promises to be one of increased transparency, collaboration, and, ultimately, improved patient outcomes. The development of robust medical XAI models is often hindered by the limited availability of labeled datasets, particularly for rare diseases or specific imaging modalities. Generative AI techniques [18] offer promising solutions to address this challenge, enabling the creation of synthetic medical data across various modalities—text, images, signals, etc. LLMs like GPT-3 excel at generating synthetic patient records that closely mimic the structure, terminology, and statistical distributions of real clinical notes. Generative Adversarial

Networks (GANs) have revolutionized medical image synthesis, producing realistic X-rays, CT scans, and MRI scans that can incorporate specific anatomical variations or pathologies. This not only increases the size of datasets but allows researchers to simulate diverse disease presentations. Additionally, GANs can augment datasets by introducing realistic variations to existing images or even by translating images between modalities (e.g., generating a potential CT scan from an MRI scan), reducing the reliance on acquiring data in multiple ways. Generative models are able to synthesize physiological signals like ECGs, EEGs, and others, replicating real-world patterns and anomalies. This can be leveraged to expand training data for diagnostic models and facilitate the development of monitoring devices. Furthermore, generative AI can create hybrid signals with controlled variations, simulating complex or unusual health conditions to diversify training data.

Finally, we can potentially leverage contrastive learning [19] to address the challenge of data-labeling exhaustion in the development of LLM-based tools similar to medGemini or BiomedGPT. Contrastive learning is a technique within self-supervised learning where a machine learning model learns by creating its own supervisory signals from unlabeled data. The primary goal of this is to train a model to recognize similarities and differences between data samples: the model learns an embedding space (a kind of representational map) where similar items are clustered together and dissimilar items are far apart. Meanwhile, deep active learning could prove to be capable of supporting and automating the process of labeling medical data.

5. Conclusions

In conclusion, this Special Issue underscores the transformative potential of artificial intelligence (AI) and machine learning (ML) in medical diagnostics, as evidenced by the collection of cutting-edge research it presents. While these advancements offer unprecedented accuracy and efficiency in disease detection and treatment planning across various medical fields, including radiology, pathology, genomics, and personalized medicine, several limitations persist. These limitations include the potential lack of generalizability, methodological constraints, and the need for the careful interpretation of results. Looking ahead, the integration of large language models (LLMs) like Gemini and medGemini holds promise for further revolutionizing healthcare diagnosis through empowering healthcare professionals with detailed textual explanations and visual heat maps to enhance the accuracy and efficiency of diagnosis. Despite these advancements, however, ongoing considerations for data privacy, security, and human oversight are essential. Ultimately, the future of healthcare diagnosis promises increased transparency, collaboration, and improved patient outcomes as AI technology continues to evolve and move forward to address current challenges.

Funding: This research received no external funding.

Acknowledgments: We would like to thank the National Research Foundation of Korea (NRF) for providing support by means of a wide range of research grants from the Korean government (MSIT) (No. RS-2022-00166402 and RS-2023-00256517) for improving AI-based medical diagnostics. These two projects were awarded funding to enhance the explainability of medical image diagnosis through visual and textual explanations using Explainable AI (XAI) techniques.

Conflicts of Interest: The authors declare no conflicts of interest.

Abbreviations

AI	artificial intelligence;
XAI	explainable artificial intelligence;
ML	machine learning;
LLM	large language model;
LLMM	large language and multimodal model;
GAN	Generative Adversarial Network;
MRI	Magnetic Resonance Imaging;
CT	Computed Tomography.

References

1. Al-Antari, M.A. Artificial intelligence for medical diagnostics—Existing and future aI technology! *Diagnostics* **2023**, *13*, 688. [CrossRef] [PubMed]
2. Gil-Rios, M.-A.; Cruz-Aceves, I.; Hernandez-Aguirre, A.; Moya-Albor, E.; Brieva, J.; Hernandez-Gonzalez, M.-A.; Solorio-Meza, S.-E. High-Dimensional Feature Selection for Automatic Classification of Coronary Stenosis Using an Evolutionary Algorithm. *Diagnostics* **2024**, *14*, 268. [CrossRef] [PubMed]
3. Ogunpola, A.; Saeed, F.; Basurra, S.; Albarrak, A.M.; Qasem, S.N. Machine Learning-Based Predictive Models for Detection of Cardiovascular Diseases. *Diagnostics* **2024**, *14*, 144. [CrossRef] [PubMed]
4. Lee, K.H.; Lee, R.W.; Kwon, Y.E. Validation of a Deep Learning Chest X-ray Interpretation Model: Integrating Large-Scale AI and Large Language Models for Comparative Analysis with ChatGPT. *Diagnostics* **2023**, *14*, 90. [CrossRef] [PubMed]
5. Marquez, E.; Barrón-Palma, E.V.; Rodríguez, K.; Savage, J.; Sanchez-Sandoval, A.L. Supervised Machine Learning Methods for Seasonal Influenza Diagnosis. *Diagnostics* **2023**, *13*, 3352. [CrossRef] [PubMed]
6. Ali, A.M.; Salim, F.; Saeed, F. Parkinson's Disease Detection Using Filter Feature Selection and a Genetic Algorithm with Ensemble Learning. *Diagnostics* **2023**, *13*, 2816. [CrossRef] [PubMed]
7. Zakareya, S.; Izadkhah, H.; Karimpour, J. A new deep-learning-based model for breast cancer diagnosis from medical images. *Diagnostics* **2023**, *13*, 1944. [CrossRef] [PubMed]
8. Al-Rimy, B.A.S.; Saeed, F.; Al-Sarem, M.; Albarrak, A.M.; Qasem, S.N. An adaptive early stopping technique for densenet169-based knee osteoarthritis detection model. *Diagnostics* **2023**, *13*, 1903. [CrossRef] [PubMed]
9. Uysal, F. Detection of monkeypox disease from human skin images with a hybrid deep learning model. *Diagnostics* **2023**, *13*, 1772. [CrossRef] [PubMed]
10. Al-Haidri, W.; Matveev, I.; Al-Antari, M.A.; Zubkov, M. A Deep Learning Framework for Cardiac MR Under-Sampled Image Reconstruction with a Hybrid Spatial and k-Space Loss Function. *Diagnostics* **2023**, *13*, 1120. [CrossRef] [PubMed]
11. Bhakar, S.; Sinwar, D.; Pradhan, N.; Dhaka, V.S.; Cherrez-Ojeda, I.; Parveen, A.; Hassan, M.U. Computational Intelligence-Based Disease Severity Identification: A Review of Multidisciplinary Domains. *Diagnostics* **2023**, *13*, 1212. [CrossRef] [PubMed]
12. Alnashwan, R.; Alhakbani, N.; Al-Nafjan, A.; Almudhi, A.; Al-Nuwaiser, W. Computational Intelligence-Based Stuttering Detection: A Systematic Review. *Diagnostics* **2023**, *13*, 3537. [CrossRef] [PubMed]
13. Team, G.; Anil, R.; Borgeaud, S.; Wu, Y.; Alayrac, J.-B.; Yu, J.; Soricut, R.; Schalkwyk, J.; Dai, A.M.; Hauth, A. Gemini: A family of highly capable multimodal models. *arXiv* **2023**, arXiv:2312.11805.
14. Yang, L.; Xu, S.; Sellergren, A.; Kohlberger, T.; Zhou, Y.; Ktena, I.; Kiraly, A.; Ahmed, F.; Hormozdiari, F.; Jaroensri, T.; et al. Advancing Multimodal Medical Capabilities of Gemini. *arXiv* **2024**, arXiv:2405.03162.
15. Zhang, K.; Yu, J.; Yan, Z.; Liu, Y.; Adhikarla, E.; Fu, S.; Chen, X.; Chen, C.; Zhou, Y.; Li, X. BiomedGPT: A Unified and Generalist Biomedical Generative Pre-trained Transformer for Vision, Language, and Multimodal Tasks. *arXiv* **2023**, arXiv:2305.17100.
16. Ukwuoma, C.C.; Cai, D.; Heyat, M.B.B.; Bamisile, O.; Adun, H.; Al-Huda, Z.; Al-Antari, M.A. Deep Learning Framework for Rapid and Accurate Respiratory COVID-19 Prediction Using Chest X-ray Images. *J. King Saud Univ.-Comput. Inf. Sci.* **2023**, *35*, 101596. [CrossRef] [PubMed]
17. Ukwuoma, C.C.; Qin, Z.; Belal Bin Heyat, M.; Akhtar, F.; Bamisile, O.; Muaad, A.Y.; Addo, D.; Al-antari, M.A. A hybrid explainable ensemble transformer encoder for pneumonia identification from chest X-ray images. *J. Adv. Res.* **2023**, *48*, 191–211. [CrossRef] [PubMed]
18. Reddy, S. Generative AI in healthcare: An implementation science informed translational path on application, integration and governance. *Implement. Sci.* **2024**, *19*, 27. [CrossRef]
19. Chuang, C.-Y.; Robinson, J.; Lin, Y.-C.; Torralba, A.; Jegelka, S. Debiased contrastive learning. *Adv. Neural Inf. Process. Syst.* **2020**, *33*, 8765–8775.

MDPI

Editorial

Artificial Intelligence for Medical Diagnostics—Existing and Future AI Technology!

Mugahed A. Al-Antari

Daeyang AI Center, Department of Artificial Intelligence, College of Software & Convergence Technology, Sejong University, Seoul 05006, Republic of Korea; en.mualshz@sejong.ac.kr

Citation: Al-Antari, M.A. Artificial Intelligence for Medical Diagnostics—Existing and Future AI Technology! *Diagnostics* **2023**, *13*, 688. https://doi.org/10.3390/diagnostics13040688

Received: 10 February 2023
Accepted: 10 February 2023
Published: 12 February 2023

We would like to express our gratitude to all authors who contributed to the Special Issue of *"Artificial Intelligence Advances for Medical Computer-Aided Diagnosis"* by providing their excellent and recent research findings for AI-based medical diagnosis. Furthermore, special thanks are extended to all reviewers who helped us to process an article in this Special Issue. Finally, we would like to express our deep and warm gratitude and respect to the editorial members working day and night on this Special Issue, providing the recent AI-based research studies to enrich the AI medical knowledge for the fourth industrial revolution.

Medical diagnostics is the process of evaluating medical conditions or diseases by analyzing symptoms, medical history, and test results. The goal of medical diagnostics is to determine the cause of a medical problem and make an accurate diagnosis to provide effective treatment. This can involve various diagnostic tests, such as imaging tests (e.g., X-rays, MRI, CT scans), blood tests, and biopsy procedures. The results of these tests help healthcare providers determine the best course of treatment for their patients. In addition to helping diagnose medical conditions, medical diagnostics can also be used to monitor the progress of a condition, assess the effectiveness of treatment, and detect potential health problems before they become serious. With the recent AI revolution, medical diagnostics could be improved to revolutionize the field of medical diagnostics by improving the prediction accuracy, speed, and efficiency of the diagnostic process. AI algorithms can analyze medical images (e.g., X-rays, MRIs, ultrasounds, CT scans, and DXAs) and assist healthcare providers in identifying and diagnosing diseases more accurately and quickly. AI can analyze large amounts of patient data, including medical 2D/3D imaging, bio-signals (e.g., ECG, EEG, EMG, and EHR), vital signs (e.g., body temperature, pulse rate, respiration rate, and blood pressure), demographic information, medical history, and laboratory test results. This could support decision making and provide accurate prediction results. This can help healthcare providers make more informed decisions about patient care. The diversity of the patient's data in terms of multimodal data is an optimal smart solution that could provide better diagnostic decisions based on multiple findings in images, signals, text representation, etc. By integrating multiple data sources, healthcare providers can gain a more comprehensive understanding of a patient's health and the underlying causes of their symptoms. The combination of multiple data sources can provide a more complete picture of a patient's health, reducing the chance of misdiagnosis and improving the accuracy of diagnosis. Multimodal data can help healthcare providers monitor the progression of a condition over time, allowing for more effective treatment and management of chronic diseases. Meanwhile, using multimodal medical data, Explainable XAI-based healthcare providers can detect potential health problems earlier, before they become serious and potentially life-threatening [1]. Moreover, AI-powered Clinical Decision Support Systems (CDSSs) could provide real-time assistance and support to make more informed decisions about patient care. XAI tools can automate routine tasks, freeing healthcare providers to focus on more complex patient care.

The future of AI-based medical diagnostics is likely to be characterized by continued growth and development as OpenAI [2]. More advanced AI technologies are being introduced into the research domain, such as quantum AI (QAI), to speed up the conventional training process and provide rapid diagnostics models [3]. Quantum computers have significantly more processing power than classical computers, and this could allow quantum AI algorithms to analyze vast amounts of medical data in real-time, leading to more accurate and efficient diagnoses. Quantum optimization algorithms can optimize decision-making processes in medical diagnostics, such as choosing the best course of treatment for a patient based on their medical history and other factors. Another concept is GAI or general AI, which is being used by different projects and companies, such as OpenAI's DeepQA, IBM's Watson, and Google's DeepMind. The goal of GAI for medical diagnostics is to improve the accuracy, speed, and efficiency of medical diagnoses, as well as provide healthcare providers with valuable insights and support in the diagnosis and treatment of patients. By using AI algorithms to analyze vast amounts of medical data and identify patterns and relationships, general AI for medical diagnostics can transform the field of medicine, leading to improved patient outcomes and a more efficient and effective healthcare system. However, the development and deployment of AI in medical diagnostics are still in the early stages, and there are several technical, regulatory, and ethical challenges that must be overcome for the technology to reach its full potential. The first challenge is due to medical data quality and availability, where AI algorithms require large amounts of high-quality labeled data to be effective, and this can be a challenge in the medical field, where data are often fragmented, incomplete, unlabeled, or unavailable. Meanwhile, AI algorithms can be biased if they are trained on data that is not representative of the population they are intended to serve, leading to incorrect or unfair diagnoses. Another issue is about the use of GAI in medical diagnostics of a private and sensitive dataset, which raises some ethical questions, including data privacy, algorithmic transparency, and accountability for decisions made by AI algorithms. Even though some solutions with federated learning have recently been presented to solve such issues, the tool still needs more investigation to approve its capability for the medical research area. In addition, AI-based medical diagnostic tools are often developed by different companies and organizations, and there is a need for interoperability standards and protocols to ensure that these tools can work together effectively. AI-based techniques can analyze a patient's medical history, genetics, and other factors to create personalized treatment plans, and this trend will likely continue to be developed in the future. However, AI-based medical diagnostics is an open research domain, and we highly recommend that researchers continue research to improve the final prediction accuracy and expedite the learning process. This will support the medical staff in hospitals and healthcare centers and even assist the industrial sector by providing novel smart solutions against epidemics or pandemics that suddenly appear and devastate communities worldwide.

Funding: This research received no external funding.

Institutional Review Board Statement: Not applicable.

Informed Consent Statement: Not applicable.

Data Availability Statement: Not applicable.

Acknowledgments: We would like to thank the National Research Foundation of Korea (NRF) for the support of a wide range of research grants by the Korean government (MSIT) (No. RS-2022-00166402) to improve AI-based medical diagnostics.

Conflicts of Interest: The authors declare no conflict of interest.

Abbreviations

GAI	General Artificial Intelligence.
XAI	Explainable Artificial Intelligence.
QAI	Quantum Artificial Intelligence.
ECG	Electrocardiogram.
EEG	Electroencephalogram.
EMG	Electromyography.
EHR	Electronic healthcare records.
MRI	Magnetic resonance imaging.
CT	Computed tomography.
DXA	Dual-energy X-ray absorptiometry.

References

1. Ukwuoma, C.C.; Qin, Z.; Heyat, M.B.B.; Akhtar, F.; Bamisile, O.; Muaad, A.Y.; Addo, D.; Al-Antari, M.A. A hybrid explainable ensemble transformer encoder for pneumonia identification from chest X-ray images. *J. Adv. Res.* 2022, *in press*. [CrossRef] [PubMed]
2. OpenAI. Available online: https://openai.com/dall-e-2/ (accessed on 10 January 2023).
3. Alonso Calafell, I.; Cox, J.D.; Radonjić, M.; Saavedra, J.R.M.; García de Abajo, F.J.; Rozema, L.A.; Walther, P. Quantum computing with graphene plasmons. *npj Quantum Inf.* **2019**, *5*, 37. [CrossRef]

 *diagnostics*

Article

A Deep Learning Framework for Cardiac MR Under-Sampled Image Reconstruction with a Hybrid Spatial and k-Space Loss Function

Walid Al-Haidri [1], **Igor Matveev** [1], **Mugahed A. Al-antari** [2,*] and **Mikhail Zubkov** [1]

1 School of Physics and Engineering, ITMO University, Saint Petersburg 191002, Russia; m.zubkov@metalab.ifmo.ru (M.Z.)
2 Department of Artificial Intelligence, College of Software & Convergence Technology, Daeyang AI Center, Sejong University, Seoul 05006, Republic of Korea
* Correspondence: en.mualshz@sejong.ac.kr

Abstract: Magnetic resonance imaging (MRI) is an efficient, non-invasive diagnostic imaging tool for a variety of disorders. In modern MRI systems, the scanning procedure is time-consuming, which leads to problems with patient comfort and causes motion artifacts. Accelerated or parallel MRI has the potential to minimize patient stress as well as reduce scanning time and medical costs. In this paper, a new deep learning MR image reconstruction framework is proposed to provide more accurate reconstructed MR images when under-sampled or aliased images are generated. The proposed reconstruction model is designed based on the conditional generative adversarial networks (CGANs) where the generator network is designed in a form of an encoder–decoder U-Net network. A hybrid spatial and k-space loss function is also proposed to improve the reconstructed image quality by minimizing the L1-distance considering both spatial and frequency domains simultaneously. The proposed reconstruction framework is directly compared when CGAN and U-Net are adopted and used individually based on the proposed hybrid loss function against the conventional L1-norm. Finally, the proposed reconstruction framework with the extended loss function is evaluated and compared against the traditional SENSE reconstruction technique using the evaluation metrics of structural similarity (SSIM) and peak signal to noise ratio (PSNR). To fine-tune and evaluate the proposed methodology, the public Multi-Coil k-Space OCMR dataset for cardiovascular MR imaging is used. The proposed framework achieves a better image reconstruction quality compared to SENSE in terms of PSNR by 6.84 and 9.57 when U-Net and CGAN are used, respectively. Similarly, it demonstrates SSIM of the reconstructed MR images comparable to the one provided by the SENSE algorithm when U-Net and CGAN are used. Comparing cases where the proposed hybrid loss function is used against the cases with the simple L1-norm, the reconstruction performance can be noticed to improve by 6.84 and 9.57 for U-Net and CGAN, respectively. To conclude this, the proposed framework using CGAN provides the best reconstruction performance compared with U-Net or the conventional SENSE reconstruction techniques. The proposed framework seems to be useful for the practical reconstruction of cardiac images since it can provide better image quality in terms of SSIM and PSNR.

Keywords: magnetic resonance imaging (MRI); medical image reconstruction; deep learning; conditional generative adversarial networks (CGANs); parallel imaging; hybrid spatial and k-space loss function

Citation: Al-Haidri, W.; Matveev, I.; Al-antari, M.A.; Zubkov, M. A Deep Learning Framework for Cardiac MR Under-Sampled Image Reconstruction with a Hybrid Spatial and k-Space Loss Function. *Diagnostics* **2023**, *13*, 1120. https://doi.org/10.3390/diagnostics13061120

Academic Editor: Ahsan Khandoker

Received: 11 February 2023
Revised: 4 March 2023
Accepted: 10 March 2023
Published: 15 March 2023

1. Introduction

Magnetic resonance imaging (MRI) is a safe and non-invasive diagnostic technique that does not use ionizing radiation to produce the images [1]. MRI is known to provide clear and detailed images of soft-tissue structures. These advantages make MRI a very effective diagnostic tool for a variety of disorders. However, the scanning procedure in modern MRI systems is very time-consuming. This leads to problems with patient comfort and causes motion artifacts [2,3]. The MRI system encodes the spatial information via the use of the magnetic field gradient in the scanned area, which results in assigning specific resonant frequency and local intensity to a specific anatomical area of the patient during the acquisition stage of MR imaging. The inverse Fourier transform is then applied to the encoded signals, which represent the phase-time space or the k-space, to produce an image [4]. Accelerated or parallel MRI has the potential to reduce medical costs and minimize patient stress. It is performed by reducing the number of gradient encoding steps during image acquisition and uses the information about the spatial locations and sensitivity profiles of the employed receiving coils to reconstruct the desired image. This allows for a shorter patient stay in the MRI scanner, which in turn decreases motion artefacts risk. However, such a reduction in the encoding step number results in the aliasing effect due to the violation of the Nyquist sampling requirements. In order to restore the desired image, parallel imaging techniques employ extra post-processing steps [5]. GRAPPA and SENSE are the dominant approaches to accelerated MRI, allowing for the effects of incomplete encoding in the phase-time domain and image domain to be mitigated via using the sensitivity maps of the receiving coils [6,7]. The SENSE technique first uses the inverse Fourier transform over the under-sampled k-space, followed by the restoration of the MR image, while GRAPPA estimates the fully sampled phase-time data and then uses Fourier transform to obtain an image [8]. Nevertheless, some information is lost in the under-sampling process, causing the signal to noise ratio (SNR) to drop and the reconstruction of artifacts.

The motivation of this work is to improve the quality of MR under-sampled image reconstruction by applying deep learning techniques. We hypothesize this would allow us to overcome the drawbacks of GRAPPA and SENSE, as the precision of MR image reconstruction has been shown to significantly increase when the latter employs recent deep learning developments. A crucial part of implementing a deep learning model to a particular task is finding a suitable loss function, sensitive to the difference between the generated and the target image. In this study, to enhance the quality of MR image reconstruction, we proposed a new hybrid spatial and k-space loss function which combines two loss functions in the spatial and frequency domains. The results of MR image reconstruction show that the developed hybrid loss could increase the precision of investigated models compared to the conventional L1-norm loss function. The developed models were trained and evaluated using the public Multi-Coil k-Space Dataset for Cardiovascular Magnetic Resonance Imaging, called OCMR [9]. Assessing the quality of cardiac MRI reconstruction was performed using the evaluation metrics of the structural similarity (SSIM) [10,11] and the peak signal to noise ratio (PSNR) [12]. Our deep learning approach allowed us to avoid the well-known parallel imaging effect of the SNR reduction, as the square root of the acceleration factor helped to mitigate this problem. The suggested algorithms and network architectures can therefore be used in applications where both fast acquisition and a high SNR is critical (with cardiac imaging being one of these areas) as a substitution for the classic parallel imaging algorithms, which do not conserve the SNR. Moreover, we investigated two state-of-the-art U-Net and CGAN networks and compared the reconstruction results using the proposed hybrid loss function against the conventional ones. The latter include L1-norm and GAN loss functions. The performance was evaluated on the OCMR [9] dataset using SSIM and PSNR metrics. The major objectives and contributions of this work are summarized as follows:

- A new AI-based accurate parallel imaging reconstruction framework was proposed for better CMR image reconstruction.
- A new hybrid spatial and k-space loss function was proposed, which improves the SNR by taking into account the difference between the target (ground-truth, GT) and the reconstructed images in both spatial and frequency domains.
- Comprehensive reconstruction experimental studies were conducted with the aim to select the best AI model for the proposed framework. The model search additionally

included the direct implementation of the conventional loss functions as well as their comparison with the proposed hybrid loss function.

2. Related Work

In this section, we summarize the recent deep-learning-based reconstruction methods for various MR imaging techniques. Based on the classical architecture of the deep network U-Net, Chang Min Hyun et al. [13] developed an algorithm for MR image reconstruction. Cardiac and brain images, obtained using the classical full k-space sampling technique, were used. The key point of their work was to combine the U-Net model with the following k-space data correction. The U-Net takes a folded image obtained from under-sampled zero-padded k-space data as input and recovers the zero-padded part of the k-space data. Then, the unpadded parts are replaced by the original k-space data to preserve the original measured data. Finally, inverse Fourier transform is performed to obtain the final reconstructed MR images. The developed algorithm showed decent results, with an average 0.90 SSIM. However, the complexity of the algorithm led to significant memory constraints when generating high-resolution images, which severely limited the output image resolution. Ghodrati et al. [14] investigated two CNN architectures: a simplified version of U-Net and the residual network (ResNet) for cardiac MR image reconstruction. The effect of four loss functions was investigated: pixel-wise L1 and L2, patch-wise structural dissimilarity (DSSIM), and feature-wise perceptual loss. According to the 57th quartile of SSIM score (0.88), U-Net–DSSIM (U-Net with DSSIM loss) performed significantly better than ResNet with different combinations of loss functions. However, U-Net has ten times the number of trainable parameters compared to ResNet, which results in increases in its computational complexity and computational time.

A GAN-based algorithm was developed for knee joint image restoration from the reduced k-space without a reference fully sampled image [15]. The authors changed the concept of the classical GAN structure, where the generator network output is compared to the fully sampled data, by letting the generator serve as a seed for imitating the imaging process via subjecting the generator output to coil sensitivity map multiplication, FFT, and a randomized under-sampling mask. As a result, this produced sparsely sampled k-space data, which could be compared to the experimentally acquired sparse data. The proposed unsupervised GAN had superior PSNR, normalized root mean-square error (NRMSE), and SSIM compared to the common compressed sensing reconstruction. The unsupervised GAN only had 0.78% worse PSNR, 4.17% worse NRMSE, and equal SSIM compared to the supervised GAN. Reinforcement learning (RL) has also found an application in the field of medical image analysis, particularly in the reconstruction of brain and knee MR images [16]. The approach differs from the classical deep learning (DL) techniques in that MRI reconstruction is formulated as a Markov Decision Process—with discrete actions and continuous action parameters. An agent in such a process is a separate neural network that is assigned to each pixel of the MR image and processes it according to the reward received at each step of the algorithm training. The reward is formed as the difference between the values of the processed pixels at step s and $(s - 1)$. The results on fastMRI data using a random 40% under-sampling mask were PSNR of 30.3 dB and SSIM of 88.0%.

The examples above, as well as this work, utilize the deep learning architecture network called U-Net which was used for biomedical image processing at the beginning of 2015 and has since shown the most powerful results. The U-Net structure is symmetrical and is divided into two main sections: the left half is called the encoder or contracting path and is made up of the basic convolutional processes, while the right section is known as the encoder or expansive path and is made up of transposed 2D convolutional layers [17]. Another candidate architecture considered in this work is the conditional generative adversarial network (CGAN), a modification of the conventional GAN. The GAN is one of the best neural network architectures for image processing and analysis, particularly for image synthesis and reconstruction [18]. Along with the complexity of some of the considered algorithms, a common drawback of the above methods is the low value of the signal to noise ratio. We assume that developing a custom hybrid loss function which calculates the difference between target and reconstructed images in both spatial and frequency domains will allow the model to overcome the drawback in recent works, i.e., the increase in the SNR, and lead to the high structural similarity of constructed images.

3. Materials and Methods

3.1. The Proposed RecCGAN Framework: End-to-End Execution Scenario

In this section, we describe the abstract view of the proposed AI-based framework for cardiac MR under-sampled image reconstruction using the hybrid spatial and frequency loss function. The proposed end-to-end workflow is presented in Figure 1 and is explained as follows:

- The fast MRI raw k-space data are collected and transformed into the spatial domain using the inverse fast Fourier transformer (IFFT).
- The MR images are resized into a fixed size of 256×256 pixels.
- After resizing, the FFT is applied to allow us to generate the under-sampled MR data in the frequency domain by removing each second column in the k-space domain (known as interleaved under-sampling).
- The IFFT is applied again to convert the under-sampled k-space data into the aliased MR images.
- All aliased images are normalized to fit all pixels within a fixed value range of $[0, 255]$ to improve the AI learning process, and hence, the reconstruction performance. More detail about the data preparation can be found in Algorithm 1 (Section 3.2.2).
- The prepared aliased MR images are randomly split into 70% training, 10% validation, and 20% testing sets.
- To increase the number of training MR images, the augmentation strategy is applied to avoid any overfitting or bias, assist in better hyper-parameters' optimization, and improve the reconstruction performance.
- For reconstruction purposes, two well-known deep learning architectures of U-Net and CGAN are adopted and used. The CGAN structure is adopted by using U-Net in an encoder–decoder fashion to build the generator network. However, we test and investigate the reconstruction performance of both U-Net and CGAN separately.
- The hybrid spatial and frequency loss function is proposed in order to improve the reconstructed image quality over conventional loss functions acting only in the spatial domain, such as L1-norm and GAN loss as a discriminator classification loss function.
- Finally, the proposed framework is evaluated using the individual U-Net and CGAN against the widely used conventional SENSE reconstruction algorithm. A direct, fair comparison is conducted using the same dataset and training environment settings.

3.2. Dataset

3.2.1. Dataset Description

To build, train, and validate the proposed AI framework, the public Multi-Coil k-Space OCMR dataset for cardiovascular magnetic resonance imaging [9] was used. The dataset is available online at https://ocmr.info/download (accessed on 20 December 2022). The OCMR dataset consists of 53 fully sampled scans and 212 under-sampled scans. The fully sampled scans comprise 81 slices, while the under-sampled scans comprise 842 slices. These slices were collected from three different planes: 2-chamber, 4-chamber, and short-axis. To build and train the proposed AI reconstruction framework, the fully sampled scan data were used. The different available cine-frames were used as separate images, resulting in a total of 1383 multi-channel (from 15 to 35) full k-space data entries used for network training and testing.

3.2.2. Dataset Preparation

Algorithm 1 shows the data preparation scenario for all 1383 multi-channel k-space data.

Algorithm 1 Dataset preparation for parallel imaging simulation.

Start:
Input: Fully sampled *k*-space cardiac MRI data
Step 1: load data
k-space data = {'kx' 'ky' 'kz' 'coil' 'phase' 'set' 'slice' 'rep' *'avg'*} ← {read
k-space data in *.h5 format}
ISMRMRD ← ISMRMRD; Python toolbox for MR image reconstruction [19]
Step 2: Average the *k*-space data accumulations (kData) if *'avg'* > 1
k-space data ← numpy.mean(kData, axis= −1)
Step 3: Apply IFFT to transform the *k*-space averaged data into the spatial domain
ImageSpaceData ← transform data from k-space into image space

Step •   Resize the MR image tensor for real and imaginary parts separately
4:   *Resized_image_tensor ← complex_mri_resize (ImageSpaceData, new_size)*

    •   Create a copy of Resized_image_tensor for under-sampling track

    copy_resized_image_tensor ← numpy.copy(Resized_image_tensor)

Step Generate aliased MR images
5:   1.   Transform the copy of resized image tensor back to the *k*-space

    resized_kspace_tensor←transform_image_to_kspace(copy_resized_image_tensor])

     2.   Generate Cartesian binary sampling mask *Binary_mask ←*
         numpy.zerose_like(resized_kspace_tensor) Binary_mask[:..:2] = 1

     3.   Under-sample the resized_kspace by removing every second column based on the
         designed binary sampling mask (Step 5: 2) *US_kspace ← resized_kspace_tensor * *
         Binary_mask

     4.   Apply IFFT to get the under-sampled MR images *US_MR_image_tensor ←*
         {transform_kspace_to_image(US_k-space)}

     5.   Merge the different channels via the sum-of-squares procedure *im_sos_full =*
         *numpy.sqrt(np.sum(np.abs(US_MR image_tensor) ** 2, 3))*

     6.   Remove singleton dimensions *Aliased_MR image_tensor ← numpy.squeeze(im_sos_full)*

Step Generate fully sampled ground-truth (GT) MR images
6:   •   Merge the different channels (Step 4) via the sum-of-squares procedure *im_sos_full =*
         *numpy.sqrt(np.sum(np.abs(resized_image_tensor) ** 2, 3))*

    •   Remove singleton dimensions *Fully sampled GT images ← numpy.squeeze(im_sos_full)*

END

Figure 2 shows the qualitative process of the MR data generation in terms of the reference or ground-truth (GT) images and aliased images. The IFFT was used twice to generate the GT and aliased MR images using the fully and under-sampled *k*-space data. To down-sample the fully sampled *k*-space data, the interleaved down-sampling strategy was used to generate the binary mask. Once the binary mask was generated with the same size as the *k*-space data, we multiplied it in the frequency domain with the original fully sampled data to generate the under-sampled *k*-space data.

3.2.3. MR Data Splitting and Augmentation

Once the MR images were prepared in the spatial domain, the whole OCMR dataset was randomly split into 70% for training (887 *k*-spaces), 10% for validation (99 *k*-spaces), and 30% for testing (415 *k*-spaces). The 10% validation set was randomly picked from the training set. Table 1 shows the OCMR data distribution used to reach the goal of this study. For augmentation, we used random rotation, vertical and horizontal flipping, and cropping.

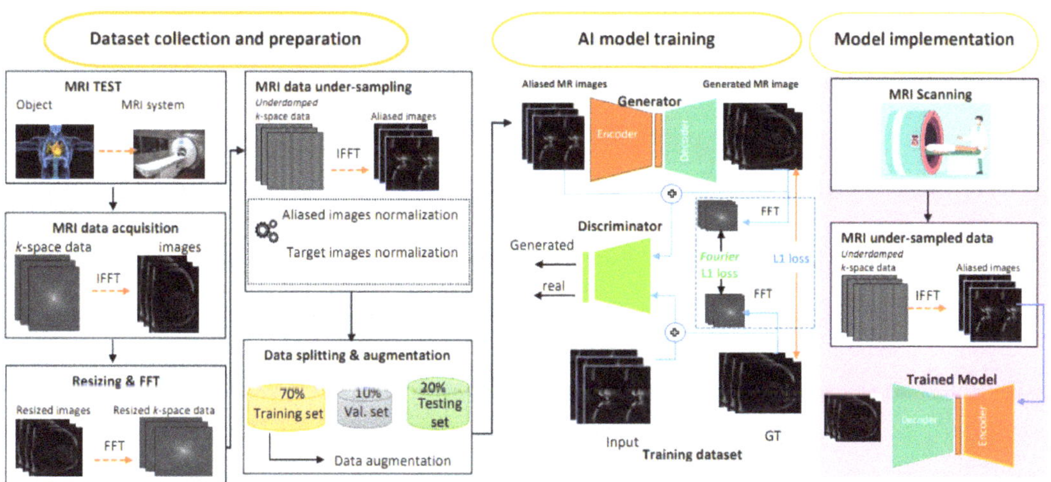

Figure 1. The proposed AI-based MRI reconstruction framework.

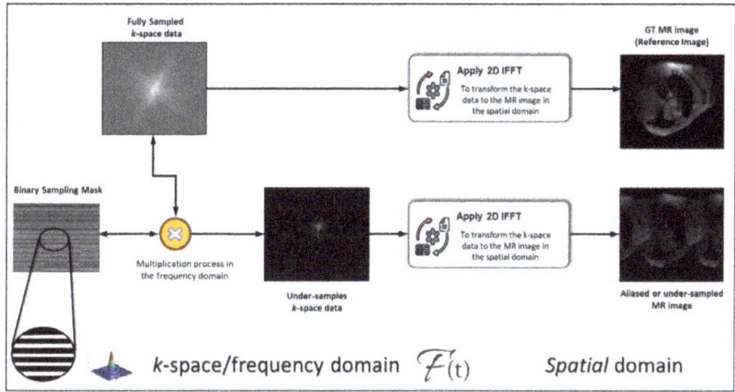

Figure 2. Generating the reference ground-truth (GT) and aliased MR images using the fully and under-sampled *k*-space data, respectively.

Table 1. MCOR data distribution.

	Training (70%)	Validation (10%)	Testing (30%)
Original Dataset	887	99	415
Augmented Dataset	3548		

3.3. Deep Learning Network Architecture and Training Details

3.3.1. U-Net with Hybrid Loss

We decided to investigate the U-Net model for MR image reconstruction due to its high efficiency in biomedical image processing. As mentioned above, the U-Net architecture comprises two parts [17]: the left part is the contracting path, in which a 3×3 convolution with zero-padding is applied for feature extraction from the input image. The rectified linear function is used as an activation function. Image down-sampling down to the bottleneck layer is conducted via a max-pooling layer with a 2×2 stride. The right section following the bottleneck layer is known as the expansive path and is made up of transposed 2D convolutional layers. To increase the reconstruction precision, the architecture suggests concatenating the up-sampled layer output on the expansive path with the corresponding feature tensor from the contracting path.

As the performance of the deep learning model depends not only on the network architecture, but also on the loss function, which plays an important role in minimizing the model error, a number of loss functions were assessed. The initial approach was to train the model with L1 loss function [13] and L2 regularization.

$$J(\theta) = \frac{1}{N} \sum_{s=1}^{N} \| \mathcal{H}_{\mathcal{L}}^{(s)} - \hat{\mathcal{H}}_{\mathcal{L}}^{(s)} \| + \lambda \cdot \| \Theta^2 \|, \tag{1}$$

where $\hat{\mathcal{H}}_{\mathcal{L}}^{(s)}$, $\mathcal{H}_{\mathcal{L}}^{(s)}$ are the model output and reference image, respectively, Θ is the tensor of trainable parameters, and λ is the regularization parameter. N denotes the batch size. To improve the model efficiency, we later extended the loss function by additionally taking into account the difference between target and reconstructed images in the frequency domain (i.e., in the k-space). The new loss term, which we call Fourier loss, is then provided by

$$\mathcal{L}_1^{\mathcal{F}} = \frac{1}{N} \sum_{s=1}^{N} \| \mathcal{F}\{\mathcal{H}_{\mathcal{L}}^{(s)}\} - \mathcal{F}\{\hat{\mathcal{H}}_{\mathcal{L}}^{(s)}\} \|, \tag{2}$$

where $\mathcal{F}\{\mathcal{H}_{\mathcal{L}}^{(s)}\}$, $\mathcal{F}\{\hat{\mathcal{H}}_{\mathcal{L}}^{(s)}\}$ are the Fourier transform of the reconstructed and the reference images, respectively. The suggested loss function was tested as a part of an extended loss function obtained by adding the Fourier loss (2) to Equation (1)

$$J(\theta) = \frac{1}{N} \sum_{s=1}^{N} \| \mathcal{H}_{\mathcal{L}}^{(s)} - \hat{\mathcal{H}}_{\mathcal{L}}^{(s)} \| + \alpha \cdot \frac{1}{N} \sum_{s=1}^{N} \| \mathcal{F}\{\mathcal{H}_{\mathcal{L}}^{(s)}\} - \mathcal{F}\{\hat{\mathcal{H}}_{\mathcal{L}}^{(s)}\} \| + \lambda \cdot \| \Theta^2 \|, \tag{3}$$

where $\alpha = 0.1$.

3.3.2. CGAN with Hybrid Loss

The GAN is another neural network architecture well-suited for image processing and analysis, particularly for image synthesis and reconstruction. GANs consist of two competing networks: the first, the generator, is the network that transforms the random noise to generate 'fake' but realistic-looking images. The second network, called the discriminator, is a different network, trained to classify whether the images generated by the generator are real or 'fake'. Here, we implemented the GAN with the U-Net architecture as a generator and a convolutional network as a discriminator. The discriminator network comprised six consecutive convolutional layers, four out six followed by batch normalization and all followed by an activation function (ReLU in the first five and sigmoid in the last layer). Another adopted modification to the classic GAN architecture fed the generator not with random noise, but with an MR image obtained from the reduced k-space [18]. This modification allowed such a model to be called image-conditional GAN (CGAN), as shown in Figure 3.

Training the generator comprises finding the minimum of the objective function. The objective specific to the CGAN is minimizing the loss function through the generator network and maximizing it through the discriminator.

$$\min_{G} \max_{D} \mathcal{L}_{CGAN}(D, G) = \mathbb{E}_{x,y}[\log D(x, y)] + \mathbb{E}_{x, G(x)}[\log(1 - D(x, G(x)))] \tag{4}$$

where $\mathbb{E}[arg]$ is the mean of arg, and $D(x, y)$ and $G(x)$ are the discriminator and the generator functions, respectively. It is also recommended to add L1 or L2 distance between the reference and the generated image to the CGAN objective [18] to increase the method accuracy. We used the L1 distance here because it encourages less blurring [18]. Adding L1 distance changes the objective to

$$Loss_{CGAN} = \min_{G}\max_{D}\mathcal{L}_{CGAN}(D,G) + \sigma\cdot\mathbb{E}_{x,y}[\|y - G(x)\|_1], \tag{5}$$

where the coefficient σ is chosen empirically. As in the case of the standalone U-Net architecture, we have found it beneficial to use Fourier loss in the objective function of the CGAN:

$$\mathcal{L}_1^{\mathcal{F}} = \mathbb{E}_{\mathcal{F}(x),\,\mathcal{F}(G(x))}[\|\mathcal{F}(y) - \mathcal{F}(G(x))\|_1]. \tag{6}$$

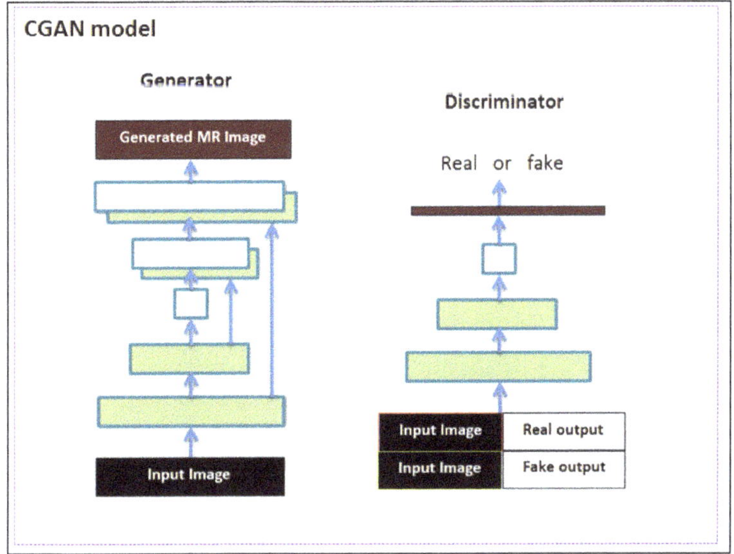

Figure 3. The CGAN model architecture. The generator employs the U-Net architecture, whereas the discriminator is a convolutional network.

Thus, identically to the bare U-Net case, two options for the loss function improvement were explored: the L1 norm and the Fourier loss as a part of the combined loss. The latter is given by

$$\begin{aligned}Loss^*_{CGAN} &= \mathbb{E}_{x,y}[\log D(x,y)] + \mathbb{E}_{x,G(x)}[\log(1 - D(x,G(x)))] + \sigma\cdot\mathbb{E}_{x,y}[\|y - G(x)\|_1]\\ &\quad + \alpha\cdot\mathbb{E}_{\mathcal{F}(x),\,\mathcal{F}(G(x))}[\|\mathcal{F}(y) - \mathcal{F}(G(x))\|_1].\end{aligned} \tag{7}$$

The discriminator model is trained separately on fake data (pairs of images acquired from the reduced k-space and the corresponding CGAN-generated images) and real data (pairs of images acquired from the reduced k-space and the reference image). The two input images are concatenated together to create one $256 \times 256 \times 2$ input to the first hidden convolutional layer. The discriminator training strategy is illustrated in Figure 4.

In the training process, the discriminator model can be updated directly, whereas the generator model must be updated via the discriminator model. This can be achieved by creating a new composite model that connects the generator model's output to the discriminator model's input. The discriminator model can then predict whether a generated image is real or fake. To prevent a misleading update of the discriminator when employing a discriminator to update the generator, the discriminator weights are specified as not trainable [18,20]. Figure 5 illustrates the training strategy for the complete CGAN model with Fourier loss.

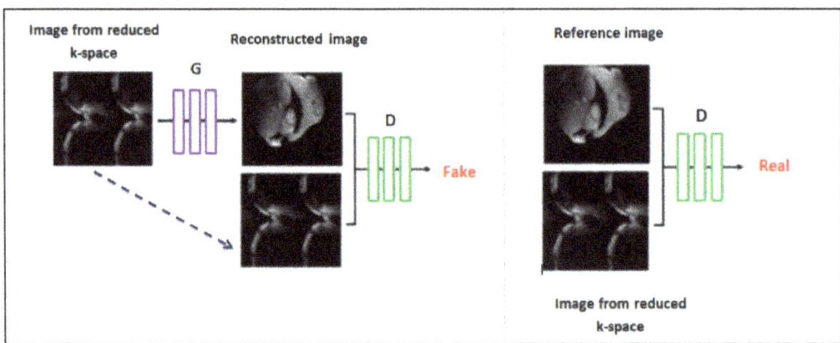

Figure 4. The CGAN model architecture. The generator employs the U-Net architecture, whereas the discriminator is a convolutional network.

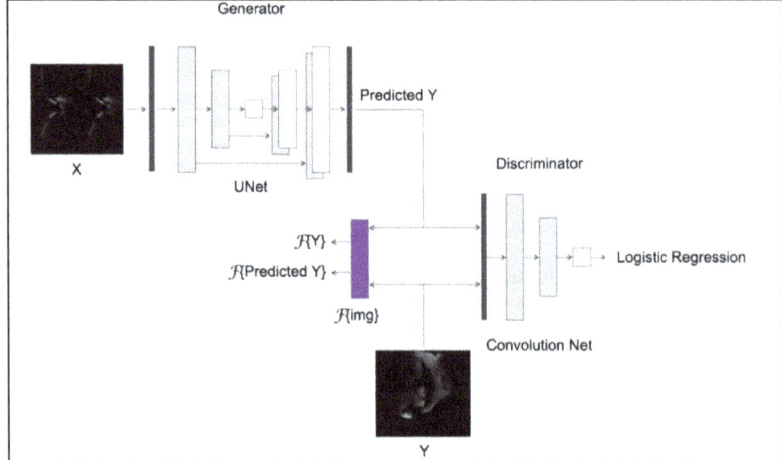

Figure 5. The training strategy for the CAGN model with Fourier loss.

To evaluate the efficiency of the deep learning approach, we compared the deep-learning-based results with the SENSE reconstruction, which is one of the most widely used parallel imaging methods, offered by the majority of MR scanner vendors. As the coil sensitivity profiles are not the part of the OCMR dataset, they were estimated for the SENSE reconstruction from fully sampled MR data using the algorithm presented in [21].

3.4. Evaluation Strategy

Two metrics were used to assess the quality of the cardiac MRI reconstruction in both the bare U-Net and CGAN cases. The first was the peak signal to noise ratio. The *PSNR* is a metric for assessing the degree of pixels' distortion caused by compression and noise [12]. It is defined as

$$PSNR = 20 \cdot \log_{10}\left(\frac{MAX_I}{\sqrt{MSE}}\right), \tag{8}$$

where MAX_I is the maximum possible pixel value of the image, and MSE is the mean squared deviation.

Structure similarity (*SSIM*) is the second metric, which is more complex and more informative than the *PSNR*. It comprises an assessment of three characteristics of the investigated images: intensity, contrast, and structural difference [11].

$$SSIM(I,K)_{local} = \frac{(2\mu_I\mu_K + c_1)(2\sigma_{I,K} + c_2)}{(\mu_I^2 + \mu_K^2 + c_1)(\sigma_I^2 + \sigma_K^2 + c_2)} \tag{9}$$

$$SSIM(I,K) = \frac{1}{M}\sum_{i=1}^{M} SSIM(I,K)_{local}$$

where $SSIM(I,K) \in [0,1]$ is the structural similarity between the target image I and the generated image K. μ and σ are the average and standard deviation, respectively. c is the constant.

The SSIM value ranges from 0 to 1. If SSIM is 1, then the images are identical. As a rule, SSIM is not used for the entire image at once, but it is used locally with a sliding window and then averaged. The local calculation of SSIM metrics allows the variability in statistical characteristics and spatial heterogeneity of the image structure to be taken into account. Here, the kernel size was chosen to be (8×8) according to [10].

3.5. Experimental Setup

End-to-end training was used for the proposed AI model. In this study, we employed a learning rate of 0.0002 with Adam optimizer. We trained all AI models using 100 epochs with Random Normal weight initialization (stddev = 0.02) and a batch size of 8. The input and output image size was fixed to the dimensions of 256×256. On the encoder side, the activation function of LeakyReLU (alpha = 0.2) was used, while ReLU was used for the decoder side.

3.6. Execution Development Environment

A computer with the following specifications was used to carry out the experiments: AMD Ryzen 7 5800X 8-Core Processor 3.80 GHz 32 GB RAM with RTX 3060 (8 GB) GRAPH-ICS CARD. Python 3.10 running on Windows 10 along with the Keras and TensorFlow backend libraries were utilized to conduct the experiments that were analyzed in this study.

4. Experimental Results

The considered models were evaluated on the test MR data subset using the PSNR and the SSIM evaluation metrics. The latter employed the image reconstructed from the under-sampled k-space with deep learning models as I in (9) and the reference images obtained from the fully sampled data as K in (9). Figure 6 and Table 2 display comparisons of SSIM and PSNR metrics of the reconstructed test MR images using: the U-Net model with the L1-loss (*U-Net_L1*), the U-Net model with the combination of the L1- and Fourier loss (*U-Net_Hybrid_Loss*), the CGAN model with L1-loss (*GAN_L1*), and the CGAN model with the combination of the L1 and Fourier losses (*GAN_ Hybrid_Loss*). As an addition, the abovementioned models were compared to the SENSE parallel MR imaging algorithm reconstruction.

Table 2. Reconstruction performance evaluation of the proposed methodology against SENSE over the test dataset.

AI Model	SSIM		PSNR		No. of Trainable Parameters (Million)	Training Time/Epoch (s)	Testing Time/Image (s)
	Mean $\pm$ SD	Median	Mean $\pm$ SD	Median			
U-Net_L1_Loss	0.857 $\pm$ 0.059	0.87	31.834 $\pm$ 2.66	32.77	54.41	1.23	0.1
GAN_L1_Loss	0.880 $\pm$ 0.054	0.89	33.678 $\pm$ 2.6	33.91	61.38	4	0.15
U-Net_Hybrid_Loss	0.876 $\pm$ 0.054	0.89	33.112 $\pm$ 2.56	33.79	54.41	1.23	0.1
GAN_ Hybrid_Loss	**0.903 $\pm$ 0.050**	**0.92**	**35.683 $\pm$ 2.77**	**36.11**	61.38	4	0.15
SENSE	0.902 $\pm$ 0.058	0.92	26.288 $\pm$ 5.48	26.70	-	-	0.29

4.1. Quality of the MR Image Reconstruction against Different AI Architectures

Our first goal was to study the impact of the deep network architecture type on the quality of the MR image reconstruction. For this purpose, the U-Net and the CGAN architectures were investigated. According to the median values of the evaluation metrics, the CGAN architecture exceeded the U-net by 2% in terms of the SSIM score, in which they reached the values of 0.89 and 0.87, respectively. The CGAN architecture also showed a better median PSNR score (33.91) when compared to the median PSNR of images reconstructed with the U-Net (32.77). Thus, the CGAN model showed overall better performance according to the SSIM and PSNR metrics.

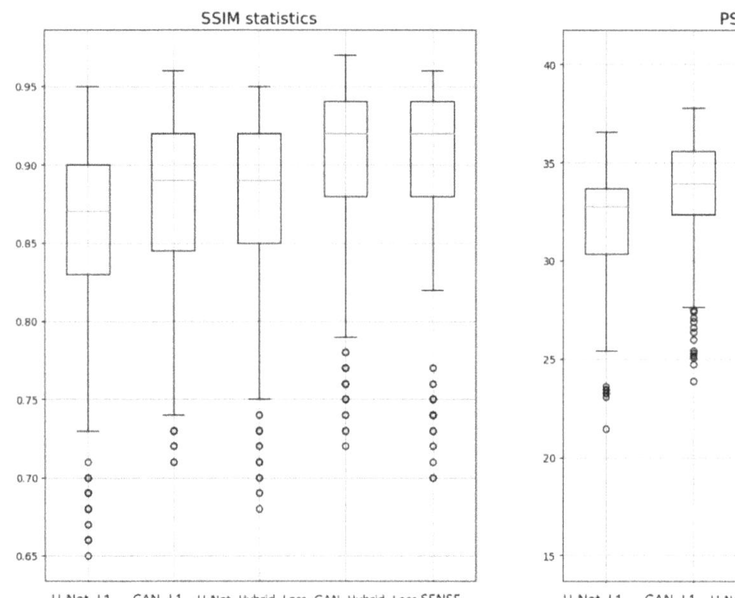
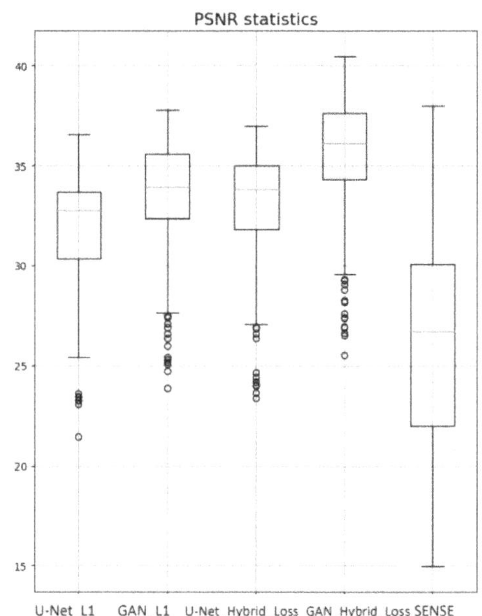

Figure 6. Comparison of the SSIM and PSNR metrics of the reconstructed test images for different algorithms: U-Net model with the L1-loss (*U-Net_L1*), U-Net model with the combination of the L1- and Fourier loss (*U-Net_L1_fft*), CGAN model with the L1-loss (*GAN_L1*), CGAN model with the combination of the L1 and Fourier losses (*GAN_L1_fft*), and the classic SENSE reconstruction.

4.2. Quality of the MR Image Reconstruction against the Proposed Hybrid Loss Function

Our second goal was to explore the impact of the introduced modified loss function. The results in Table 2 and Figure 6 show the contribution of the proposed Fourier loss function, which was designed to take into account the difference between target and reconstructed images in the frequency domain. Adding Fourier loss to the U-Net and the CGAN model resulted in increases in the SSIM score by 2% and 3%, respectively, whereas the PSNR score increased by about 1.02 and 2.2 for the two network architectures. Thus, we can see that the proposed Fourier hybrid loss helped to enhance the image reconstruction quality. Making the model minimize the error in the k-space resulted in minimizing the reconstruction error.

5. Discussion

5.1. Deep Learning Approach against Classical Algorithms of MR Image Reconstruction

The results above show the ability of the deep learning algorithms to reconstruct MRI images as an alternative to classical algorithms such as SENSE. Different deep network architectures (U-Net and CGAN) achieve better image reconstruction quality against SENSE in terms of the PSNR by 6.84 and 9.57, respectively. Similarly, the deep network architectures studied in this paper display comparable SSIM results of the reconstructed MR images. The advantage of the developed deep learning approach is that there is no need for coil sensitivity maps compared to the SENSE algorithms of the MRI image reconstruction. The higher value of PSNR, comparable SSIM metric, and the absence of the need for a coil sensitivity map, along with other possibilities, open up promising prospects for the development of deep learning approaches in MRI image reconstruction problems.

L1 distance in image space forces the model to enhance the structural characteristics of generated (reconstructed) images. Using Fourier (k-space) L1 loss, we encouraged the model to take important frequency components into account; thus, we achieved an increase in the quality of image reconstruction for both networks. Adding Fourier (k-space) loss to the U-Net and the CGAN model resulted in increasing the SSIM score by 2% and 3%, respectively, whereas the PSNR score increased by about 1.02 and 2.2 for the two network architectures. Thus, we can see that the proposed hybrid loss helps to enhance the image reconstruction quality.

5.2. Statistical Significance of the Results

After the PSNR and SSIM metrics were calculated using the test dataset for every architecture, the metrics distributions were tested for statistically significant differences. The choice of the statistical significance test depends critically on the type of the data distribution. As a rule, if the samples have a normal distribution, then a t-test is used. If the data distribution does not meet the requirements for normality, then other approaches are undertaken, the most commonly known being the Mann–Whitney U-test. It is therefore necessary to first conduct a test for data normality and then evaluate the statistical significance. Normality tests were carried out using a qualitative histogram evaluation as well as a Shapiro–Wilk normality test. The essence of normality verification is to put forward the null hypotheses that the data are distributed normally with the error probability of 0.05. Thus, H0 is "the data come from the normal distributions (accepted if $p > 0.05$)"; otherwise, H1 (rejected), meaning that data do not come from the normal distribution. The results of the Shapiro–Wilk normality test are presented in Table 3.

Table 3. Shapiro–Wilk normality test for SSIM and PSNR of studied AI models.

AI Model	SSIM			PSNR		
	Statistics	df	p-Value	Statistics	df	p-Value
U-Net_L1_Loss	0.885		5.3×10^{-17}	0.925		1.61×10^{-13}
GAN_L1_Loss	0.921		7×10^{-14}	0.943		1.66×10^{-11}
U-Net_Hybrid_Loss	0.882	414	2.8×10^{-17}	0.905	414	2.30×10^{-15}
SENSE	0.819		5.75×10^{-17}	0.954		1.11×10^{-13}
GAN_Hybrid_Loss	0.868		3×10^{-18}	0.941		9.46×10^{-12}

Since for all the architectures in Table 3, the p-values for SSIM and PSNR distributions were much less than the alpha (p-value 0.05), we rejected all the null hypotheses and concluded that none of the samples came from normal distributions. These results were also confirmed by the graphic evaluation. It can be seen in Figure 7 that the data were not distributed normally.

Due to the non-normality of the SSIM and PSNR distributions, we could not use a t-test to study statistical significance, and thus, a nonparametric Mann–Whitney U-test was used. We compared the SSIM and PSNR distributions provided by the final architecture (GAN_Hybrid_Loss) with the rest of the reconstruction strategies. The results of the Mann–Whitney U-test are presented in Table 4.

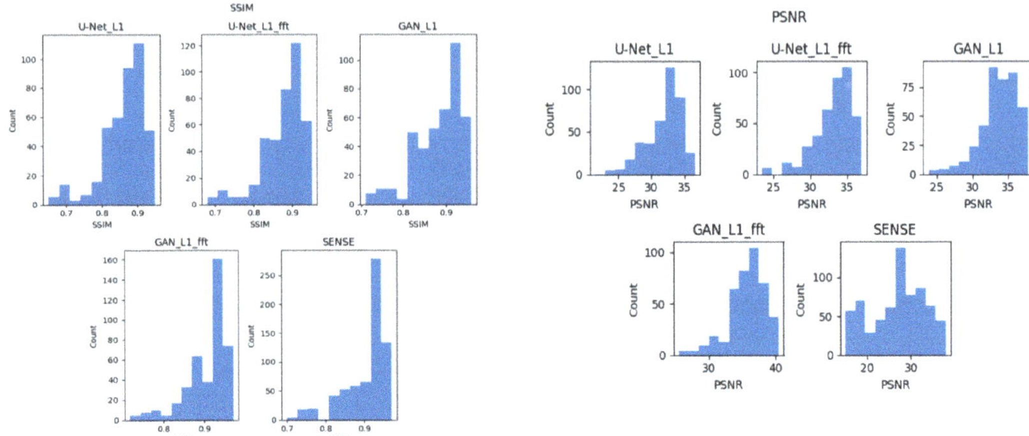

Figure 7. Graphical evaluation of SSIM and PSNR for studied models.

Table 4. The results of Mann–Whitney U test for SSIM and PSNR of the studied AI models against the model with the highest SSIM and PSNR (GAN_Hybrid_Loss).

AI Model	SSIM p-Value	PSNR p-Value
U-Net_L1_Loss	3.26×10^{-38}	3.20×10^{-74}
GAN_L1_Loss	3.83×10^{-12}	2.02×10^{-28}
U-Net_Hybrid_Loss	2.02×10^{-18}	2.12×10^{-43}
SENSE	**0.092**	2.40×10^{-116}

Using the Mann–Whitney U paired test with a significance level of a 0.05 p-value, the AI models were investigated. Assuming the null hypothesis, there was no significant performance difference between our proposed model (GAN_Hybrid_Loss) and others, whereas the alternative hypothesis intended to show if the proposed GAN_ Hybrid_Loss model provided metrics that were statistically different from the other approaches. According to the obtained p-values of the SSIM metric, shown in Table 4, we could infer that statistical differences existed between GAN_ Hybrid_Loss and other models (p-value much smaller than 0.05), except SENSE (p-value = 0.09 > 0.05). This shows the differences between the GAN_Hybrid_Loss model and other models (seen in Figure 6 and Table 2) were statistically significant differences in all cases except SENSE, which thus can be concluded to have had a comparable performance. On the other hand, the p-values of the PNSR distribution tests for all models were smaller than the 0.05 threshold, which confirms the improvement in the performance of our GAN_Hybrid_Loss model against other models, including the SENSE algorithm, to be statistically significant.

5.3. Comparison Results against the Recent Research Works

During this study, we conducted a comparison of the proposed algorithm with the latest AI research works for MRI image reconstruction. Table 5 shows the used models, implemented loss functions, and some quantitative results of studied deep learning algorithms for MR image construction. The analysis of these approaches shows that regardless of how good the SSIM metrics are, the PSNR is still close to the mean PSNR of the classical SENSE algorithm. In [9], some k-space correction was used, but it was employed as a post-processing step, so the model itself did not learn this correction in the k-space. Unfortunately, the it did not provide any information about the PSNR, so it is difficult to evaluate the contribution of this post-processing procedure. However, our approach, thanks to the use of hybrid spatial and *k*-space loss, overcame the presented models in the PSNR metric (mean PSNR = 35.68). Thus, we can conclude that in MR image reconstruction, it is important to pay attention to the difference between reconstructed and target images not only in the special space, but also in the *k*-space. This will guarantee the achievement of the more accurate quality of the image reconstruction.

Table 5. Comparison of the evaluation results against the latest AI research works for MRI image reconstruction.

Reference	Model	Loss Function	SSIM	PSNR
Hyun CM et al. (2017), [9]	U-net with *k*-space correction	L2-norm	0.903	-
Ghodrati V et al. (2019), [10]	Resnet-L1	L1-norm	0.81	26.39
Ghodrati V et al. (2019), [10]	Unet–Dssim	Structural dissimilarity	0.86	27.04
Cole, Elizabeth et al. (2020), [11]	Unsupervised GAN	GAN loss	0.88	29
The proposed, (U-Net_Hybrid_Loss)	U-Net	Hybrid Loss function	0.876 ± 0.03	33.11 ± 2.56
The proposed, (GAN_Hybrid_Loss)	**CGAN**	**Hybridd Loss**	**0.903 ± 0.05**	**35.68 ± 2.77**

6. Conclusions

In this work, we developed a deep learning approach to reconstruct cardiac MR images from under-sampled *k*-space data. Two deep network architectures were considered: the U-Net and the CGAN. The results showed that the CGAN model outperformed the U-Net model by 2% in terms of the SSIM score. To enhance the model efficiency, we extended the loss function by additionally taking into account the difference between target and reconstructed images in the frequency domain. The proposed loss, referred to as the Fourier loss, was shown to increase the SSIM by another 2% for the U-Net model and by 3% for the CGAN. The PSNR score was also improved by employing the Fourier loss by 1 for the U-net model and by 2.2 for the CGAN model.

Because the GAN model with the combination of L1 and Fourier losses (*GAN_Hybrid_Loss*) yielded the best results among the other studied deep learning models, we also compared it with the reconstruction employing the SENSE algorithm. According to the SSIM metric, the results of *GAN_Hybrid_Loss* are comparable to the SENSE results. However, the PSNR of *GAN_Hybrid_Loss* was greater than that of the SENSE by 8.7 (36.11 and 27.40, respectively). The latter could have resulted from the known effect of SNR reduction in parallel imaging as the square root of the acceleration factor, while the deep learning algorithms do seem to help mitigate this problem. The suggested algorithms and network architectures can therefore be used in applications in which both fast acquisition and a high SNR is critical (cardiac imaging being one of these areas) as a substitution for the classic parallel imaging algorithms, which do not conserve the SNR.

Author Contributions: Conceptualization, W.A.-H., I.M. and M.Z.; data curation, I.M. and M.Z.; formal analysis, M.A.A.-a.; investigation, W.A.-H., M.Z. and M.A.A.-a.; methodology, W.A.-H., I.M. and M.Z.; project administration, M.A.A.-a.; resources, M.Z.; software, W.A.-H. and I.M.; supervision, M.Z.; validation, I.M. and M.Z.; visualization, W.A.-H. and M.A.A.-a.; writing—original draft, W.A.-H. and M.Z.; writing—review and editing, M.Z. and M.A.A.-a. All authors were informed about each step of manuscript processing, including submission, revision, revision reminder, etc., via emails from our system or assigned Assistant Editor. All authors have read and agreed to the published version of the manuscript.

Funding: The work was supported by the Ministry of Science and Higher Education of the Russian Federation (075-15-2021-592).

Institutional Review Board Statement: Not applicable.

Informed Consent Statement: Not applicable.

Data Availability Statement: The datasets used in this paper are publicly available at: https://ocmr.info/download/ (accessed on 20 December 2022).

Acknowledgments: The work was supported by the Ministry of Science and Higher Education of the Russian Federation (075-15-2021-592). Mugahed A. Al-antari acknowledges the support of the National Research Foundation of Korea (NRF) grant funded by the Korean government (MSIT) (No. RS-2022-00166402).

Conflicts of Interest: The authors declare no conflict of interest.

Abbreviations

AI	Artificial Intelligence
GAN	Generative Adversarial Network
CGAN	Conditional Generative Adversarial Network
MRI	Magnetic Resonance Imaging
PSNR	Peak Signal to Noise Ratio
SSIM	Structure Similarity
SENSE	Sensitivity Encoding
GRAPPA	Generalized Autocalibrating Partial Parallel Acquisition
FFT	Fast Fourier Transform
IFFT	Inverse Fast Fourier Transform

References

1. Sobol, W.T. Recent advances in MRI technology: Implications for image quality and patient safety. *Saudi J. Ophthalmol.* **2012**, *26*, 393–399. [CrossRef] [PubMed]
2. Krupa, K.; Bekiesińska-Figatowska, M. Artifacts in Magnetic Resonance Imaging. *Pol. J. Radiol.* **2015**, *80*, 93–106. [CrossRef] [PubMed]
3. Zbontar, J.; Knoll, F.; Sriram, A.; Murrell, T.; Huang, Z.; Muckley, M.J.; Defazio, A.; Stern, R.; Johnson, P.; Bruno, M.; et al. fastMRI: An Open Dataset and Benchmarks for Accelerated MRI. *arXiv* **2018**, arXiv:1811.08839.
4. Martí-Bonmatí, L. MR Image Acquisition: From 2D to 3D. In *3D Image Processing: Techniques and Clinical Applications*; Caramella, D., Bartolozzi, C., Eds.; Springer: Berlin/Heidelberg, Germany, 2002; pp. 21–295.
5. Yanasak, N.; Clarke, G.; Stafford, R.J.; Goerner, F.; Steckner, M.; Bercha, I.; Och, J.; Amurao, M. *Parallel Imaging in MRI: Technology, Applications, and Quality Control*; American Association of Physicists in Medicine: Alexandria, VA, USA, 2015. [CrossRef]
6. Blaimer, M.; Breuer, F.; Mueller, M.; Heidemann, R.M.; Griswold, M.A.; Jakob, P.M. SMASH, SENSE, PILS, GRAPPA: How to choose the optimal method. *Top Magn. Reason. Imaging TMRI* **2004**, *15*, 223–236. [CrossRef] [PubMed]
7. Pruessmann, K.P.; Weiger, M.; Scheidegger, M.B.; Boesiger, P. SENSE: Sensitivity encoding for fast MRI. *Magn. Reason. Med.* **1999**, *42*, 952–962. [CrossRef]
8. Hoge, W.S.; Brooks, D.H. On the complimentarity of SENSE and GRAPPA in parallel MR imaging. In Proceedings of the 2006 International Conference of the IEEE Engineering in Medicine and Biology Society, New York, NY, USA, 30 August–3 September 2006; pp. 755–758. [CrossRef]
9. Chen, C.; Liu, Y.; Schniter, P.; Tong, M.; Zareba, K.; Simonetti, O.; Potter, L.; Ahmad, R. OCMR (v1.0)–Open-Access Multi-Coil k-Space Dataset for Cardiovascular Magnetic Resonance Imaging. *arXiv* **2020**, arXiv:2008.03410.
10. Nilsson, J.; Akenine-Möller, T. Understanding SSIM. *arXiv* **2020**, arXiv:2006.13846.
11. Wang, Z.; Bovik, A.C.; Sheikh, H.R.; Simoncelli, E.P. Image quality assessment: From error visibility to structural similarity. *IEEE Trans. Image Process* **2004**, *13*, 600–612. [CrossRef] [PubMed]
12. Tao, D.; Di, S.; Liang, X.; Chen, Z.; Cappello, F. Fixed-PSNR Lossy Compression for Scientific Data. In Proceedings of the 2018 IEEE International Conference on Cluster Computing (CLUSTER), Belfast, UK, 10–13 September 2018; pp. 314–318.
13. Hyun, C.M.; Kim, H.P.; Lee, S.M.; Lee, S.; Seo, J.K. Deep learning for undersampled MRI reconstruction. *Phys. Med. Biol.* **2017**, *63*, 135007. [CrossRef] [PubMed]
14. Ghodrati, V.; Shao, J.; Bydder, M.; Zhou, Z.; Yin, W.; Nguyen, K.L.; Yang, Y.; Hu, P. MR image reconstruction using deep learning: Evaluation of network structure and loss functions. *Quant. Imaging Med. Surg.* **2019**, *9*, 1516–1527. [CrossRef] [PubMed]
15. Cole, E.K.; Pauly, J.M.; Vasanawala, S.S.; Ong, F. Unsupervised MRI Reconstruction with Generative Adversarial Networks. *arXiv* **2020**, arXiv:2008.13065. [CrossRef]
16. Li, W.; Feng, X.; An, H.; Ng, X.Y.; Zhang, Y.J. MRI Reconstruction with Interpretable Pixel-Wise Operations Using Reinforcement Learning. *Proc. AAAI Conf. Artif. Intell.* **2020**, *34*, 792–799. [CrossRef]
17. Ronneberger, O.; Fischer, P.; Brox, T. U-Net: Convolutional Networks for Biomedical Image Segmentation. In *Medical Image Computing and Computer-Assisted—MICCAI*; Navab, N., Hornegger, J., Wells, W., Frangi, A., Eds.; Lecture Notes in Computer Science; Springer: Cham, Switzerland, 2015; Volume 9351. [CrossRef]

18. Isola, P.; Zhu, J.-Y.; Zhou, T.; Efros, A.A. Image-to-Image Translation with Conditional Adversarial Networks. In Proceedings of the 2017 IEEE Conference on Computer Vision and Pattern Recognition (CVPR), Honolulu, HI, USA, 21–26 July 2017; pp. 5967–5976.
19. Available online: https://github.com/ismrmrd/ismrmrd-python (accessed on 25 February 2023).
20. Hossain, K.F.; Kamran, S.A.; Tavakkoli, A.; Pan, L.; Ma, X.; Rajasegarar, S.; Karmaker, C. ECG-Adv-GAN: Detecting ECG Adversarial Examples with Conditional Generative Adversarial Networks. In Proceedings of the 2021 20th IEEE International Conference on Machine Learning and Applications (ICMLA), Online, 13–15 December 2021; pp. 50–56.
21. Bernstein, M.A.; King, K.F.; Zhou, Z.J. *Handbook of MRI Pulse Sequences*; Academic Press: Amsterdam, The Netherlands, 2004.

 *diagnostics*

Article

High-Dimensional Feature Selection for Automatic Classification of Coronary Stenosis Using an Evolutionary Algorithm

Miguel-Angel Gil-Rios [1], Ivan Cruz-Aceves [2,*], Arturo Hernandez-Aguirre [3], Ernesto Moya-Albor [4], Jorge Brieva [4], Martha-Alicia Hernandez-Gonzalez [5] and Sergio-Eduardo Solorio-Meza [6]

[1] Tecnologías de Información, Universidad Tecnológica de León, Blvd. Universidad Tecnológica 225, Col. San Carlos, León 37670, Mexico; mgil@utleon.edu.mx
[2] CONACYT, Centro de Investigación en Matemáticas (CIMAT), A.C., Jalisco S/N, Col. Valenciana, Guanajuato 36000, Mexico
[3] Departamento de Computación, Centro de Investigación en Matemáticas (CIMAT), A.C., Jalisco S/N, Col. Valenciana, Guanajuato 36000, Mexico; artha@cimat.mx
[4] Facultad de Ingeniería, Universidad Panamericana, Augusto Rodin 498, Ciudad de México 03920, Mexico; emoya@up.edu.mx (E.M.-A.); jbrieva@up.edu.mx (J.B.)
[5] Unidad Médica de Alta Especialidad (UMAE), Hospital de Especialidades No. 1. Centro Médico Nacional del Bajio, IMSS, Blvd. Adolfo López Mateos esquina Paseo de los Insurgentes S/N, Col. Los Paraisos, León 37320, Mexico; martha.hernandez@imss.gob.mx
[6] División Ciencias de la Salud, Universidad Tecnológica de México, Campus León, Blvd. Juan Alonso de Torres 1041, Col. San José del Consuelo, León 37200, Mexico; sergio_solorio@my.unitec.edu.mx
* Correspondence: ivan.cruz@cimat.mx

Citation: Gil-Rios, M.-A.; Cruz-Aceves, I.; Hernandez-Aguirre, A.; Moya-Albor, E.; Brieva, J.; Hernandez-Gonzalez, M.-A.; Solorio-Meza, S.-E. High-Dimensional Feature Selection for Automatic Classification of Coronary Stenosis Using an Evolutionary Algorithm. *Diagnostics* **2024**, *14*, 268. https://doi.org/10.3390/diagnostics14030268

Academic Editor: Mugahed A. Al-antari

Received: 18 December 2023
Revised: 11 January 2024
Accepted: 23 January 2024
Published: 26 January 2024

Abstract: In this paper, a novel strategy to perform high-dimensional feature selection using an evolutionary algorithm for the automatic classification of coronary stenosis is introduced. The method involves a feature extraction stage to form a bank of 473 features considering different types such as intensity, texture and shape. The feature selection task is carried out on a high-dimensional feature bank, where the search space is denoted by $O(2^n)$ and $n = 473$. The proposed evolutionary search strategy was compared in terms of the Jaccard coefficient and accuracy classification with different state-of-the-art methods. The highest feature selection rate, along with the best classification performance, was obtained with a subset of four features, representing a 99% discrimination rate. In the last stage, the feature subset was used as input to train a support vector machine using an independent testing set. The classification of coronary stenosis cases involves a binary classification type by considering positive and negative classes. The highest classification performance was obtained with the four-feature subset in terms of accuracy (0.86) and Jaccard coefficient (0.75) metrics. In addition, a second dataset containing 2788 instances was formed from a public image database, obtaining an accuracy of 0.89 and a Jaccard Coefficient of 0.80. Finally, based on the performance achieved with the four-feature subset, they can be suitable for use in a clinical decision support system.

Keywords: bank of features; coronary angiograms; evolutionary algorithm; feature selection; K-nearest neighbor; stenosis classification

1. Introduction

Coronary artery disease (CAD) stands as a leading cause of mortality in the majority of developed countries [1]. According to the British Heart Foundation (BHF) [2], coronary heart disease was the main cause of death in the year 2021 around the world. In Figure 1, a comparative chart of different death causes with data extracted from the BHF is illustrated.

According to Figure 1, coronary heart disease presents the highest rate of cases, with 9.2 million registered cases around the world.

In coronary artery disease, atherosclerosis leads to the development of coronary stenosis at various locations [3]. Nowadays, X-ray coronary angiograms are the gold standard for the detection of coronary stenosis in clinical practice. Consequently, a cardiology specialist must exhaustively examine the entire angiogram, and according to their expertise, all regions in which coronary stenosis cases can occur are labeled by hand. Figure 2 presents X-ray coronary angiogram samples with stenosis regions labeled by a specialist in cardiology.

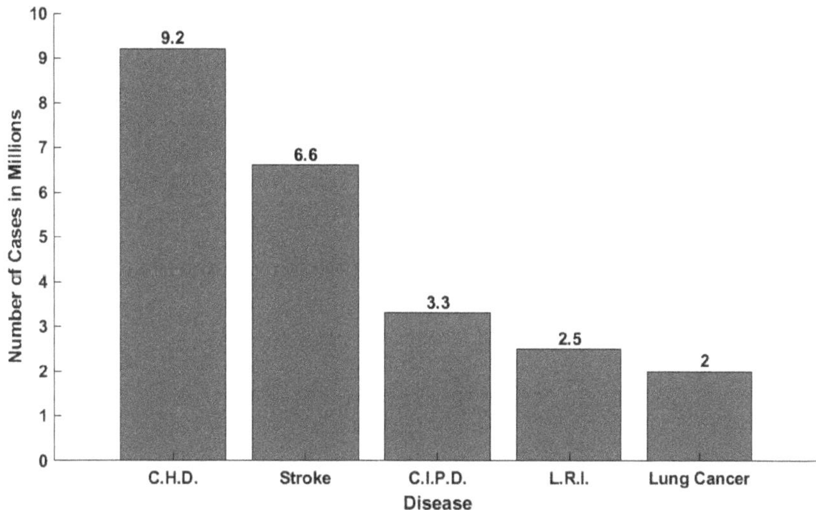

Figure 1. Comparison of five main diseases around the world: coronary heart disease (C.H.D.), stroke, chronic inflammatory pulmonary disease (C.I.P.D.), lower respiratory infections (L.R.I.) and lung cancer.

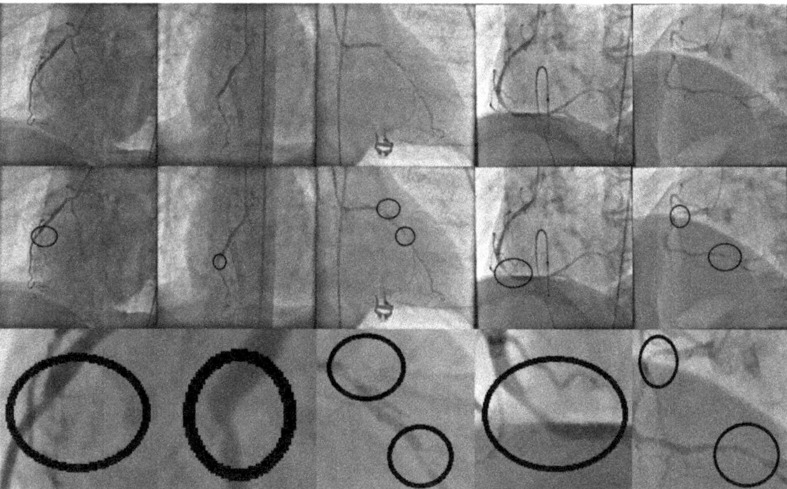

Figure 2. Coronary angiograms with their respective coronary stenosis regions labeled by the specialist.

The coronary stenosis problem has been studied in digital image processing, in which several challenging issues must be addressed, such as the presence of noise and weak contrast. In the literature, the method proposed by Saad [4] requires a previous vessel segmentation of a coronary angiogram to identify atherosclerosis using a vessel-width variation measure. A coronary stenosis measure grading method was proposed by Kishore and Jayanthi [5], using the pixel intensities of a previously enhanced image. Alternative methods, such as Brieva et al. [6], used a Hessian-based approach to extract different texture and shape vessel features to classify positive and negative stenosis cases.

Moreover, machine learning-based techniques have been used to address the stenosis classification problem. A naive Bayes classifier was used by Taki et al. [7] to classify calcified and non-calcified coronary artery plaques. The method of Cruz–Aceves et al. [8] uses a $3D$ intensity feature vector computed from the histogram of an image to detect a specific type of stenosis using a Bayesian-based classifier. Giannoglou et al. [9] proposed a fuzzy criterion for a feature selection process in atherosclerotic plaques. Chen et al. [10] use a $6D$ vector of shape features for the detection of coronary artery disease.

The main disadvantage of previous methods is the use of a fixed threshold measurement in order to classify coronary stenosis cases. Since image datasets of coronary angiograms have different size, contrast, and noise levels, the artery feature values present considerable variations. In consequence, the obtained results are highly dependent on the applied vessel enhancement method. In the proposed method, a number of spatial and frequency domain filters have been adopted for vessel enhancement in order to capture relevant information from different domains.

The method proposed by Antczak and Liberadzki [11], generates synthetic coronary stenosis and non-stenosis patches aiming to enhance the performance of a convolutional neural network (CNN) from scratch. Data augmentation techniques [12] are also used to generate a large number of instances that are used in the training and testing steps of a CNN. In addition, using the *explainable AI* concept, it is difficult to identify what features are really useful for correct classification and what they represent, which is the main drawback of the CNN [13]. Convolutional neural networks are still considered "black box" systems that do not offer any insight or explanations on how the decision is obtained, limiting the clinical applicability due to their lack of transparency and interpretability. Since the proposed method works with identifiable features, it is possible to know their relevance in the coronary stenosis classification by performance employing statistical analysis.

In the present paper, a novel strategy to perform high-dimensional feature selection using an evolutionary algorithm for the classification of coronary stenosis is proposed. The method involves a binary classification in order to identify positive and negative coronary stenosis cases. An automatic feature selection step is driven by a hybrid-evolutionary algorithm over a high-dimensional bank of features. The accuracy metric was used as a fitness function, while the discrimination rate was also maximized. Shape feature values such as vessel length, bifurcation points and others are dependent on a vessel enhancement technique. Therefore, different enhancement methods were applied to the original images. This strategy allows the formation of a feature bank of 473 different features. To obtain an optimal subset of features, an automatic feature selection stage is performed using a K-nearest neighbor classifier (KNN) to select relevant features. The classification of positive and negative stenosis cases is performed adequately in terms of the accuracy and Jaccard coefficients, which are useful to evaluate the rate of true-positive coronary stenosis cases and avoid the rate of true-negative cases. Since the problem involves a high-dimensional search space, which can be expressed as $O(2^{473})$, a hybrid-evolutionary algorithm is appropriate for addressing the feature selection optimization problem. An evolutionary algorithm is a high-dimensional optimization technique for working in discrete and continuous domains using the Darwinian theory about the evolution of biological organisms throughout various generations. The hybrid-evolutionary strategy achieved a discrimination rate of 0.99, obtaining a subset of four features from the extracted bank of 473. In the experiments, a database containing 608 images was used for training of the proposed method (508 for training and 100 for testing). In addition, a second public domain database [11], of 2788 coronary patch images was used for testing.

The rest of this paper is structured as follows. The background methods of vessel enhancement methods, along with intensity, texture, and shape features, are presented in Section 2. In Section 3, the bank of 473 features, the hybrid-evolutionary algorithm, and the performance metrics of the proposed method are introduced. Experimental results are described in Section 4, and conclusions are presented in Section 5.

2. Materials and Methods

2.1. Experiment Materials

For the experiments, two distinct banks of images were used. The first image database was formed from a bank of images provided by the Mexican Institute for Social Healthcare and authorized by an institutional review board only for research purposes under the reference R-2019-1001-078. The bank contains 180 digital images of coronary angiograms, which are 512×512 pixels and have a grayscale color scheme. It also included the corresponding ground-truth images with the coronary stenosis cases labeled by a cardiology specialist. From the provided image bank, 304 patches of size 64×64 pixels containing coronary stenosis cases were extracted. In addition, 304 additional patches with non-stenosis cases were extracted in order to form a balanced database. A total set of 608 image patches were extracted to form the first database. The size of the patches corresponds to the area enclosing the stenosis cases and was labeled and validated by the specialist.

In order to assess the obtained results, a second image database was formed from the Antczak [11] image database, which is in the public domain. Each image corresponds to a coronary patch of size 32×32 pixels in a grayscale color scheme. The original bank is formed by 122 natural coronary patches containing a stenosis case. In addition, it contains 1394 natural coronary patches with non-coronary stenosis cases. Since the proportion of positive and negative stenosis cases is unbalanced, 1272 additional synthetic patches with a coronary stenosis case were added, which were taken from the same image bank. Consequently, the second image database consists of 2788 images with a balance of positive and negative coronary stenosis cases.

2.2. Feature Extraction

The description and measurement of objects of interest, properties of an image, or a specific region is commonly known as feature extraction [14]. It is possible to extract distinct feature types, as described in the literature [15,16]. According to their nature, features can be classified into texture, intensity, and shape.

2.2.1. Intensity Features

Intensity features are relevant in digital image processing because they are related to the corresponding value for each pixel in the image. Here, the five minimum, maximum, median, average, and standard deviation statistical measures of the pixel intensity have been extracted.

2.2.2. Texture Features

Texture features are relevant in different cardiovascular problems [17,18]. One of the most used approaches for the extraction of texture-related features over images is the gray-level co-occurrence matrix (GLCM) [19]. The GLCM computes the frequency of variation between the intensity levels of a pixel. It is expressed as a matrix whose rows and columns correspond to the entire image's pixel intensities. The frequencies of intensity variations are computed in a specific spatial relationship denoted by $(\Delta x, \Delta y)$ between two different pixels with intensity levels i and j as follows:

$$C_{\Delta x, \Delta y}(i,j) = \sum_{x=1}^{n}\sum_{y=1}^{m} \begin{cases} 1, \text{if } I(x,y) = i \text{ and } I(x+\Delta x, y+\Delta y) = j \\ 0, \text{otherwise} \end{cases}, \tag{1}$$

where $C_{\Delta x, \Delta y}(i,j)$ is the frequency in which two pixels with intensities i and j at a specific offset $(\Delta x, \Delta y)$ occur, and n and m represent the height and width of the image.

Additionally, it is also possible to extract texture features from the output of alternative representation methods, such as the Radon transform [20]. The Radon transform is the projection of the image intensity along with a radial line oriented at some specific angle, which can be computed as follows:

$$R(\rho,\theta) = \int_{-\infty}^{\infty} \int_{-\infty}^{\infty} f(x,y)\delta(\rho - x\cos\theta - y\sin\theta)dxdy, \tag{2}$$

where $R(\rho,\theta)$ is the Radon Transform of a function $f(x,y)$ at an angle θ, $\delta(r)$ is the Dirac delta function, and $\delta(\rho - x\cos\theta - y\sin\theta)$ forces the integration of $f(x,y)$ along the line $\rho - x\cos\theta - y\sin\theta = 0$.

2.2.3. Shape Features

Shape-based features enable the extraction of quantifiable information concerning various aspects related to the artery shape; for example, a segment length, the tortuosity level present in a determined arterial section, or the number of bifurcations present on it and the vessel width. Nevertheless, to extract shape-based features, it is necessary to perform a previous vessel enhancement process over the original image in which useless information such as noise and background are identified. Consequently, the shape measure values are highly dependent on the applied filtering method. The use of the Hessian matrix and eigenvalues methodology [21] has proved to be adequate for vessel enhancement. However, an automatic thresholding strategy such as the Otsu method [22] is necessary to separate vessel and non-vessel pixels.

The vessel skeleton has been used to extract vessel-shape-related information. To obtain the corresponding vessel skeleton from a binary segmented image, the medial-axis transform method has been commonly applied [23].

2.3. Vessel Enhancement Methods

Vessel enhancement methods are useful for discriminating irrelevant non-vessel information on coronary angiograms. In the literature, spatial and frequency domain filters have been applied to the vessel enhancement problem, achieving suitable results. In the experiments, eight state-of-the-art vessel enhancement methods have been adopted, which are mentioned below.

1. Spatial domain filters

    (a) Hessian-based methods.

      i. Vesselness measure [24]. The Frangi method computes a vesselness measure using the eigenvalues of a Hessian matrix.

      ii. Hessian matrix and clustering [25]. The Hessian matrix is also used by Salem et al. for vessel enhancement by computing the largest eigenvalue and vessel orientation over all scales.

    (b) Morphological top-hat filter [26]. In mathematical morphology, the top-hat filter is useful to enhance images with non-uniform illumination. Because of this property, the top-hat operator has been used to enhance vessel-like structures [27,28].

    (c) Multi-scale line detection [29]. An alternative approach that has been used for artery enhancement is the linear matched filter. This method works under the assumption that blood vessels can be modeled by linear segments that share the same orientation and length.

    (d) Gaussian matched filter

      i. Single-scale Gaussian filter (GMF). In this approach, a gray-scale template is formed from a Gaussian distribution, which is convolved with the input image.

      ii. Multi-scale Gaussian filter. The main limitation of GMF is the use of a fixed vessel diameter represented by the σ parameter in which non-corresponding vessel diameters will be distinguished. In order to overcome this disadvantage, a multi-scale Gaussian matched filter was proposed by Cruz–Aceves et al. [30] considering different vessel width scales.

2. Frequency domain filters

(a) Gabor filter

i. Single-scale Gabor filter [31]. The Gabor filter is a Gaussian curve modulated by a sinusoidal function, which is useful for the detection of directional features. In addition, Rangayyan et al. [32] simplified the matching template equation so it is governed by only two parameters.

ii. Multi-scale Gabor filter [33]. Similar to the GMF, the use of a fixed vessel diameter represented by the τ parameter will only detect the main artery tree and, as a consequence, discriminate vessels with diameters lower than τ. In order to overcome this disadvantage, Rangayyan et al. proposed a multi-scale Gabor filter for retinal vessels.

2.4. Metaheuristics

The classification techniques learn and predict by classifying instances defined by their features. Therefore, the classification accuracy performance is highly dependent on the used feature set since not all of the used features could be relevant for the classification process. In this context, a feature selection task is necessary after the feature extraction stage is performed. However, the feature selection task is turned into a high-dimensional complexity problem when the number of involved features is elevated because the number of different combinations that are required to find the most suitable feature subset is denoted by 2^n, where n is the number of involved features. Consequently, the use of high-dimensional optimization algorithms is appropriate to address the feature selection problem.

2.4.1. Simulated Annealing

Simulated annealing (SA) is a metaheuristic that was abstracted from an industrial process. In the annealing process, the material is exposed to a certain high temperature, and after, a controlled cooling process is performed. Simultaneously, care must be taken in order to preserve certain molecular alignments in the material to ensure their quality. This process was adapted as a computational search technique by Kirpatrick et al. [34] to solve combinatorial and continuous optimization problems. The algorithm is governed by the T_{min}, T_{max} and T_{step} parameters, which refer to the minimum, maximum, and changing-step temperatures, respectively. At each iteration step, a new solution is generated based on a computed probability that involves the current temperature and the decreasing parameter ΔE, which represents the objective function response. The probability is calculated from the Boltzmann distribution as follows:

$$P(\Delta E, T) = \frac{f(s') - f(s)}{T},\tag{3}$$

where $P(\Delta E, T)$ is the probability computed from the Boltzmann distribution, and $f(s')$ and $f(s)$ denote the objective function value obtained with the current and the previous SA solution, respectively.

2.4.2. Boltzmann Univariate Marginal Distribution Algorithm (BUMDA)

BUMDA [35] is a population-based method that uses the estimation of distribution to generate new individuals. The main idea of BUMDA is the use of a distribution probability computed from the best solutions of the current generation in order to generate the new one [36]. The Boltzmann probability distribution used by BUMDA is calculated as follows:

$$\mu = \sum_j W(X_j)x_j, \text{ where } W(X_j) = \frac{g(X_j)}{\sum_{X_j} g(X_j)},\tag{4}$$

$$v = \sum_j W'(X_j)(X_j - \mu)^2, \text{ where } W'(X_j) = \frac{g(X_j)}{\sum_{X_j} g(X_j) + 1},\tag{5}$$

where μ is the objective function average, v is the objective function variance that was obtained from the population. $g(X_j)$ corresponds to the value of the objective function obtained by the individual j^{th}, which is an individual of the population X. Consequently, in order to generate the next population, a fraction of the current one that contains the best individuals is used to produce the new generation, as follows ($npop$ is the population size):

$$\theta^{t+1} = \begin{cases} f(x_{\text{npop}}) & \text{if } t = 1, \\ f(x_{\frac{\text{npop}}{2}}) & \text{if } f(x_{\frac{\text{npop}}{2}}) >= \theta^t, \\ f(x_i) & \text{when } f(x_i) >= \theta^t \end{cases} \Bigg|_{i=\frac{\text{npop}}{2}+1}^{\text{npop}}, \tag{6}$$

2.5. Machine Learning-Based Classifiers

Classifiers are useful for deciding if a specific instance belongs to part of one class or another. For the coronary stenosis classification problems, they are useful for determining if an image or a region over it corresponds to a positive stenosis case or a negative one.

2.5.1. K-Nearest Neighbor

K-nearest neighbor (KNN) represents a fast classification method that was first proposed by Evelyn Fix and Joseph Hodges in 1952 [37]. Later, in 1967, Thomas Cover and Peter E. Hart expanded the initial proposal by introducing the concept of *nearest neighbor* [38]. The KNN is governed only by the k parameter, which is a positive integer and indicates the number of associated nearest neighbors that a new instance will have in order to measure its probability of membership to different classes. The KNN inputs are labeled vectors in a multidimensional feature space. In the first stage, the training of the KNN model consists only of the storage of the feature vectors and their corresponding class. In the second stage, the KNN classifies new instances by measuring their frequency among the k nearest instances to determine the label of the new instance. In addition, the "nearest" term is associated with the similarity concept in which measurement is commonly based on a distance metric (commonly, Euclidean distance).

2.5.2. Support Vector Machine

Originally conceived as a linear separator for binary classification in supervised learning, the Support Vector Machine (SVM) faces challenges when dealing with instances characterized by significant data overlaps, making linear separability unattainable [39]. To address this issue, SVM has the capability of projecting instances from their original representation space into higher-dimensional orders, enabling successful classification [40]. To execute these projections, SVM leverages instances situated on both sides of the separation boundary, whether it be a line, plane, or hyperplane. The SVM is then formulated as follows [41]:

$$f(x) = W^T \phi(X) + b, \tag{7}$$

where W is the weight vector and normal to the hyperplane, ϕ is the projection function or kernel, b is the bias or threshold, and X is the data point to be classified.

3. Proposed Method

The proposed strategy consists of three stages. The first stage corresponds to the feature extraction in order to form a bank of 473 features involving intensity, shape, and texture types. For shape features, 8 vessel enhancement methods from the state-of-the-art were used. In the second stage, a subset of features is selected (feature selection) by using a hybrid-evolutionary algorithm in order to maximize the classification accuracy in training data while minimizing the number of features. In the final stage, the selected subset is tested for the classification of coronary stenosis using an independent test set of angiograms. In Figure 3, the steps of the proposed strategy are illustrated.

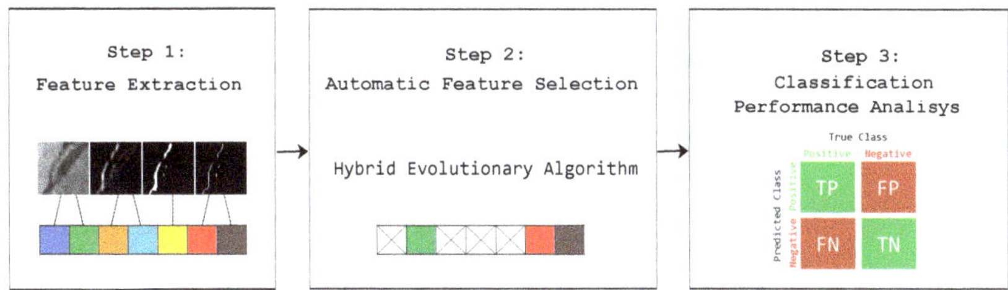

Figure 3. Proposed feature extraction, selection, and classification methods to classify coronary stenosis.

The bank of 473 features is formed as follows. Intensity-based features, such as minimum, maximum, median, mean, and standard deviation of the pixel intensities, are present in the original image. In addition, these features were also extracted from the responses of the different enhancement methods. In texture features, the Haralik [42] methodology was applied in order to obtain 14 distinct texture-related features. Consequently, 50 different shape features were extracted, including those used by Welikala [43]. Since shape feature values depend on the previous image enhancement process used, different methods were applied, such as those by Frangi et al. and Salem et al., as well as single and multi-scale Gaussian matched filtering, single and multi-scale Gabor filtering, linear multi-scale, and a multi-scale top-hat operator. All extracted features are described below.

Intensity-based Features

The intensity features correspond to the statistical measures of standard deviation, minimum, maximum, average, and median of the pixel intensities. Those features were computed from the original image and from the filter response of the 8 enhancement methods described previously. Consequently, 45 intensity-based features were extracted.

Texture Features

From the original image, the set of 14 texture features proposed by Haralik [42] were computed. In addition, this set of features was also applied to the Radon transform response. The total number of texture features is 28.

Shape Features

For shape features, 14 of them were extracted as described in the Welikala methodology [43], as follows:

1. The total number of vessel pixels.
2. The total number of vessel segments.
3. Vessel density.
4. Tortuosity.
5. The minimum vessel length.
6. The maximum vessel length.
7. The median vessel length.
8. The mean vessel length.
9. The standard deviation length.
10. The number of bifurcation points.
11. Gray level coefficient of variation.
13. Gradient mean.
14. Gradient coefficient of variation.

In addition, in the study by Gil et al. [23], 5 shape features were extracted considering continuous arterial sections and their corresponding segments delimited by tortuosity as follows:

1. The minimum standard deviation of the segments in length pixels considering all arterial sections. Since each arterial section is composed of continuous segments, it is possible to measure the length of each segment and compute the standard deviation

for each section. Therefore, if several arterial sections are present in the image, it is possible to obtain statistical measures over the arterial sections.

2. The maximum standard deviation of segments in length pixels considering all arterial sections.
3. The median standard deviation of the segments in length pixels considering all arterial sections.
4. The average of standard deviations of the segments in length pixels considering all arterial sections.
5. The variance of the standard deviations of the segments in length pixels considering all arterial sections.

In addition, 25 shape features were computed as following:

1. Minimum perimeter. The perimeter of an arterial section is the length of its boundary.
2. Maximum perimeter.
3. Median perimeter.
4. Mean perimeter.
5. Standard deviation of the perimeters.
6. Minimum compactness. It can be computed as follows:

$$\text{Compactness} = \frac{\text{Perimeter}^2}{\text{Area}}. \tag{8}$$

7. Maximum compactness.
8. Median compactness.
9. Mean compactness.
10. Standard deviation of compactness.
11. Minimum circularity ratio. It can be computed as follows:

$$\text{Circularity Ratio} = \frac{4 \cdot \pi \cdot \text{Area}}{\text{Perimeter}^2}. \tag{9}$$

Similar to previous measures, for images containing several arterial sections, it is possible to compute circularity for each section and obtain statistical measurements.
12. Maximum circularity ratio.
13. Median circularity ratio.
14. Mean circularity ratio.
15. Standard deviations of the circularity ratios.
16. Minimum rectangularity. It can be computed as follows:

$$\text{Rectangularity} = \frac{\text{Area of Arterial Region}}{\text{Area of Bounding Rectangle of Arterial Region}}. \tag{10}$$

17. Maximum rectangularity.
18. Median rectangularity.
19. Mean rectangularity.
20. Standard deviation of rectangularities.
21. Minimum elongatedness. It can be computed as follows:

$$\text{Elongatedness} = \frac{l}{w}, \tag{11}$$

where l is the arterial section length in pixels, and w represents the vessel width in pixels.
22. Maximum elongatedness.
23. Median elongatedness.
24. Mean elongatedness.
25. Standard deviation of elongatedness.

Finally, 6 shape-density features were also extracted.

1. Minimum vessel pixel density of all arterial sections present in the patch.
2. Maximum vessel pixel density.
3. Median vessel pixel density.
4. Mean vessel pixel density.
5. Standard deviation of the vessel pixel densities.
6. Sum of the vessel pixel densities of all arterial sections.

A set with 50 distinct shape-based features was described previously. Since shape feature values are dependent on the applied enhancement method, they were extracted from the 8 responses corresponding to each applied enhancement method in order to extract a set with 400 shape-related features.

After the feature extraction process is concluded, a numeric feature dataset is generated and partitioned randomly into training and testing instances in a balanced manner. The feature selection task is performed on the training dataset, and it is turned into a search process that is conducted by the hybrid-evolutionary algorithm involving the BUMDA and SA metaheuristics.

Since the total number of extracted features is 473, the identification of an optimal feature subset using an exhaustive search process involves a computational cost of $O(2^{473})$, which is highly difficult to perform. By involving a single search evolutionary method, the problem can be solved partially. However, due to the high-dimensional complexity of the problem, it is possible to improve the solution achieved by the evolutionary method at each iteration, applying a refined search. This will lead to the use of a hybrid-evolutionary method in which the main goal of feature selection is to identify an optimal subset that improves classification accuracy and reduces model complexity.

The proposed method for the automatic feature selection task is formed by the BUMDA and the SA strategies. The use of BUMDA is adequate because the use of the Boltzmann distribution produces spread populations in comparison with the UMDA and other related techniques [35], which also decreases the risk of falling into local-optima solutions. In addition, since the SA algorithm is a single-solution search method, it is suitable to improve the best solution produced by the BUMDA at each iteration in a refined search step. The combination of these metaheuristics produces a hybrid-evolutionary method focused on the search for the best suitable feature subset by considering the minimization of its size and the maximization of the accuracy classification performance.

Since the feature selection task involves the evaluation of each feature subset produced by the search techniques, a classification model must be trained. By considering that BUMDA is governed by the population size (ps) and the number of generations (ng), and the SA is governed by the number of iterations (ni) computed from the initial, final, and step temperature, the total number of classification models to train is computed as $ps \times ng \times ni$. For instance, if $ps = 100$, $ng = 1000$, and $ni = 1000$, the total number of different classifiers to be evaluated is 1×10^8. Consequently, a fast-convergent classification method such as the K-nearest neighbor is suitable in this stage. Subsequently, when the feature selection stage is completed, a more complex classification technique, such as the SVM, can be used in a testing stage to improve the classification performance.

Figure 4 illustrates the flowchart corresponding to the proposed hybrid-evolutionary method.

To evaluate the feature selection performance of the proposed method, the feature decreasing rate (FDR) metric was used and it can be computed as follows:

$$\text{FDR} = 1 - \frac{\text{Number of Selected Features}}{\text{Total Number of Features}}. \tag{12}$$

The FDR metric is used as a way to measure the feature selection performance. When the number of selected features decreases, the FDR increases, allowing the hybrid-evolutionary method to conduct the search for an optimal feature subset by maximizing the FDR.

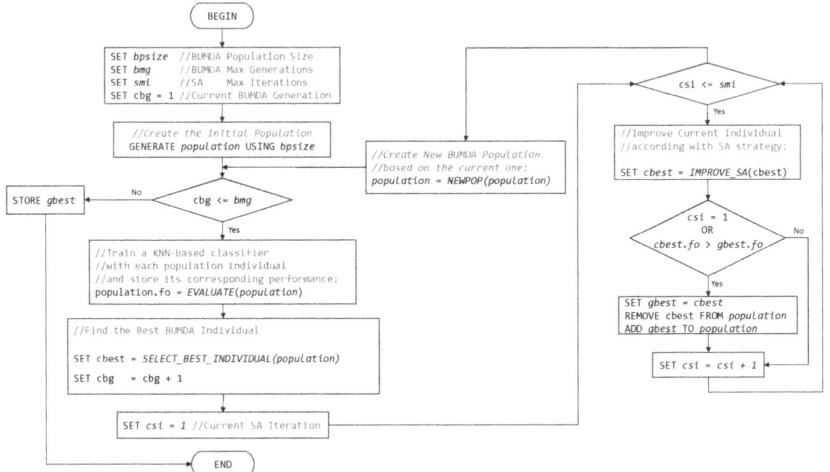

Figure 4. Flowchart of the proposed hybrid method to perform automatic feature selection.

The final step of the proposed method is the classification of an independent test set using the subset of selected features. In the experiments, the first database of 608 images was divided into training (508 images) and testing (100 images) sets. Both datasets use a balance of positive and negative cases. Each patch is in a grayscale color scheme and their corresponding size is 64×64 pixels. In Figure 5, a subset of sample images, along with its filter response, are presented.

In order to evaluate the obtained results, a dataset was formed from the second image database described in Section 2.1. The dataset was divided into the training set with $\approx 65\%$ (1828 instances) and testing set (960 instances). In Figure 6, sample patches corresponding to the second dataset, are presented.

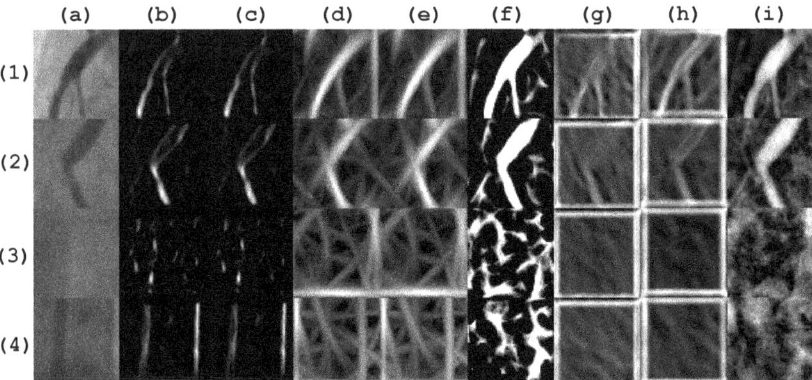

Figure 5. Patch samples from a database of 608 images. Rows (**1**) and (**2**) correspond to positive stenosis cases. Rows (**3**) and (**4**) correspond to negative stenosis cases. Column (**a**) corresponds to the original patch image. Columns (**b**–**i**) correspond to the different vessel enhancement method responses, as follows: Frangi, Salem, simple-scale Gabor, multi-scale Gabor, multi-scale linear, multi-scale matched filter, single-scale matched filter, and top-hat operator.

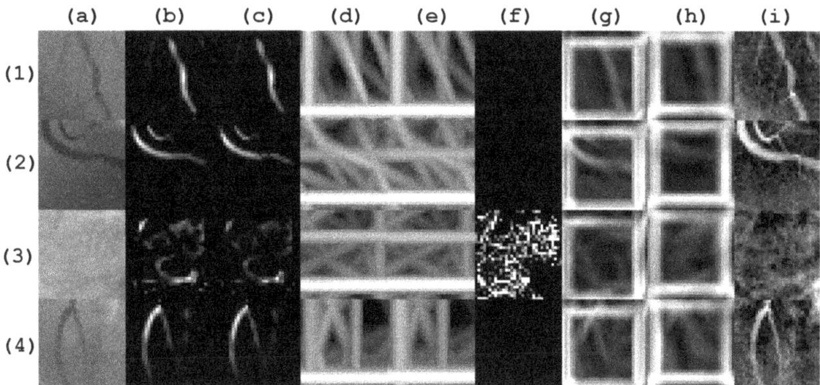

Figure 6. Patch samples from the second database of 2788 images. Rows (**1**) and (**2**) correspond to positive stenosis cases. Rows (**3**) and (**4**) correspond to negative stenosis cases. Column (**a**) corresponds to the original patch image. Columns (**b–i**) correspond to the different enhancement method responses as follows: Frangi, Salem, simple-scale Gabor, multi-scale Gabor, multi-scale linear, multi-scale matched filter, single-scale matched filter, and top-hat operator.

To determine the classification performance, the accuracy and the Jaccard Coefficient metrics were used. The accuracy metric can be calculated as follows:

$$Acc = \frac{TP + TN}{TP + TN + FP + FN},\tag{13}$$

where Acc is the classification accuracy, TP is the fraction of positive cases classified correctly, TN is the fraction of negative cases classified correctly, FP is the fraction of negative cases classified as positive, and FN represents the fraction of positive cases classified as negative. Moreover, the Jaccard coefficient (JC) is calculated as follows:

$$JC = \frac{TP}{TP + FP + FN},\tag{14}$$

The accuracy and the JC metrics are convenient for measuring binary classification performance. The accuracy metric allows us to know how well the classification of positive and negative stenosis cases is performed. However, in coronary stenosis classification, it is important to know the performance of positive coronary stenosis classification, which is evaluated by the JC metric. These two metrics are the most commonly used in image binary classification problems.

4. Results and Discussion

In this section, the results achieved in each stage of the proposed method are presented and discussed. Moreover, they are compared with other methods from the literature. All the experiments were implemented in MATLAB software version 2018 and executed on a computer with an Intel Core i7 processor and 8 GB of RAM.

In the first stage, a bank of 473 features was extracted. This stage involved the original images as well as the responses of the different vessel enhancement methods applied to them. In Table 1, a summary of the distinct feature types extracted is described.

After the feature extraction step was concluded, an optimal subset of features was obtained by applying a hybrid-evolutionary algorithm. The feature subset was evaluated in terms of the training accuracy and FDR metrics. Furthermore, the obtained results were compared with other search methods from the literature. In Table 2, different search methods are described with their corresponding parameter settings, including a statistical analysis of the FDR for each of them.

Table 1. Summary of the different feature types extracted from the original images and their corresponding vessel-enhancement responses.

Feature Type	Require the Original Image	Require Vessel-Enhancement	Quantity
Intensity	Yes	Yes	45
Texture	Yes	No	28
Shape	Yes	Yes	400
		Total Extracted Features:	**473**

Table 2. Parameter settings and statistical analysis of the feature decreasing rate (FDR) for the feature selection performance of distinct search metaheuristics of 30 independent trials.

Method	Pop. Size	Max. Gens.	Min. FDR	Max. FDR	Median FDR	Mean FDR	Std. Dev. FDR
UMDA	100	1000	0.48	0.98	0.92	0.83	0.17
BUMDA	100	1000	0.57	0.50	0.57	0.53	0.02
GA	100	1000	0.94	0.91	0.94	0.93	0.01
SA [1]	-	10000	0.87	0.97	0.92	0.92	0.03
Hybrid Evolutionary [2]	100	100/5000 [1]	0.92	0.99	0.97	0.97	0.03

[1] Simulated annealing ($T_{max} = 1, T_{min} = 0, T_{step} = 0.0001$). [2] BUMDA was configured with 100 maximum number of generations. Subsequently, the inner SA strategy was configured with ($T_{max} = 1, T_{min} = 0, T_{step} = 0.005$). In total, the hybrid-evolutionary strategy iterated 5×10^5 times.

By considering the results described in Table 2, the highest FDR was obtained using the hybrid-evolutionary strategy, which uses the smallest feature subset of four shape features, as follows:

- Mean Intensity extracted from the Frangi method response.
- Average standard deviation of the segments in length pixels for all arterial sections, extracted from the Frangi filter response.
- Gradient mean extracted from the linear multi-scale method response.
- Gradient coefficient of variation extracted from the top-hat method response.

In addition, the training classification accuracy performance was also considered. For almost all techniques, the FDR variation was small, which means that the search behavior was stable. By involving the FDR and the training classification accuracy, a performance evaluation was conducted. Figure 7 illustrates the performance chart of the best trial for each compared strategy considering the FDR and the training accuracy.

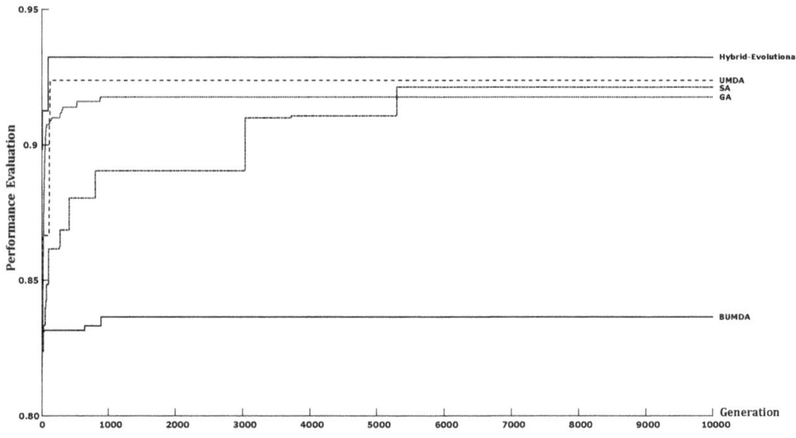

Figure 7. Performance evaluation chart considering the FDR and the training accuracy for selected metaheuristics: UMDA, BUMDA, GA, SA, and the hybrid-evolutionary method.

The classification performance of each method was measured using the testing set. In Table 3, the performance obtained using the testing dataset, which corresponds to the first database, is described in terms of the accuracy and Jaccard coefficient metrics. In addition to the KNN-based classifier, an SVM was also trained in order to measure the performance of the selected features under different classifiers. It was established that there was a maximum number of 1000 iterations for KNN and the SVM and a cross-validation with $k = 10$. The values for those parameters were chosen as a tradeoff between an optimal result and the computational time.

Table 3. Description of the test results for each compared strategy using the KNN and the SVM classifiers, including the number of selected features (NSF) with their corresponding feature decreasing rate (FDR), the achieved classification accuracy, and Jaccard coefficient (JC), using the first image database containing 100 balanced instances.

Method	NSF	FDR	Classifier	Accuracy	JC
GLNet [44]	–	–	–	0.85	**0.76**
UNet [45]	–	–	–	0.85	0.75
CNN-16C [11]	–	–	–	0.86	**0.76**
UMDA [46]	10	0.98	KNN	0.81	0.75
			SVM	0.80	0.67
BUMDA	205	0.57	KNN	0.81	0.70
			SVM	0.82	0.72
GA	29	0.94	KNN	0.79	0.65
			SVM	0.79	0.66
SA	16	0.97	KNN	0.81	0.65
			SVM	0.85	0.75
Hybrid-Evolutionary	4	0.99	KNN	**0.87**	**0.76**
			SVM	**0.86**	0.75

The results described in Table 3 shows that the highest rate, in terms of the accuracy and Jaccard coefficient was obtained using the subset of 4 features selected by the hybrid-evolutionary method along with the KNN and the SVM classifiers. Correspondingly, using only 4 of the 473 features represents a discrimination rate of 0.99% over the initial bank of features. Moreover, the obtained results are similar to those achieved using the Deep Learning CNN-16C and GLNet architectures, which exhibits the robustness of the proposed strategy.

To evaluate the efficiency of the previously discussed results, an additional dataset formed from the publically available database [11], was used for training and testing a KNN and an SVM with the bank of 4 features described in Table 4.

Table 4 shows that the highest classification rate in terms of the accuracy and Jaccard coefficient metrics was also achieved by the proposed method, along with the SVM-based classifier. This result is relevant because it shows that the SVM classification performance is competitive using the bank of four features. Similarly, results obtained using the KNN-based classifier were close to those achieved by the SVM, also using the $4D$ feature vector. Consequently, the accuracy and Jaccard coefficient rates show how competitive the feature subset was in classifying stenosis cases. An analysis of the frequency for each of the selected features is described in Figure 8.

Table 4. Description of the testing results for each compared strategy using the KNN and the SVM classifiers, including the number of selected features (NSF) with their corresponding feature decreasing rate (FDR), the achieved classification accuracy and Jaccard coefficient, using a dataset formed from the Antczak image database, containing 960 balanced instances.

Method	NSF	FDR	Classifier	Accuracy	JC
GLNet [44]	–	–	–	0.72	0.63
UNet [45]	–	–	–	0.76	0.72
CNN-16C [11]	–	–	–	0.86	0.74
UMDA [46]	10	0.98	KNN	0.85	0.75
			SVM	0.83	0.71
BUMDA	205	0.57	KNN	0.86	0.77
			SVM	0.74	0.56
GA	29	0.94	KNN	0.85	0.75
			SVM	0.87	0.78
SA	16	0.97	KNN	0.85	0.75
			SVM	0.85	0.75
Hybrid Evolutionary	4	0.99	KNN	0.84	0.74
			SVM	**0.89**	**0.80**

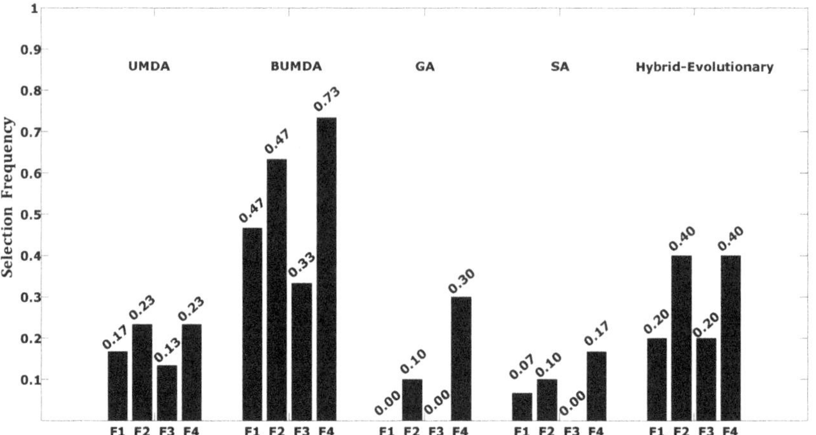

Figure 8. Frequencies for each selected feature in the best solution achieved by the hybrid-evolutionary method contrasted with the frequency for that same feature in the best solution achieved by the other methods. Selected features are as follows: Frangi filter—mean intensity (F1), Frangi filter—mean std. dev. of segments length in all arterial sections (F2), linear multi-scale filter—gradient mean (F3), top-hat—gradient coefficient of variation (F4).

According to Figure 8, almost all selected features by the hybrid-evolutionary method have a presence in the other contrasted search methods, which provides evidence of their relevance in the classification process.

Based on Figure 9, it is relevant that two of the four features that were selected by the hybrid-evolutionary strategy are in the group of features with the highest selection probabilities considering all trials of all compared techniques. This finding is important because it statistically validates the relevance of the final selected 4D feature vector for the automatic classification of coronary stenosis cases. In addition, the mean time required for the classification of a single testing instance was $\approx$0.02 s.

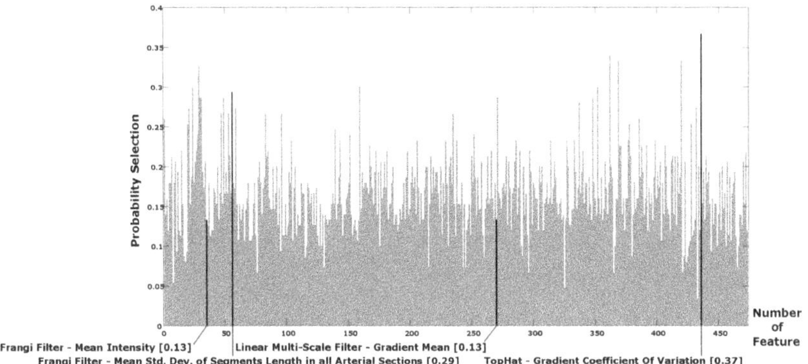

Figure 9. Selection probability for all 473 features based on the frequency in which each of them was selected in all trials considering all contrasted strategies.

Even when the achieved results were competitive, the weak vessel contrast present in almost all coronary angiograms and the continuous heart movement decreased the response accuracy of vessel enhancement methods, leading to classification errors. Since the human body is composed of a variety of molecular substances, the reaction of the contrast medium to the X-rays is not enough to produce an optimal artery distinction in the majority of the coronary angiograms. However, although technological innovations have been produced to improve the generation of coronary angiograms [47], the elevated cost of medical imaging devices makes it difficult to adopt in the short term.

One of the strengths of the proposed methods is the identification of a subset with relevant features, which allows performing a positive and negative classification of coronary stenosis cases with a high accuracy rate. However, the conformation of the feature bank can be a limitation in the proposed method and also a robust discrete optimization technique for high-dimensional problems.

After the testing stage was finished, the obtained results with the proposed method achieved a high classification performance in terms of the accuracy and the JC metrics using only a $4D$ feature vector. For instance, using the proposed method with the first testing set, accuracy and JC rates of 0.87 and 0.76 were achieved, respectively. It is relevant that the obtained accuracy was slightly higher than that achieved by the deep learning model proposed by Antczak et al., which was 0.86. In addition, the hybrid-evolutionary strategy results also overcome to those obtained using single search metaheuristics. The single BUMDA and SA methods achieved a classification accuracy performance of 0.82 and 0.85 in their best result. Accordingly, the single SA technique was close to the best accuracy result. However, the single SA found a $15D$ feature vector with respect to the $4D$ feature vector found using the hybrid-evolutionary strategy, which shows how the hybrid approach was relevant to producing an optimal feature vector in terms of size and classification performance. Furthermore, when the second testing dataset was used to measure the classification performance, accuracy and JC rates of 0.89 and 0.80 were achieved, respectively, which surpassed those obtained by all the compared methods, including the results obtained with deep learning techniques (0.86 and 0.74 for accuracy and JC metrics) and single metaheuristics.

5. Conclusions

In this paper, a novel strategy consisting of three stages was presented for the automatic classification of positive and negative coronary stenosis cases. In the first step, a bank of 473 features was formed, which represents a computational search complexity of $O(2^{473})$, which was explored using different search strategies from the literature. Results achieved in this stage were relevant since only 4 of the 473 features (decreasing rate of 0.99) were selected using the hybrid-evolutionary method across all of the other methods compared. It was also relevant that three of the four features are of shape type: Frangi filter—average of the standard deviations of the segments length across all arterial sections, the linear multi-scale filter—gradient mean, and the tophat—gradient coefficient of variation. The remaining feature is intensity type and corresponds to the mean intensity, which is also extracted from the response of the Frangi enhancement method. It is relevant that none of the selected features were obtained from the original image directly. In the testing stage, the proposed method achieved the highest classification performance in terms of accuracy and JC metrics. For the first testing set, the highest accuracy and JC rates were 0.87 and 0.86, respectively. Correspondingly, with the second dataset, the highest performance in terms of accuracy and JC was 0.84 and 0.89, respectively. Additionally, taking into account the computational time and classification accuracy of the proposed method based on four selected features, it can be a potential method as part of a computer-aided diagnosis system in cardiology. Finally, future work can be conducted in fast convergence of feature selection for high-dimensional spaces in order to improve the computational time without decreasing the classification performance rate.

Author Contributions: Conceptualization, M.-A.G.-R., I.C.-A., M.-A.H.-G. and S.-E.S.-M.; formal analysis, M.-A.H.-G.; investigation, S.-E.S.-M.; methodology, M.-A.G.-R., I.C.-A., E.M.-A. and J.B.; project administration, I.C.-A. and A.H.-A.; software, M.-A.G.-R. and A.H.-A.; validation, M.-A.G.-R., I.C.-A., M.-A.H.-G. and S.-E.S.-M.; visualization, E.M.-A. and J.B.; writing—original draft, M.-A.G.-R. and I.C.-A. All authors have read and agreed to the published version of the manuscript.

Funding: This work was supported by CONACyT under Project IxM-CONACyT No. 3150-3097. Ernesto Moya-Albor and Jorge Brieva would like to thank Facultad de Ingeniería of Universidad Panamericana for supporting this work. The APC was partially funded by Facultad de Ingeniería, Universidad Panamericana.

Institutional Review Board Statement: Not applicable.

Informed Consent Statement: Not applicable.

Data Availability Statement: The first database presented in this article are not readily available because ethic and legal restrictions. The second database is not our authorship. It is publicly available at: https://github.com/KarolAntczak/DeepStenosisDetection.

Conflicts of Interest: The authors declare no conflicts of interest.

Abbreviations

The following abbreviations are used in this manuscript:

BUMDA	Boltzmann univariate marginal distribution algorithm
FDR	feature decreasing rate
GA	genetic algorithm
KNN	K-nearest neighbor
NSF	number of selected features
SA	simulated annealing
SVM	support vector machine
UMDA	univariate marginal distribution algorithm

References

1. Duggan, J.P.; Peters, A.S.; Trachiotis, G.D.; Antevil, J.L. Epidemiology of Coronary Artery Disease. *Surg. Clin.* **2022**, *102*, 499–516. [CrossRef] [PubMed]
2. British-Heart-Foundation. Global Hearth and Cirsculatory Diseases Factsheet. Available online: https://www.bhf.org.uk/-/media/files/for-professionals/research/heart-statistics/bhf-cvd-statistics-global-factsheet.pdf?rev=f323972183254ca0a10436 83a9707a01&hash=5AA21565EEE5D85691D37157B31E4AAA (accessed on 8 January 2024).
3. Frąk, W.; Wojtasińska, A.; Lisińska, W.; Młynarska, E.; Franczyk, B.; Rysz, J. Pathophysiology of Cardiovascular Diseases: New Insights into Molecular Mechanisms of Atherosclerosis, Arterial Hypertension, and Coronary Artery Disease. *Biomedicines* **2022**, *10*, 1938. [CrossRef] [PubMed]

4. Saad, I.A. Segmentation of Coronary Artery Images and Detection of Atherosclerosis. *J. Eng. Appl. Sci.* **2018**, *13*, 7381–7387. [CrossRef]

5. Kishore, A.N.; Jayanthi, V. Automatic stenosis grading system for diagnosing coronary artery disease using coronary angiogram. *Int. J. Biomed. Eng. Technol.* **2019**, *31*, 260–277. [CrossRef]

6. Brieva, J.; Gálvez, M.; Toumoulin, C. Coronary extraction and stenosis quantification in X-ray angiographic imaging. In Proceedings of the The 26th Annual International Conference of the IEEE Engineering in Medicine and Biology Society, San Francisco, CA, USA, 1–5 September 2004; Volume 26, pp. 1714–1717.

7. Taki, A.; Roodaki, A.; Setahredan, S.K.; Zoroofi, R.A.; Konig, A.; Navab, N. Automatic segmentation of calcified plaques and vessel borders in IVUS images. *Int. J. Comput. Assist. Radiol. Surg.* **2008**, *2008*, 347–354. [CrossRef]

8. Cruz-Aceves, I.; Cervantes-Sanchez, F.; Hernandez-Aguirre, A. Automatic Detection of Coronary Artery Stenosis Using Bayesian Classification and Gaussian Filters Based on Differential Evolution. In *Hybrid Intelligence for Image Analysis and Understanding*; John Wiley & Sons, Ltd.: Hoboken, NJ, USA, 2017; Chapter 16; pp. 369–390. [CrossRef]

9. Giannoglou, V.G.; Stavrakoudis, D.G.; Theocharis, J.B. IVUS-based characterization of atherosclerotic plaques using feature selection and SVM classification. In Proceedings of the 2012 IEEE 12th International Conference on Bioinformatics & Bioengineering (BIBE), Larnaca, Cyprus, 11–13 November 2012; pp. 715–720. [CrossRef]

10. Chen, X.; Fu, Y.; Lin, J.; Ji, Y.; Fang, Y.; Wu, J. Coronary Artery Disease Detection by Machine Learning with Coronary Bifurcation Features. *Appl. Sci.* **2020**, *10*, 7656. [CrossRef]

11. Antczak, K.; Liberadzki, Ł. Stenosis Detection with Deep Convolutional Neural Networks. *MATEC Web Conf.* **2018**, *210*, 04001. [CrossRef]

12. Garcea, F.; Serra, A.; Lamberti, F.; Morra, L. Data augmentation for medical imaging: A systematic literature review. *Comput. Biol. Med.* **2023**, *152*, 106391. [CrossRef]

13. Bas, H.; Hugo, J.K.; Gilhuijs, K.G.; Viergever, M.A. Explainable artificial intelligence (XAI) in deep learning-based medical image analysis. *Med. Image Anal.* **2022**, *79*, 102470. [CrossRef]

14. Li, Y.; Wang, S.; Tian, Q.; Ding, X. A survey of recent advances in visual feature detection. *Neurocomputing* **2015**, *149*, 736–751. [CrossRef]

15. Tessmann, M.; Vega-Higuera, F.; Fritz, D.; Scheuering, M.; Greiner, G. Multi-scale feature extraction for learning-based classification of coronary artery stenosis. In Proceedings of the Medical Imaging 2009: Computer-Aided Diagnosis, Lake Buena Vista, FL, USA, 10–12 February 2009; Karssemeijer, N., Giger, M.L., Eds.; International Society for Optics and Photonics, SPIE: Bellingham, WA, USA, 2009; Volume 7260, p. 726002. [CrossRef]

16. Fazlali, H.R.; Karimi, N.; Soroushmehr, S.M.R.; Sinha, S.; Samavi, S.; Nallamothu, B.; Najarian, K. Vessel region detection in coronary X-ray angiograms. In Proceedings of the 2015 IEEE International Conference on Image Processing (ICIP), Quebec City, QC, Canada, 27–30 September 2015; pp. 1493–1497. [CrossRef]

17. Acharya, U.R.; Sree, S.V.; Krishnan, M.M.R.; Molinari, F.; Saba, L.; Ho, S.Y.S.; Ahuja, A.T.; Ho, S.C.; Nicolaides, A.; Suri, J.S. Atherosclerotic Risk Stratification Strategy for Carotid Arteries Using Texture-Based Features. *Ultrasound Med. Biol.* **2022**, *38*, 899–915. [CrossRef] [PubMed]

18. Ricciardi, C.; Valente, A.S.; Edmunds, K.; Cantoni, V.; Green, R.; Fiorillo, A.; Picone, I.; Santini, S.; Cesarelli, M. Linear discriminant analysis and principal component analysis to predict coronary artery disease. *Health Inform. J.* **2020**, *26*, 2181–2192. [CrossRef] [PubMed]

19. Barburiceanu, S.; Terebes, R.; Meza, S. 3D Texture Feature Extraction and Classification Using GLCM and LBP-Based Descriptors. *Appl. Sci.* **2021**, *11*, 2332. [CrossRef]

20. Murphy, L.M. Linear feature detection and enhancement in noisy images via the Radon transform. *Pattern Recognit. Lett.* **1986**, *4*, 279–284. [CrossRef]

21. Cruz-Aceves, I.; Oloumi, F.; Rangayyan, R.M.; Aviña-Cervantes, J.G.; Hernandez-Aguirre, A. Automatic segmentation of coronary arteries using Gabor filters and thresholding based on multiobjective optimization. *Biomed. Signal Process. Control* **2016**, *25*, 76–85. [CrossRef]

22. Otsu, N. A Threshold Selection Method from Gray-Level Histograms. *IEEE Trans. Syst. Man Cybern.* **1979**, *9*, 62–66. Available online: https://cw.fel.cvut.cz/b201/_media/courses/a6m33bio/otsu.pdf (accessed on 8 January 2024). [CrossRef]

23. Gil-Rios, M.A.; Chalopin, C.; Cruz-Aceves, I.; Lopez-Hernandez, J.M.; Hernandez-Gonzalez, M.A.; Solorio-Meza, S.E. Automatic Classification of Coronary Stenosis Using Feature Selection and a Hybrid Evolutionary Algorithm. *Axioms* **2023**, *12*, 462. [CrossRef]

24. Frangi, A.; Nielsen, W.; Vincken, K.; Viergever, M. Multiscale vessel enhancement filtering. In Proceedings of the Medical Image Computing and Computer-Assisted Intervention—MICCAI'98: First International Conference, Cambridge, MA, USA, 11–13 October 1998; Volume 1496, pp. 130–137. [CrossRef]

25. Salem, N.M.; Salem, S.A.; Nandi, A.K. Segmentation of retinal blood vessels based on analysis of the hessian matrix and Clustering Algorithm. In Proceedings of the 2007 15th European Signal Processing Conference, Poznan, Poland, 3–7 September 2007; pp. 428–432.

26. Eiho, S.; Qian, Y. Detection of coronary artery tree using morphological operator. In Proceedings of the Computers in Cardiology 1997, Lund, Sweden, 7–10 September 1997; pp. 525–528. [CrossRef]

27. Qian, Y.; Eiho, S.; Sugimoto, N.; Fujita, M. Automatic extraction of coronary artery tree on coronary angiograms by morphological operators. In Proceedings of the Computers in Cardiology 1998. Vol. 25 (Cat. No.98CH36292), Cleveland, OH, USA, 13–16 September 1998; pp. 765–768. [CrossRef]

28. Kang, W.; Kang, W.; Li, Y.; Wang, Q. The segmentation method of degree-based fusion algorithm for coronary angiograms. In Proceedings of the 2013 2nd International Conference on Measurement, Information and Control, Harbin, China, 16–18 August 2013; Volume 01, pp. 696–699. [CrossRef]

29. Ricci, E.; Perfetti, R. Retinal Blood Vessel Segmentation Using Line Operators and Support Vector Classification. *IEEE Trans. Med. Imaging* **2007**, *26*, 1357–1365. [CrossRef]

30. Cruz-Aceves, I.; Cervantes-Sanchez, F.; Avila-Garcia, M.S. A Novel Multiscale Gaussian-Matched Filter Using Neural Networks for the Segmentation of X-Ray Coronary Angiograms. *J. Healthc. Eng.* **2018**, *2018*, 5812059. [CrossRef] [PubMed]

31. Gabor, D. Theory of communication. Part 1: The analysis of information. *J. Inst. Electr.-Eng. Part Iii Radio Commun. Eng.* **1946**, *93*, 429–441. [CrossRef]

32. Rangayyan, R.M.; Oloumi, F.; Oloumi, F.; Eshghzadeh-Zanjani, P.; Ayres, F.J. Detection of Blood Vessels in the Retina Using Gabor Filters. In Proceedings of the 2007 Canadian Conference on Electrical and Computer Engineering, Vancouver, BC, Canada, 22–26 April 2007; pp. 717–720. [CrossRef]

33. Rangayyan, R.M.; Ayres, F.J.; Oloumi, F.; Oloumi, F.; Eshghzadeh-Zanjani, P. Detection of blood vessels in the retina with multiscale Gabor filters. *J. Electron. Imaging* **2008**, *17*, 023018. [CrossRef]

34. Kirkpatrick, S.; Gelatt, C.D.; Vecchi, M.P. Optimization by simulated annealing. *Science* **1983**, *220*, 671–680. [CrossRef]

35. Valdez-Peña, S.I.; Hernández, A.; Botello, S. A Boltzmann based estimation of distribution algorithm. *Inf. Sci.* **2013**, *236*, 126–137. [CrossRef]

36. Dang, D.C.; Lehre, P.K.; Nguyen, P.T.H. Level-Based Analysis of the Univariate Marginal Distribution Algorithm. *Algorithmica* **2017**, *81*, 668–702. [CrossRef]

37. Fix, E.; Hodges, J.J.L. *Discriminatory Analysis—Nonparametric Discrimination: Small Sample Performance*; Technical Report; Air University, USAF School of Aviation Medecine: San Antonio, TX, USA, 1952; pp. 13–37.

38. Cover, T.; Hart, P. Nearest neighbor pattern classification. *IEEE Trans. Inf. Theory* **1967**, *13*, 21–27. [CrossRef]

39. Cortes, C.; Vapnik, V. Support-Vector Networks. *Mach. Learn.* **1995**, *20*, 273–297. [CrossRef]

40. Noble, W.S. What is a support vector machine? *Nat. Biotechnol.* **2006**, *24*, 1565–1567. [CrossRef] [PubMed]

41. Hari, B.N.; Salankar, S.S.; Bora, V.R. MRI brain cancer classification using Support Vector Machine. In Proceedings of the 2014 IEEE Students' Conference on Electrical, Electronics and Computer Science, Bhopal, India, 1–2 March 2014; pp. 1–6. [CrossRef]

42. Haralick, R.; Shanmugam, K.; Dinstein, I. Textural Features for Image Classification. *IEEE Trans. Syst. Man Cybern.* **1973**, *SMC-3*, 610–621. [CrossRef]

43. Welikala, R.; Fraz, M.; Dehmeshki, J.; Hoppe, A.; Tah, V.; Mann, S.; Williamson, T.; Barman, S. Genetic algorithm based feature selection combined with dual classification for the automated detection of proliferative diabetic retinopathy. *Comput. Med. Imaging Graph.* **2015**, *43*, 64–77. [CrossRef] [PubMed]

44. Szegedy, C.; Liu, W.; Jia, Y.; Sermanet, P.; Reed, S.; Anguelov, D.; Erhan, D.; Vanhoucke, V.; Rabinovich, A. Going Deeper with Convolutions. *arXiv* **2014**, arXiv:1409.4842 [CrossRef]

45. Harouni, A.; Karargyris, A.; Negahdar, M.; Beymer, D.; Syeda-Mahmood, T. Universal multi-modal deep network for classification and segmentation of medical images. In Proceedings of the 2018 IEEE 15th International Symposium on Biomedical Imaging (ISBI 2018), Washington, DC, USA, 4–7 April 2018; pp. 872–876. [CrossRef]

46. Gil-Rios, M.A.; Guryev, I.V.; Cruz-Aceves, I.; Avina-Cervantes, J.G.; Hernandez-Gonzalez, M.A.; Solorio-Meza, S.E.; Lopez-Hernandez, J.M. Automatic Feature Selection for Stenosis Detection in X-ray Coronary Angiograms. *Mathematics* **2021**, *9*, 2471. [CrossRef]

47. Hajhosseiny, R.; Bustin, A.; Munoz, C.; Rashid, I.; Cruz, G.; Manning, W.J.; Prieto, C.; Botnar, R.M. Coronary Magnetic Resonance Angiography: Technical Innovations Leading Us to the Promised Land? *JACC Cardiovasc. Imaging* **2020**, *13*, 2653–2672. [CrossRef] [PubMed]

Article

Machine Learning-Based Predictive Models for Detection of Cardiovascular Diseases

Adedayo Ogunpola [1], Faisal Saeed [1,*], Shadi Basurra [1], Abdullah M. Albarrak [2] and Sultan Noman Qasem [2]

[1] DAAI Research Group, College of Computing and Digital Technology, Birmingham City University, Birmingham B4 7XG, UK; adedayo.ogunpola@mail.bcu.ac.uk (A.O.); shadi.basurra@bcu.ac.uk (S.B.)

[2] Computer Science Department, College of Computer and Information Sciences, Imam Mohammad Ibn Saud Islamic University (IMSIU), Riyadh 11432, Saudi Arabia; amsbarrak@imamu.edu.sa (A.M.A.); snmohammed@imamu.edu.sa (S.N.Q.)

* Correspondence: faisal.saeed@bcu.ac.uk

Abstract: Cardiovascular diseases present a significant global health challenge that emphasizes the critical need for developing accurate and more effective detection methods. Several studies have contributed valuable insights in this field, but it is still necessary to advance the predictive models and address the gaps in the existing detection approaches. For instance, some of the previous studies have not considered the challenge of imbalanced datasets, which can lead to biased predictions, especially when the datasets include minority classes. This study's primary focus is the early detection of heart diseases, particularly myocardial infarction, using machine learning techniques. It tackles the challenge of imbalanced datasets by conducting a comprehensive literature review to identify effective strategies. Seven machine learning and deep learning classifiers, including K-Nearest Neighbors, Support Vector Machine, Logistic Regression, Convolutional Neural Network, Gradient Boost, XGBoost, and Random Forest, were deployed to enhance the accuracy of heart disease predictions. The research explores different classifiers and their performance, providing valuable insights for developing robust prediction models for myocardial infarction. The study's outcomes emphasize the effectiveness of meticulously fine-tuning an XGBoost model for cardiovascular diseases. This optimization yields remarkable results: 98.50% accuracy, 99.14% precision, 98.29% recall, and a 98.71% F1 score. Such optimization significantly enhances the model's diagnostic accuracy for heart disease.

Keywords: cardiovascular diseases; deep learning; disease detection; heart diseases; machine learning; ensemble learning; XGBoost

Citation: Ogunpola, A.; Saeed, F.; Basurra, S.; Albarrak, A.M.; Qasem, S.N. Machine Learning-Based Predictive Models for Detection of Cardiovascular Diseases. *Diagnostics* **2024**, *14*, 144. https://doi.org/10.3390/diagnostics14020144

Academic Editor: Mugahed A. Al-antari

Received: 27 November 2023
Revised: 21 December 2023
Accepted: 25 December 2023
Published: 8 January 2024

1. Introduction

The heart plays a crucial role in sustaining life by effectively pumping oxygenated blood and regulating important hormones to maintain optimal blood pressure levels. Any deviation from its functioning can lead to the development of heart conditions, collectively known as cardiovascular diseases (CVD). CVD includes a range of disorders that affect both the heart and blood vessels, such as cerebrovascular problems, congenital anomalies, pulmonary embolisms, irregular heart rhythms (arrhythmias), peripheral arterial issues, coronary artery disease (CAD), rheumatic heart ailments, coronary heart disease (CHD), and cardiomyopathies that affect the heart muscle.

Notably, CHD is the subtype among cardiovascular diseases, accounting for a significant 64% of all cases. While it primarily affects men, women are also susceptible to its impact. Within the realm of CVDs, CAD is particularly concerning due to its association with global mortality rates. According to the World Health Organization (WHO) [1], the consequences of CVDs are profound, with staggering statistics indicating an estimated 17.9 million deaths annually are attributed to these diseases worldwide. These alarming numbers highlight the significance of research efforts and medical advancements dedicated to combatting and lessening the impact of cardiovascular diseases worldwide. There are risk factors that contribute to the development of CVDs, including blood pressure, excess body weight and obesity, abnormal lipid profiles, glucose irregularities or diabetes conditions, tobacco usage or smoking habits, physical inactivity or sedentary lifestyle, alcohol consumption, and cholesterol levels. The WHO predicts that CVD will remain a cause of mortality, silently posing a substantial threat to human life for the foreseeable future, possibly even beyond 2030.

Machine learning, as highlighted by Ramesh et al. [2], enjoys major transformative capability within the healthcare industry. Its outstanding advancements can be ascribed to its exceptional data processing abilities, which are far superior to those of humans. Consequently, the field of healthcare has observed the development of several AI applications that leverage machine learning's speed and accuracy, paving the way for revolutionary solutions to diverse healthcare challenges. Several machine learning methods have been applied for the purpose of detecting cardiovascular diseases. However, there is still a need to enhance the predictive models and address the research gaps in the existing detection approaches, such as the challenge of imbalanced datasets, which can lead to biased predictions.

By investigating the effectiveness of hybrid models combining different techniques, various researchers have explored diverse methodologies, including neural networks and various machine learning methods, to enhance prediction accuracy [3–12]. While these studies provide valuable insights, the variability in datasets, models, and outcomes underscores the complexity of the predictive task. Despite the advancements, there remains a pressing need for further investigations to refine existing models and improve the overall performance of cardiovascular disease prediction. The diverse landscape of machine learning applications in this domain emphasizes the importance of continued research to enhance the accuracy, reliability, and generalizability of predictive models, ultimately contributing to more effective clinical interventions and patient care.

In this paper, we have explored the strengths and limitations of the existing machine learning (ML) techniques in the context of heart disease analysis. Then, we investigated and applied seven machine learning-driven predictive models that can enhance the detection of cardiovascular and cerebrovascular diseases; these models include K-Nearest Neighbors, Support Vector Machine, Logistic Regression, Convolutional Neural Network, Gradient Boost, XGBoost, and Random Forest. Two datasets were used in this study, which were pre-processed using different techniques such as oversampling, feature scaling, normalization, and dimensionality reduction to optimize data for effective machine learning analysis. Finally, we evaluated and compared the efficacy of different machine learning (ML) techniques for analyzing heart diseases within the healthcare sector.

2. Related Works

In this paper, we present a concise technical background and review pertinent literature related to research studies conducted on the early forecast of heart disease utilizing machine learning and deep learning techniques. We highlight the different methods that have been employed in these studies to foretell heart disease at an initial stage.

2.1. Machine Learning Approach

Machine learning remains a rapidly advancing discipline of computational algorithms that try to imitate human intelligence by learning through data and the surrounding environment. These algorithms play a crucial role in processing and analyzing large-scale data, often referred to as "big data". Machine learning techniques have demonstrated their effectiveness in various domains, including pattern recognition, computer vision, spacecraft engineering, as well as biomedical and medical applications. Their versatility and success have made them indispensable tools in addressing complex challenges and extracting valuable insights from diverse datasets [13].

Machine learning is a specialized approach that automates the process of model building. Using algorithms, machines can discover hidden patterns and insights within datasets. Importantly, in machine learning, we do not particularly instruct machines on where to explore for insights; instead, the algorithms enable the machines to learn and adapt their techniques and outputs as they uncover new-found data and scenarios. This iterative nature of machine learning allows for continuous improvement and adaptation, making it a powerful tool for processing and analyzing complex datasets [14].

There exist two main approaches in machine learning: supervised learning and unsupervised learning. In one approach, supervised learning, algorithms are trained using specific examples. The machine is provided with input data along with their corresponding correct outputs. Learning takes place by comparing the machine's experimental outcomes with the accurate outputs to discover blunders. This sort of learning is suitable after previous data has been utilized to foretell future occurrences [15].

The other approach, unsupervised learning, involves the machine exploring the records and attempting to discover patterns or structures on its own. It needs to create models commencing from scratch and is not provided with any precise outputs to guide its learning process. Unsupervised learning is commonly employed to detect and distinguish outliers in the data. This approach is particularly useful when there is limited or no labeled data available for training [14]. Researchers worldwide have made significant efforts to combat cardiovascular disease (CVD) and improve patient outcomes [16]. These efforts include enhancing clinical decision support systems to achieve precise early detection and enable effective treatment. Machine learning (ML) and artificial intelligence (AI) techniques have played a pivotal role in the early detection and diagnosis of CVD.

CVD detection encompasses different distinct approaches. The first approach involves utilizing AI models that analyze various test reports to distinguish between CVD patients and healthy citizens. The second approach utilizes signals such as electrocardiogram (ECG) and heart sound signals as vital information for ML models to classify individuals as either healthy or having CVD [16].

2.2. Deep Learning Approach

In recent years, there has been remarkable progress in the field of deep learning, with a primary focus on developing intelligent automated systems that aid doctors in predicting and diagnosing diseases through the utilization of the Internet of Things (IoT). While conventional machine learning techniques were often restricted by their dependency on single datasets, the advent of deep learning has brought significant enhancements to the accuracy of existing algorithms. Deep learning leverages artificial neural networks, which consist of multiple hidden layers organized in a cascading pattern. This architecture enables the processing of non-linear datasets, allowing for more complex patterns and relationships to be captured and learned by the model. As a result, deep learning has emerged as a powerful tool in medical applications, providing improved predictive capabilities and enhancing disease diagnosis through the integration of IoT devices and data sources. This approach has shown promising results, outperforming older machine learning algorithms in terms of accuracy. As accurate medical support systems for detecting hidden patterns and predicting diseases are still lacking, deep learning offers the potential to accurately predict heart diseases at an early stage, allowing for timely intervention and treatment [17].

Sudha and Kumar [18] observed that the Convolutional Neural Network (CNN) is a suitable method for diagnosing heart disease. CNN's ability to learn and represent features in a concise and conceptual manner is advantageous, especially as the network's depth increases. Additionally, they proposed a hybrid model that combines Convolutional Neural Networks (CNN) with Long Short-Term Memory (LSTM) units, which are a type of recurrent neural network (RNN). LSTM units are known for their ability to store and transmit relevant information over long sequences, making them particularly useful for time-series data such as heart disease data. By integrating CNN and LSTM, the hybrid model aimed to enhance the accuracy of heart disease classification. The CNN component is adept at capturing spatial patterns in the data, while the LSTM component excels at recognizing temporal dependencies and patterns. This combination allows the model to effectively learn complex features from the data, leading to improved classification accuracy. Experimental results from the study revealed promising outcomes, with the hybrid model achieving an accuracy of 89%, sensitivity of 81%, and specificity of 93%. These results outperformed conventional machine learning classifiers, indicating the potential of the proposed hybrid approach in advancing the accuracy of heart disease classification [18].

The healthcare sector has emerged as a prime beneficiary of the growing volume and accessibility of data [19]. Various entities, such as healthcare providers, pharmacological firms, research institutions, and government parastatals, are now accumulating vast volumes of data from diverse sources, including research, clinical trials, public health programs, and insurance data. The merging of such data holds immense potential for advancing healthcare practices and decision-making [20]. Traditionally, doctors used to diagnose and treat patients based on their symptoms alone. However, evidence-based medicine has become the prevailing approach, where physicians review extensive datasets obtained from medical trials and treatment paths on a huge scale to make decisions built on the most comprehensive and up-to-date information available. This shift towards data-driven decision-making is transforming healthcare practices, improving patient outcomes, and driving further advancements in the medical field [14].

Numerous industry and research initiatives are actively working on implementing machine learning expertise in the healthcare sector to enhance patient care and well-being globally. One such initiative is the Shah Lab, based at Stanford University [14]. The Shah Lab focuses on leveraging machine learning and data science to address critical healthcare challenges and develop innovative solutions for various medical applications. Through these initiatives, researchers and experts aim to harness the power of machine learning to analyze large-scale healthcare data, including electronic health records, medical imaging, genomics, and patient outcomes. By extracting valuable insights and patterns from this data, they aim to improve disease diagnosis, treatment prediction, personalized medicine, and overall patient management. The goal is to provide healthcare professionals with advanced tools and technologies that can assist them in making more accurate and timely clinical decisions, leading to better patient outcomes and an overall improvement in healthcare services worldwide. Table 1 below presents a summary of the performance metrics related to the existing methods under evaluation, with each entry corresponding to specific evaluation criteria.

Table 1. Summary of the performance of the existing methods.

Study	Method	Results
Mohan et al. [21]	Hybrid Random Forest with Linear Model (HRFLM)	Accuracy: 88% Sensitivity: 92.8% Specificity: 82.6%
Singh et al. [22]	SVM K-Nearest Neighbors Decision Tree Linear Regression	83% Accuracy SVM 79% (DT) 78% (LR) 87% (KNN)
Gavhane et al. [23]	Neural Network	Precision rate: 91% Recall rate: 89%
Kavitha et al. [24]	Hybrid Model (Random Forest (RF) and Decision Tree (DT))	Accuracy: 88%
Amiri and Armano [25]	Classification—CART	Accuracy: 99.14% Sensitivity: 100% Specificity: 98.28%
Liu and Kim [26]	Classifier—Long Short Term Memory (LSTM)	Accuracy: 98.4%

2.3. Datasets Collection and Preprocessing

In their study, Algarni et al. [27] utilized a dataset of coronary artery X-ray angiography images obtained from a clinical database. These images exhibited challenging characteristics, including uneven vessel thickness, complex vascular structures in the background, and the presence of noise. The dataset consisted of 130 X-ray coronary angiograms, each having a size of 300 × 300 pixels. The data was collected from the cardiology department of the Mexican Social Security Institute, and ethical approval was obtained (reference number R-2019-1001-078) for the use of this medical database in heart disease diagnosis. To train and evaluate their proposed model, called ASCARIS, the dataset was randomly divided into two parts: a training set containing 100 images and a test set comprising 30 images. The ASCARIS model was developed based on color, diameter, and shape features extracted from the angiography images.

Al Mehedi et al. [28] utilized a dataset of 299 heart failure patients obtained from the Faisalabad Institute of Cardiology and the Allied Hospital in Faisalabad. The dataset consisted of 13 attributes, including features such as Age, Anemia, High Blood Pressure, Creatinine Phosphokinase (CPK), Diabetes, Ejection Fraction, Sex, Serum Creatinine, Serum Sodium, Smoking, Time, and a target column labeled as "Death Event", which was used for binary classification. The dataset underwent preprocessing to ensure its quality and consistency. After preprocessing, the dataset was divided into separate train and test sets for model training and evaluation. Two feature selection methods were applied to the train set to identify the most relevant features for the heart failure prediction task.

Deepika and Seema [29] conducted a study on heart disease with datasets available online from the UCI Machine Learning Repository at the University of California, Irvine. They comprise 76 attributes, including the target property, but only 14 of these attributes were considered essential for analysis. The researchers used two specific datasets for their study: the Cleveland Clinic Foundation dataset, with records from 303 patients, and the Hungarian Institute of Cardiology dataset, with records from 294 patients. Various machine learning algorithms, including Naïve Bayes (NB), Support Vector Machine (SVM), Decision Tree (DT), and Artificial Neural Networks, were employed in the analysis to predict heart disease. Within the broader context, Table 2 clarifies the preprocessing approaches and predictive methodologies utilized in previous studies.

Table 2. Preprocessing and predictive methods.

Study	Dataset	Preprocessing and Modeling	Results
Algarni et al. [27]	Coronary artery X-ray angiography images obtained from a clinical database.	Training: 100 images Test: 30 images ASCARIS model (based on color, diameter, and shape features).	Accuracy: 97%
Uyar and İlhan [30]	Cleveland dataset for heart disease.	Removal of 6 instances with missing entries from the dataset and categorization of the diagnosis attribute (num) into two classes: absence (num = 0) and presence (num = 1, 2, 3, or 4) of heart disease. Recursive Fuzzy Neural Network (RFNN)	Testing set accuracy: 97.78% Overall accuracy: 96.63%
Deng et al. [31]	Fuwai ECG database and public PTB database	training phase for dynamics acquisition and a test phase for dynamics reuse Attention-based Res-BiLSTM-Net model	F1 scores ranging from 0.72 to 0.98
Das et al. [32]	UCI dataset	SAS-based software Neural Networks	Training accuracy: 86.4%, Validation accuracy: 89.011%

2.4. Discussions on the Research Limitations

The literature review involved an in-depth exploration of the existing research and knowledge pertaining to heart disease prediction using diverse machine learning and deep learning techniques. Several studies reviewed the recent advancements and limitations of applying machine learning for cardiovascular disease detection [10,33–36]. For instance, the studies [8,37–40] proposed different data mining and machine learning methods based on heartbeat segmentation and selection process, ECG images, images of carotid arteries, and others.

Numerous studies have concentrated on applying machine learning algorithms such as Decision Tree, Naïve Bayes, Random Forest, Support Vector Machine, and Logistic Regression on the Heart Disease Dataset, yielding promising accuracy rates for classification. Moreover, deep learning methods, particularly Convolutional Neural Networks (CNN), have gained significant traction for effectively handling complex tasks and unstructured data. The review also examined discussions regarding the implementation of data preprocessing techniques, feature selection methods, and performance evaluation metrics to optimize the efficiency of predictive models. Some studies underscored the importance of data quality and the relevance of specific features in enhancing the accuracy of the models.

Machine learning algorithms play a crucial role in precisely foretelling heart disease by discovering suppressed patterns in data, making predictions, and improving performance based on historical data. These programs make it possible for us to anticipate and diagnose heart disease more accurately, while deep learning, fueled by artificial neural networks, is a critical factor in handling complex computations on large volumes of data. These algorithms play an essential role in identifying key attributes and patterns in both structured and unstructured data, enhancing more efficient data analysis and processing.

Employing machine learning and deep learning approaches offers considerable potential in the field of heart disease diagnosis and treatment. These sophisticated techniques enable the integration of various data sources, such as medical records, imaging data, genetics, and lifestyle factors, to create a universal and individualized approach to healthcare. The iterative nature of machine learning acknowledges continuous learning and adaptation, resulting in progressed diagnostic and predictive models over time. This promises to enhance the accuracy and effectiveness of heart disease management, ultimately leading to better patient outcomes.

After reviewing the available literature, it is evident that there is a lack of extensive experimentation on the use of Gradient Boosting models in the detection of heart disease. However, considering the unique capabilities of Gradient Boosting models in analyzing data and capturing temporal dependencies, their potential in this domain is worth exploring.

The potential of Gradient Boosting models to progressively enhance predictive accuracy by refining weaker learners within the model positions them as promising contenders for improving the precision of heart disease detection. Consequently, there is a need for further exploration and experimentation dedicated to harnessing the capabilities of Gradient Boosting models in this context.

By embracing the use of Gradient Boosting models in heart disease detection and conducting more targeted experiments, we can unlock new possibilities for advancing healthcare interventions and ultimately enhancing patient outcomes and well-being.

3. Materials and Methods

The following methods are adapted to achieve the goals of this research. They are applied to explore and comprehend various dimensions of heart-related conditions, ultimately contributing to the creation of precise models for the diagnosis and prediction of these conditions. The general research method framework of this study is shown in Figure 1.

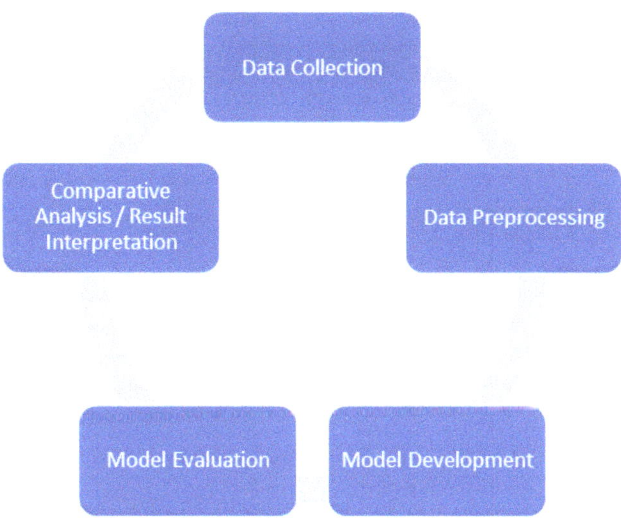

Figure 1. Research method workflow.

3.1. Datasets

To carry out this research study, two datasets were examined, namely the Cardiovascular Heart Disease Dataset, which was retrieved from the Mendeley database, and the Heart Disease Cleveland Dataset, which was retrieved from the Kaggle database. The "Cardio" and "Target" columns on both datasets refer to the column we are trying to predict with numeric values 0 (no disease) and 1 (disease). It is important to note that neither dataset has any missing values. The detailed descriptions of all these attributes are listed below:

The Cardiovascular Heart Disease Dataset (Table 3) holds significant importance within the healthcare and machine learning domains. It serves as an asset for tasks associated with the prediction and classification of cardiovascular diseases while holding data of 1000 data samples in 13 attributes, each representing a potential risk factor.

Table 3. Cardiovascular Heart Disease Dataset.

Features	Details
1. Patient Id	Individual unique identifier.
2. Age	Numeric representation of patients' age in years.
3. Gender	Binary (1, 0 (0 = female, 1 = male))
4. Chestpain	Nominal (0, 1, 2, 3 (Value 0: typical angina Value 1: atypical angina Value 2: non-anginal pain Value 3: asymptomatic))
5. restingBP	Numeric (94–200 (in mm HG))
6. serumcholestrol	Numeric (126–564 (in mg/dL))
7. fastingbloodsugar	Binary (0, 1 > 120 mg/dL (0 = false, 1 = true))
8. restingrelectro	Nominal (0, 1, 2 (Value 0: normal, Value 1: having ST-T wave abnormality (T wave inversions and/or ST elevation or depression of >0.05 mV), Value 2: showing probable or definite left ventricular hypertrophy by Estes' criteria))
9. maxheartrate	Numeric (71–202)
10. exerciseangia	Binary (0, 1 (0 = no, 1 = yes))
11. oldpeak	Numeric (0–6.2)
12. slope	Nominal (1, 2, 3 (1-upsloping, 2-flat, 3-downsloping))
13. noofmajorvessels	Numeric (0, 1, 2, 3)
14. target	Binary (0, 1 (0 = Absence of Heart Disease, 1= Presence of Heart Disease))

Shifting our focus to the Heart Disease Cleveland Dataset (Table 4), a widely recognized dataset frequently employed in the fields of machine learning and healthcare, which has been extensively used in tasks related to predicting and classifying heart disease. This dataset holds prominence for its pivotal role in assessing the effectiveness of diverse machine learning algorithms in diagnosing heart disease with 303 patients' information in 14 attributes. Its primary objective revolves around predicting whether heart disease is present or absent.

Table 4. Heart Disease Cleveland Dataset.

Features	Details
1. Age	Numeric representation of patients' age in years.
2. Sex	Categorical feature representing gender, where Male is encoded as 1 and Female as 0.
3. cp	Categorical attribute indicating the various types of chest pain felt by the patient. 0 for typical angina, 1 for atypical angina, 2 for non-anginal pain, and 3 for asymptomatic.
4. trestbps	Numerical measurement of the patient's blood pressure at rest, recorded in mm/HG.
5. chol	Numeric value indicating the serum cholesterol intensity of the patient, calculated in mg/dL.
6. fbs	Categorical representation of fasting blood sugar levels, with 1 signifying levels above 120 mg/dL and 0 indicating levels below.
7. restecg	Categorical feature describing the result of the electrocardiogram conducted at rest. 0 for normal, 1 for ST-T wave abnormalities, and 2 for indications of probable or definite left ventricular hypertrophy according to Estes' criteria.
8. thalach	Numeric representation of the heart rate realized by the patient.
9. exang	Categorical feature denoting whether exercise-induced angina is present. 0 signifies no, while 1 signifies yes.
10. oldpeak	Numeric value indicating exercise-induced ST-depression relative to the rest state.
11. slope	Categorical attribute representing the slope of the ST segment during peak exercise. It can take three values: 0 for up-sloping, 1 for flat, and 2 for down-sloping.
12. ca	Categorical feature indicating the number of major blood vessels, ranging from 0 to 3.
13 thal	Categorical representation of a blood disorder called thalassemia. 0 for NULL, 1 for normal blood flow, 2 for fixed defects (indicating no blood flow in a portion of the heart), and 3 for reversible defects (indicating abnormal but observable blood flow).
14. target	The target variable to predict heart disease, encoded as 1 for patients with heart disease and 0 for patients without heart disease.

3.2. Data Pre-Processing

Data preprocessing is an essential step within machine learning that aims to improve dataset quality and reliability before analysis and modeling. This phase tackles challenges such as missing data, inconsistencies, outliers, and skewed class distributions. Addressing missing values is crucial to ensure accurate insights by utilizing techniques such as imputation. Detecting and managing outliers is also vital, as these data points can skew results. A key concern is class distribution balance, where methods like oversampling mitigate imbalanced datasets. Considering these considerations, employing techniques such as feature scaling, normalization, and dimensionality reduction can optimize data for effective machine learning analysis.

3.3. Model Development

The conclusion of the thorough literature work brings us to the pivotal stage of model development. This section encompasses seven notable machine learning techniques:

Logistic Regression, Convolutional Neural Network, Support Vector Machine (SVM), Gradient Boosting, K-Nearest Neighbors (KNN), XGBoost, and Random Forest. Each algorithm contributes distinct characteristics to unveil predictive revelations in the analysis of cardiovascular and cerebrovascular diseases, utilizing resources such as Scikit-Learn and Keras libraries.

Each of these models possesses unique traits, spanning from linear approaches to ensemble techniques and deep learning architectures. Through thorough empirical investigations, we assessed the effectiveness of every model in terms of recall, precision, accuracy, and F1-score metrics.

3.4. Model Evaluation

Model Evaluation stands as a pivotal phase in the realm of machine learning, dedicated to thoroughly gauging how well-trained models predict outcomes. This essential step ensures that models can generalize to new data effectively, informing decisions about deployment and refinement. The following key techniques and metrics will contribute to a comprehensive evaluation of this study:

Confusion Matrix: Offering insight into true positives, true negatives, false positives, and false negatives, this matrix forms the basis for calculating vital metrics.

Accuracy: Providing an overall view of model performance by measuring correctly predicted instances against the total dataset.

$$\text{Accuracy} = (TP + TN)/(TP + FP + TN + FN) \qquad (1)$$

Precision and Recall: Precision assesses positive prediction accuracy, while recall gauges the model's ability to capture positive instances.

$$\text{Precision} = TP/(TP + FP) \qquad (2)$$

$$\text{Recall} = TP/(TP + FN) \qquad (3)$$

F1-Score: Striking a balance between precision and recall, this score is essential for harmonizing performance aspects.

$$F1 = (2 \times \text{precision} \times \text{recall})/(\text{precision} + \text{recall}) \qquad (4)$$

Cross-Validation: This technique partitions data for training and testing, guarding against overfitting.

Hyperparameter Tuning: Optimizing model parameters through techniques like GridSearch enhances performance.

4. Results

This section explores the detailed analysis of machine learning models for heart disease prediction, leveraging two distinct datasets: the Cardiovascular Heart Disease Dataset and the Heart Disease Cleveland Dataset using the Python programming language.

Our primary objective is to identify the most effective predictive models, considering both traditional tabular datasets while keeping in mind the aims of the study.

4.1. Pre-Processing Results

To harness the potential of the Cardiovascular Heart Disease Dataset and the Heart Disease Cleveland Dataset for machine learning applications, it becomes imperative to execute preliminary data preprocessing procedures. These procedures encompass a range of actions, including managing missing data, encoding categorical variables, standardizing or normalizing feature values, and partitioning the dataset into distinct training and testing subsets. Additionally, the utilization of exploratory data analysis (EDA) techniques and data visualization tools proves instrumental in gaining insights into data distributions and inter-variable relationships.

Firstly, a correlation matrix heatmap is created, as shown in Figure 2. This heatmap computes the correlation coefficients among diverse attributes in the datasets and represents them graphically. Its purpose is to facilitate the visual examination of associations between various features. Positive correlations are depicted using green hues, whereas negative correlations are represented in red. This heatmap serves the purpose of identifying the features that exhibit the most substantial correlations with the target variable, thereby revealing their impact on the presence or absence of cardiovascular disease. On the left side is the Cardiovascular Heart Disease Dataset, while on the right is the Heart Disease Cleveland Dataset.

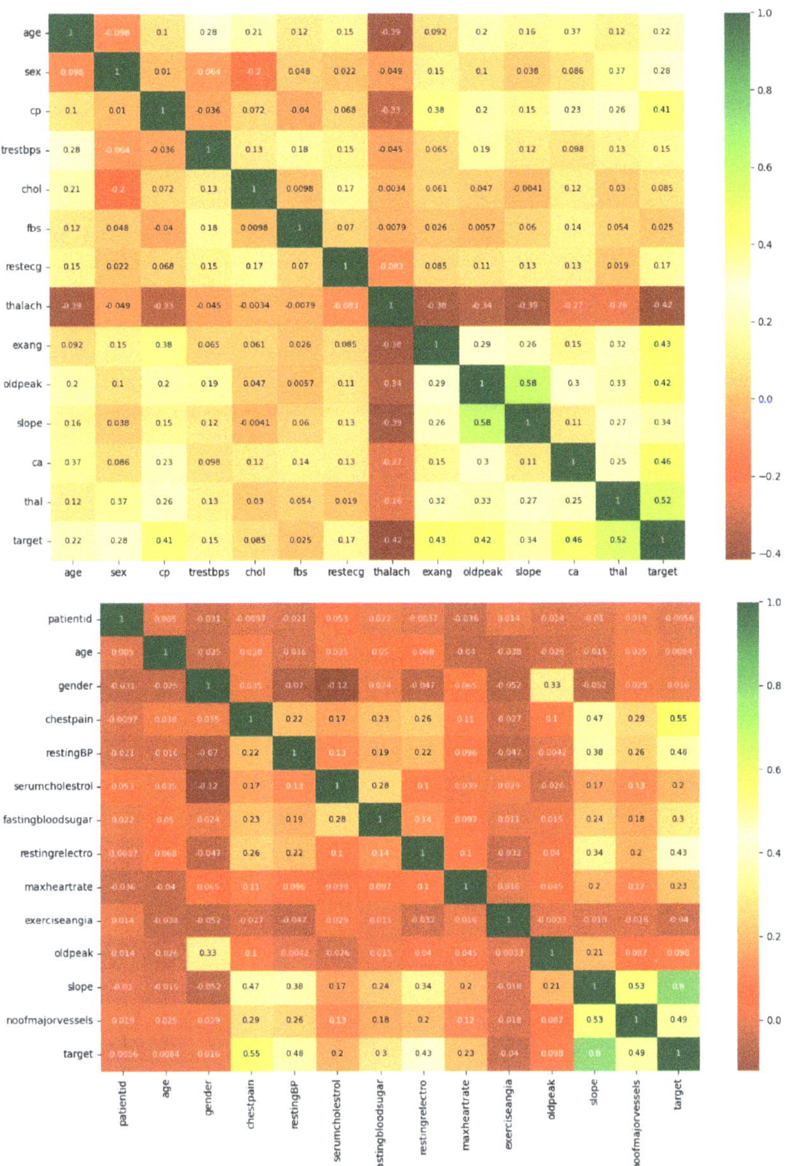

Figure 2. Heatmap distribution of the dataset features.

The histograms corresponding to individual dataset attributes provide valuable insights by allowing exploration of each feature's distribution, as shown in Figure 3. They are instrumental in the detection of potential outliers and provide a rapid overview of the characteristics and spans of these features. This visualization is a helpful tool for comprehending the overall shape and distribution of the data. The pictorial evidence of both datasets can be seen below, where the Cardiovascular Heart Disease Dataset is on the left, and the Heart Disease Cleveland Dataset can be seen on the right.

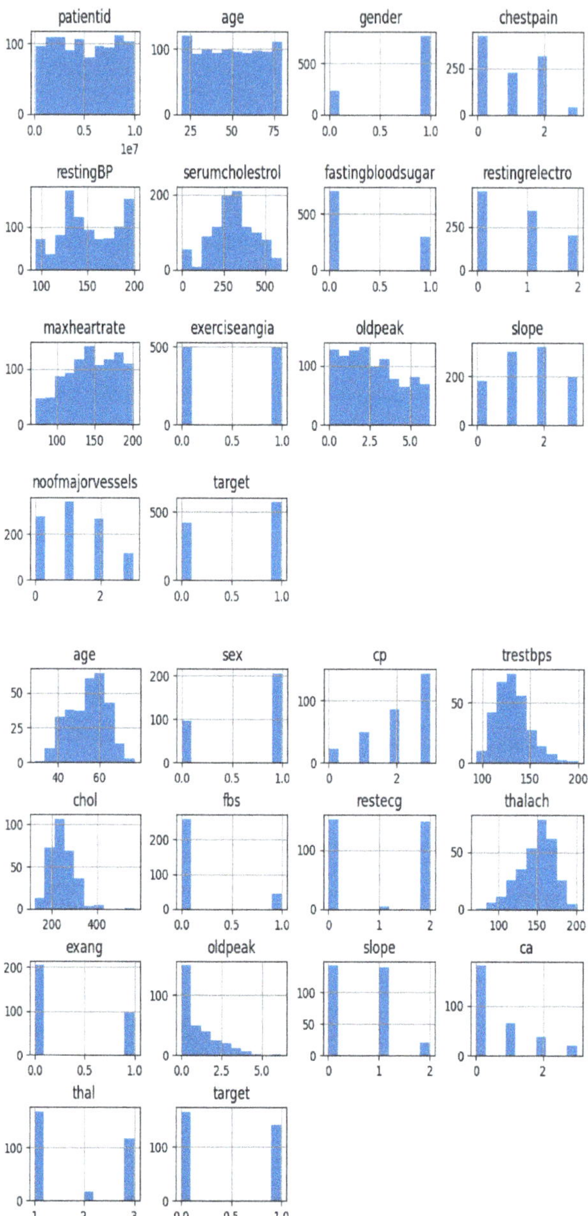

Figure 3. Histogram distribution of the dataset features.

As shown in Figure 4, the pie chart is utilized to depict the distribution of the target variable, which signifies the existence or non-existence of cardiovascular disease. The figure shows the distribution of features in the target variable, where 1 represents features with heart disease, and 0 represents features without heart disease. It enumerates the instances of each class and exhibits the proportions as percentages in the pie chart, illustrating the presence and absence of cardiovascular disease. In Figure 4, the pie chart on the right represents features of the target column distribution of the Cardiovascular Heart Disease

Dataset, while the left represents the feature of the target column distribution of the Heart Disease Cleveland Dataset.

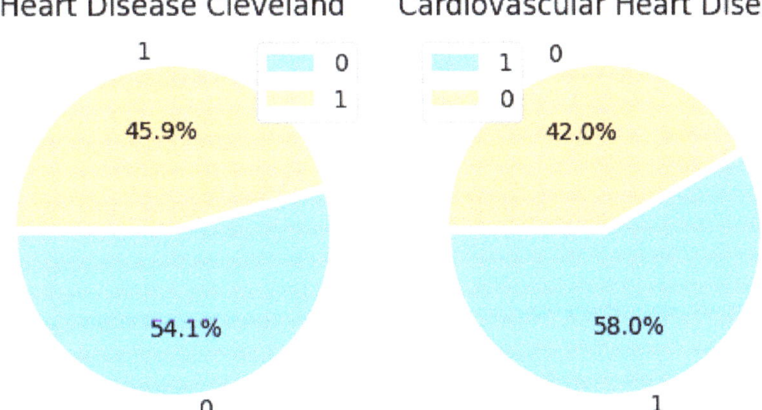

Figure 4. The distribution of features in the target variable.

After successfully preprocessing and visualizing the features of the dataset, we conducted an in-depth exploration of various machine learning models to discern their predictive efficacy.

4.2. K-Nearest Neighbors (KNN) Results

We commenced the analysis by employing the K-Nearest Neighbors (KNN) algorithm with varying 'k' values, representing the number of nearest neighbors considered during the predictions. Employing cross-validation, we computed scores for each 'k' value, ultimately discerning that 'k = 7' yielded the most favorable mean cross-validation score. This outcome underscores that configuring KNN with 'k = 7' exhibits significant promise.

As shown in Tables 5–8, the implementation of this model yielded an impressive accuracy rate of 96.50% and 91.80% on the datasets, respectively, serving as an overarching measure of the model's correctness in its predictions. Furthermore, meticulous hyperparameter tuning was carried out to guarantee optimal performance. The precision score, gauging the proportion of true positive predictions among all positive predictions, achieved a notable level, approximately 96.61% and 96.55%. Additionally, the recall, representing the proportion of true positive predictions among all actual positives, exhibited a strong value, approximately 97.44%, and 87.50%. Similarly, the F1 Score attained an impressive value, hovering around 97.02% and 91.80%. These metrics collectively affirm the exceptional performance of the KNN model within the dataset.

Table 5. Results on Precision measure.

Classification Model	Precision (in %)	
	Dataset 1	**Dataset 2**
KNN	96.50%	**96.55%**
RF	98.63%	94.44%
LR	96.55%	93.10%
GB	99.13%	90.00%
SVM	95.00%	80.65%
CNN	99.14%	87.50%
XGBoost	**99.14%**	90.00%

Table 6. Results on Recall measure.

Classification Model	Recall (in %)	
	Dataset 1	Dataset 2
KNN	97.44%	87.50%
RF	**98.97%**	85.61%
LR	95.73%	84.38%
GB	97.44%	84.38%
SVM	97.44%	78.12%
CNN	98.29%	**89.77%**
XGBoost	98.29%	84.38%

Table 7. Results on F1-Score measure.

Classification Model	F1-Score (in %)	
	Dataset 1	Dataset 2
KNN	97.02%	**91.80%**
RF	**98.80%**	89.81%
LR	96.14%	88.52%
GB	98.28%	87.10%
SVM	96.20%	79.37%
CNN	97.80%	87.50%
XGBoost	98.71%	87.10%

Table 8. Results on Accuracy measure.

Classification Model	Accuracy (in %)	
	Dataset 1	Dataset 2
KNN	96.50%	**91.80%**
RF	**98.60%**	91.09%
LR	95.50%	88.52%
GB	98.00%	86.89%
SVM	95.50%	78.69%
CNN	97.50%	86.89%
XGBoost	98.50%	86.89%

4.3. Random Forest Results

By conducting an extensive hyperparameter tuning process, we modified the number of trees (n_estimators) to 200 within the Random Forest ensemble model. As shown in Tables 5–8, the tuned model achieved an outstanding accuracy level, hovering at around 98.60% and 91.09%. The assessment of precision showed a significant enhancement, which obtained 98.63% and 94.44%.

Similarly, the F1 Score, which amalgamates precision and recall, demonstrated the model's robustness, registering a value of 98.80% and 89.81, respectively. Furthermore, the recall score, measuring the model's aptitude for recognizing genuine positive cases, reached a remarkable value of 98.97% and 85.61.

4.4. Logistic Regression (LR) Results

By implementing a custom threshold of 0.6, the model was configured to adopt a cautious approach when classifying instances as positive. To be specific, if the predicted probability of an instance belonging to the positive class (class 1) equaled or exceeded 0.6, it was categorized as positive; otherwise, it was designated as negative. This threshold selection significantly influenced how the model struck a balance between precision and recall. As shown in Tables 5–8, the model's precision score was 96.55% and 93.10%, signifying its proficiency in minimizing false positive predictions.

The recall scores stood at 95.73% and 84.38%, emphasizing the model's importance in correctly identifying all positive cases, particularly in scenarios where missing potential cases of heart disease is a critical concern. The F1 Score captured genuine positive cases at 96.14% and 88.52%. Regarding overall accuracy, the model achieved an accuracy score of 95.50% and 88.52%.

4.5. Gradient Boosting (GB) Results

Through the GridSearchCV process, we effectively fine-tuned the model's hyperparameters. The optimal hyperparameters selected encompassed a learning rate of 0.2, a maximum depth of 3 for individual trees, and 100 boosting stages (n_estimators). These hyperparameters were chosen based on their exceptional performance on the validation datasets. When tested on independent data, the refined Gradient Boosting model consistently delivered exceptional results. As shown in Tables 5–8, it attained an impressive precision score of 99.13% and 90.90%, indicative of its ability to minimize false positive predictions effectively.

Furthermore, the model exhibited a recall score of 97.44% and 84.38%, which holds paramount importance in medical applications where identifying potential cases of heart disease is critical. The F1 Score, which harmonizes precision and recall, reached an impressive value of 98.28% and 87.10.

The model's accuracy on the test dataset was consistently high, measuring 98.00%, although it achieved 86.89% on the Heart Disease Cleveland Dataset. These findings collectively underscore the Gradient Boosting model's exceptional suitability for the task of heart disease classification, highlighting its potential to accurately detect individuals with heart disease while maintaining a low rate of false positives. Such performance makes it an asset for healthcare professionals and researchers in the cardiology field.

4.6. Support Vector Machine (SVM) Results

The process of tuning hyperparameters, carried out through GridSearchCV, effectively determined the most suitable hyperparameter configuration for the SVM model. This configuration included a regularization parameter (C) set to 10, a polynomial kernel with a degree of 2, and the utilization of a linear kernel.

As shown in Tables 5–8, for post-tuning, the model achieved a precision score of 95.00% and 80.65%, a recall score of 97.44% and 78.12%, and an F1 Score of 96.20% and 79.37%.

On the test dataset, the model exhibited an accuracy of approximately 95.50% and 78.69%, affirming its consistent and accurate predictive capabilities.

4.7. Convolutional Neural Network (CNN) Results

The model architecture consists of three layers: an initial layer with 128 units employing the ReLU activation function, followed by a hidden layer featuring 64 units with ReLU activation, and ultimately, an output layer utilizing the sigmoid activation function. During model compilation, the Adam optimizer was employed alongside binary cross-entropy loss, with accuracy serving as the evaluation metric.

To mitigate the risk of overfitting, a precautionary measure known as early stopping was integrated into the training process. This involved monitoring the validation loss for a maximum of 10 epochs and restoring the model's weights to their best configuration. The training was conducted using scaled training data over a maximum of 100 epochs, employing a batch size of 64.

As shown in Tables 5–8, the model's performance on the test dataset is particularly noteworthy. Precision achieved an impressive score of 97.46% and 87.50%.

This suggests that when the model predicts an individual as having heart disease, it is highly likely to be accurate. Furthermore, the recall scores were 98.29% and 87.50%. The F1 Score demonstrates resilience at 97.87% and 87.50%. Overall accuracy, which reflects the ratio of correctly predicted cases to the total cases, stands at 97.50% and 86.89%, respectively.

4.8. XGBoost Results

Through the utilization of GridSearchCV, a highly effective process of hyperparameter tuning was carried out. This process led to the discovery of optimal hyperparameters for the XGBoost model, which included a learning rate of 0.2, a maximum tree depth of 3, 100 boosting rounds (n_estimators), and a subsample fraction of 1.0. The recall of these chosen hyperparameters was substantiated by a remarkable validation score of approximately 98.00% on the Cardiovascular Heart Disease Dataset and 84% on the Heart Disease Cleveland Dataset, respectively.

On the test dataset, the fine-tuned XGBoost model upheld its exceptional performance by achieving a precision score of 99.14% and 90.00%, signifying its adeptness in accurately categorizing positive cases. Moreover, the recall score, at 98.29% and 84.38%, holds particular significance. The F1 Score exhibits resilience at 98.71% and 87.10%. The model's overall accuracy on the test data hovers at 98.50% and 86.89%. These remarkable outcomes underscore the XGBoost model's aptness for heart disease classification.

5. Discussion

The experimental results are shown in Tables 5–8 and Figure 5. The thorough assessment of machine learning models, specifically the XGBoost and K-Nearest Neighbors models, in the context of heart disease prediction, provides valuable insights. These insights align with the research conducted by Zhang et al. [41], which underscores the effectiveness of the XGBoost algorithm in this specific domain.

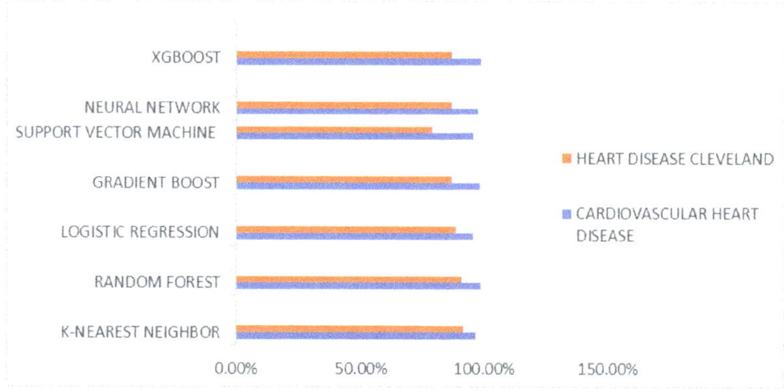

Figure 5. Accuracy of machine learning models on both datasets.

Across both datasets, these models consistently demonstrate exceptional performance, emphasizing their efficacy in heart disease prediction. Notably, the XGBoost model stands out with an impressive accuracy rate of 98.50% in the Cardiovascular Heart Disease Dataset, while the K-Nearest Neighbors (KNN) model achieves a commendable accuracy of 91.80% in the Heart Disease Cleveland Dataset. These high levels of accuracy emphasize the models' reliability, positioning them as valuable tools for diagnosing heart disease.

Precision, a critical metric in healthcare, reflects the models' ability to identify heart disease cases precisely. Both models achieve outstanding precision, with the XGBoost model leading at 99.14%, closely followed by the KNN model at 96.55%. These elevated precision levels significantly reduce the occurrence of false positive diagnoses, alleviating unnecessary concerns for patients.

Furthermore, the F1 Score, which balances precision and recall, highlights the XGBoost model's effectiveness in recognizing heart disease cases while minimizing the risk of overlooking positive instances. The model achieves F1 Scores of 98.71% and 91.80% in both datasets, showcasing its ability to strike this delicate balance effectively.

6. Conclusions and Future Scope

As we discussed the broader scope of model selection and its implications for heart disease prediction, the conducted analysis has unearthed invaluable insights. Among the array of models under scrutiny, K-Nearest Neighbors and XGBoost have consistently risen to prominence as top-performing candidates across both datasets, as shown below. These models have exhibited remarkable accuracy and recall scores, rendering them robust contenders for the precise classification of heart disease. It is noteworthy, however, that other models, including Logistic Regression, Convolutional Neural Network, Gradient Boost, Random Forest (RF), and Support Vector Machines (SVM), have showcased significant predictive capabilities once their hyperparameters were meticulously tuned. In this diverse ensemble, XGBoost emerges as a standout performer, marked by its exceptional accuracy and recall scores, coupled with a harmoniously balanced F1 Score and precision on the Cardiovascular Heart Disease Dataset. This points out XGBoost's transformative potential in the realm of heart disease prediction and diagnosis, positioning it as an invaluable tool for healthcare professionals. The model instills a high level of confidence in identifying potential cases of heart disease, firmly establishing itself as an exemplary choice within this dataset. The exceptional precision and accuracy exhibited by these models bear profound implications for the diagnosis and care of individuals with heart disease. Such precision not only enhances diagnostic accuracy but also opens new avenues for interventions and treatments that can be initiated with heightened confidence. In the quest for the most suitable model, it is imperative to align the selection with the specific requirements and constraints of the application at hand. Practical considerations such as interpretability, computational complexity, and data availability should guide the decision-making process, ensuring that the chosen model is tailored to meet the unique needs of the task. These findings culminate in a valuable resource that can empower informed decision-making within the realm of heart disease prediction, particularly in clinical settings. The potential to revolutionize heart disease diagnosis and patient care is emphasized, further cementing the significance of machine learning in the field of healthcare. In practical terms, this implies that when the model indicates an individual as having heart disease, the likelihood of accuracy is notably high, signifying a significant advancement in the landscape of medical diagnostics. Future directions for this study could involve expanding the scope by incorporating more extensive medical imaging datasets. Leveraging such data could enhance image-based heart disease prediction, potentially leading to even more accurate and robust diagnostic tools in the field of cardiovascular health. Furthermore, exploring ensemble models that merge the strengths of multiple algorithms may offer promising avenues for further improving predictive accuracy in the field of heart disease prediction. These considerations shed light on the multifaceted nature of heart disease prediction research, emphasizing the need for ongoing refinement and innovation in this critical domain. Future research directions should also prioritize the refinement of models and expansion of datasets. In contrast to [42,43], our study employs a distinct dataset, leveraging its unique characteristics to enhance the robustness and generalizability of the models. Furthermore, the selection of machine learning models in our work deviates from those used in the cited studies, contributing to the innovative aspect of our approach. Importantly, the outcomes of our models exhibit a noteworthy improvement in predictive accuracy, establishing a superior performance benchmark.

This nuanced combination of dataset, model selection, and elevated accuracy underscores the distinctive contribution of our work to the field of heart disease prediction. It positions our study as an advancement beyond existing research, offering a more refined and accurate predictive framework.

Author Contributions: Conceptualization, A.O. and F.S.; methodology, A.O. and F.S.; software, A.O.; validation, S.B., A.M.A. and S.N.Q.; formal analysis, A.O., F.S., S.B., A.M.A. and S.N.Q.; investigation, A.O., S.B., A.M.A. and S.N.Q.; resources, S.B., A.M.A. and S.N.Q.; data curation, A.O.; writing—original draft preparation, A.O. and F.S.; writing—review and editing, A.O., F.S., S.B.,

A.M.A. and S.N.Q.; visualization, A.O.; supervision, F.S.; project administration, F.S. and A.M.A.; funding acquisition, A.M.A. and S.N.Q. All authors have read and agreed to the published version of the manuscript.

Funding: This work was supported and funded by the Deanship of Scientific Research at Imam Mohammad Ibn Saud Islamic University (IMSIU) (grant number IMSIU-RG23077).

Institutional Review Board Statement: Not applicable.

Informed Consent Statement: Not applicable.

Data Availability Statement: The datasets are available online and upon request.

Acknowledgments: The authors extend their appreciation to the Deanship of Scientific Research at Imam Mohammad Ibn Saud Islamic University for funding this work through Grant Number IMSIU-RG23077.

Conflicts of Interest: The authors declare no conflicts of interest.

References

1. World Health Organization. WHO Cardiovascular Diseases. Available online: https://www.who.int/health-topics/cardiovascular-diseases#tab=tab_1 (accessed on 19 January 2022).
2. Ramesh, A.N.; Kambhampati, C.; Monson, J.R.; Drew, P.J. Artificial intelligence in medicine. *Ann. R. Coll. Surg. Engl.* **2004**, *86*, 334. [CrossRef] [PubMed]
3. Abdellatif, A.; Mubarak, H.; Abdellatef, H.; Kanesan, J.; Abdelltif, Y.; Chow, C.-O.; Chuah, J.H.; Gheni, H.M.; Kendall, G. Computational detection and interpretation of heart disease based on conditional variational auto-encoder and stacked ensemble-learning framework. *Biomed. Signal Process. Control* **2024**, *88*, 105644. [CrossRef]
4. Tartarisco, G.; Cicceri, G.; Bruschetta, R.; Tonacci, A.; Campisi, S.; Vitabile, S.; Cerasa, A.; Distefano, S.; Pellegrino, A.; Modesti, P.A.; et al. An intelligent Medical Cyber–Physical System to support heart valve disease screening and diagnosis. *Expert Syst. Appl.* **2024**, *238*, 121772. [CrossRef]
5. Cuevas-Chávez, A.; Hernández, Y.; Ortiz-Hernandez, J.; Sánchez-Jiménez, E.; Ochoa-Ruiz, G.; Pérez, J.; González-Serna, G. A Systematic Review of Machine Learning and IoT Applied to the Prediction and Monitoring of Cardiovascular Diseases. *Healthcare* **2023**, *11*, 2240. [CrossRef] [PubMed]
6. Plati, D.K.; Tripoliti, E.E.; Bechlioulis, A.; Rammos, A.; Dimou, I.; Lakkas, L.; Watson, C.; McDonald, K.; Ledwidge, M.; Pharithi, R.; et al. A Machine Learning Approach for Chronic Heart Failure Diagnosis. *Diagnostics* **2021**, *11*, 1863. [CrossRef] [PubMed]
7. Kim, J.O.; Jeong, Y.-S.; Kim, J.H.; Lee, J.-W.; Park, D.; Kim, H.-S. Machine Learning-Based Cardiovascular Disease Prediction Model: A Cohort Study on the Korean National Health Insurance Service Health Screening Database. *Diagnostics* **2021**, *11*, 943. [CrossRef]
8. Mhamdi, L.; Dammak, O.; Cottin, F.; Ben Dhaou, I. Artificial Intelligence for Cardiac Diseases Diagnosis and Prediction Using ECG Images on Embedded Systems. *Biomedicines* **2022**, *10*, 2013. [CrossRef]
9. Özbilgin, F.; Kurnaz, Ç.; Aydın, E. Prediction of Coronary Artery Disease Using Machine Learning Techniques with Iris Analysis. *Diagnostics* **2023**, *13*, 1081. [CrossRef]
10. Brites, I.S.G.; da Silva, L.M.; Barbosa, J.L.V.; Rigo, S.J.; Correia, S.D.; Leithardt, V.R.Q. Machine Learning and IoT Applied to Cardiovascular Diseases Identification through Heart Sounds: A Literature Review. Repositório Comum (Repositório Científico de Acesso Aberto de Portugal). 2021. Available online: https://www.preprints.org/manuscript/202110.0161/v1 (accessed on 15 June 2023).
11. Papandrianos, N.I.; Feleki, A.; Papageorgiou, E.I.; Martini, C. Deep Learning-Based Automated Diagnosis for Coronary Artery Disease Using SPECT-MPI Images. *J. Clin. Med.* **2022**, *11*, 3918. [CrossRef]
12. Al-Absi, H.R.H.; Islam, M.T.; Refaee, M.A.; Chowdhury, M.E.H.; Alam, T. Cardiovascular Disease Diagnosis from DXA Scan and Retinal Images Using Deep Learning. *Sensors* **2022**, *22*, 4310. [CrossRef]
13. El Naqa, I.; Murphy, M.J. *What Is Machine Learning?* Springer International Publishing: Berlin/Heidelberg, Germany, 2015; pp. 3–11.
14. Bhardwaj, R.; Nambiar, A.R.; Dutta, D. A study of machine learning in healthcare. In Proceedings of the 2017 IEEE 41st Annual Computer Software and Applications Conference (COMPSAC), Torino, Italy, 4–8 July 2017; IEEE: New York, NY, USA, 2017; Volume 2, pp. 236–241.
15. Brownlee, J. What is Machine Learning: A Tour of Authoritative Definitions and a Handy One-Liner You Can Use. Available online: www.machinelearningmastery.com (accessed on 25 November 2023).
16. Oresko, J.J.; Jin, Z.; Cheng, J.; Huang, S.; Sun, Y.; Duschl, H.; Cheng, A.C. A wearable smartphone-based platform for real-time cardiovascular disease detection via electrocardiogram processing. *IEEE Trans. Inf. Technol. Biomed.* **2010**, *14*, 734–740. [CrossRef] [PubMed]
17. Sharean, T.M.A.M.; Johncy, G. Deep learning models on Heart Disease Estimation—A review. *J. Artif. Intell.* **2022**, *4*, 122–130. [CrossRef]

18. Sudha, V.K.; Kumar, D. Hybrid CNN and LSTM network For heart disease prediction. *SN Comput. Sci.* **2023**, *4*, 172. [CrossRef]
19. Bhardwaj, R.; Sethi, A.; Nambiar, R. Big data in genomics: An overview. In Proceedings of the 2014 IEEE International Conference on Big Data (Big Data), Beijing, China, 4–7 August 2014; IEEE: New York, NY, USA, 2014; pp. 45–49.
20. Kayyali, B.; Knott, D.; Van Kuiken, S. *The Big-Data Revolution in US Health Care: Accelerating Value and Innovation*; Mc Kinsey & Company: Chicago, IL, USA, 2013; Volume 2, pp. 1–13.
21. Mohan, S.; Thirumalai, C.; Srivastava, G. Effective heart disease prediction using hybrid machine learning techniques. *IEEE Access* **2019**, *7*, 81542–81554. [CrossRef]
22. Singh, A.; Kumar, R. February. Heart disease prediction using machine learning algorithms. In Proceedings of the 2020 International Conference on Electrical and Electronics Engineering (ICE3), Gorakhpur, India, 14–15 February 2020; IEEE: New York, NY, USA, 2020; pp. 452–457.
23. Gavhane, A.; Kokkula, G.; Pandya, I.; Devadkar, K. March. Prediction of heart disease using machine learning. In Proceedings of the 2018 Second International Conference on Electronics, Communication and Aerospace Technology (ICECA), Coimbatore, India, 29–31 March 2018; IEEE: New York, NY, USA, 2018; pp. 1275–1278.
24. Kavitha, M.; Gnaneswar, G.; Dinesh, R.; Sai, Y.R.; Suraj, R.S. Heart disease prediction using hybrid machine learning model. In Proceedings of the 2021 6th International Conference on Inventive Computation Technologies (ICICT), Coimbatore, India, 20–22 January 2021; IEEE: New York, NY, USA, 2021; pp. 1329–1333.
25. Amiri, A.M.; Armano, G. Heart sound analysis for diagnosis of heart diseases in newborns. *APCBEE Procedia* **2013**, *7*, 109–116. [CrossRef]
26. Liu, M.; Kim, Y. Classification of heart diseases based on ECG signals using long short-term memory. In Proceedings of the 2018 40th Annual International Conference of the IEEE Engineering in Medicine and Biology Society (EMBC), Honolulu, HI, USA, 18–21 July 2018; IEEE: New York, NY, USA, 2018; pp. 2707–2710.
27. Algarni, M.; Al-Rezqi, A.; Saeed, F.; Alsaeedi, A.; Ghabban, F. Multi-constraints based deep learning model for automated segmentation and diagnosis of coronary artery disease in X-ray angiographic images. *PeerJ Comput. Sci.* **2022**, *8*, e993. [CrossRef] [PubMed]
28. Hasan, A.M.; Shin, J.; Das, U.; Srizon, A.Y. Identifying prognostic features for predicting heart failure by using machine learning algorithm. In Proceedings of the ICBET'21: 2021 11th International Conference on Biomedical Engineering and Technology, Tokyo, Japan, 17–20 March 2021; pp. 40–46.
29. Deepika, K.; Seema, S. Predictive analytics to prevent and control chronic diseases. In Proceedings of the 2016 2nd International Conference on Applied and Theoretical Computing and Communication Technology (iCATccT), Bangalore, India, 21–23 July 2016; IEEE: New York, NY, USA, 2016; pp. 381–386.
30. Uyar, K.; Ilhan, A. Diagnosis of heart disease using genetic algorithm based trained recurrent fuzzy neural networks. *Procedia Comput. Sci.* **2017**, *120*, 588–593. [CrossRef]
31. Deng, M.; Wang, C.; Tang, M.; Zheng, T. Extracting cardiac dynamics within ECG signal for human identification and cardiovascular diseases classification. *Neural Netw.* **2018**, *100*, 70–83. [CrossRef]
32. Das, R.; Turkoglu, I.; Sengur, A. Effective diagnosis of heart disease through neural networks ensembles. *Expert Syst. Appl.* **2009**, *36*, 7675–7680. [CrossRef]
33. Huang, J.-D.; Wang, J.; Ramsey, E.; Leavey, G.; Chico, T.J.A.; Condell, J. Applying artificial intelligence to wearable sensor data to diagnose and predict cardiovascular disease: A review. *Sensors* **2022**, *22*, 8002. [CrossRef]
34. Moshawrab, M.; Adda, M.; Bouzouane, A.; Ibrahim, H.; Raad, A. Smart Wearables for the Detection of Cardiovascular Diseases: A Systematic Literature Review. *Sensors* **2023**, *23*, 828. [CrossRef] [PubMed]
35. Alkayyali, Z.K.; Idris, S.A.B.; Abu-Naser, S.S. A Systematic Literature Review of Deep and Machine Learning Algorithms in Cardiovascular Diseases Diagnosis. *J. Theor. Appl. Inf. Technol.* **2023**, *101*, 1353–1365.
36. Jafari, M.; Shoeibi, A.; Khodatars, M.; Ghassemi, N.; Moridian, P.; Alizadehsani, R.; Khosravi, A.; Ling, S.H.; Delfan, N.; Zhang, Y.-D.; et al. Automated diagnosis of cardiovascular diseases from cardiac magnetic resonance imaging using deep learning models: A review. *Comput. Biol. Med.* **2023**, *160*, 106998. [CrossRef] [PubMed]
37. Kim, H.; Ishag, M.I.M.; Piao, M.; Kwon, T.; Ryu, K.H. A data mining approach for cardiovascular disease diagnosis using heart rate variability and images of carotid arteries. *Symmetry* **2016**, *8*, 47. [CrossRef]
38. Boulares, M.; Alotaibi, R.; AlMansour, A.; Barnawi, A. Cardiovascular disease recognition based on heartbeat segmentation and selection process. *Int. J. Environ. Res. Public Health* **2021**, *18*, 10952. [CrossRef] [PubMed]
39. Moradi, H.; Al-Hourani, A.; Concilia, G.; Khoshmanesh, F.; Nezami, F.R.; Needham, S.; Baratchi, S.; Khoshmanesh, K. Recent developments in modeling, imaging, and monitoring of cardiovascular diseases using machine learning. *Biophys. Rev.* **2023**, *15*, 19–33. [CrossRef]
40. Bhatt, C.M.; Patel, P.; Ghetia, T.; Mazzeo, P.L. Effective heart disease prediction using machine learning techniques. *Algorithms* **2023**, *16*, 88. [CrossRef]
41. Zhang, S.; Yuan, Y.; Yao, Z.; Wang, X.; Lei, Z. Improvement of the performance of models for predicting coronary artery disease based on XGBoost algorithm and feature processing technology. *Electronics* **2022**, *11*, 315. [CrossRef]
42. Hagan, R.; Gillan, C.J.; Mallett, F. Comparison of machine learning methods for the classification of cardiovascular disease. *Inform. Med. Unlocked* **2021**, *24*, 100606. [CrossRef]

43. Ghongade, O.S.; Reddy, S.K.S.; Tokala, S.; Hajarathaiah, K.; Enduri, M.K.; Anamalamudi, S. A Comparison of Neural Networks and Machine Learning Methods for Prediction of Heart Disease. In Proceedings of the 2023 3rd International Conference on Intelligent Communication and Computational Techniques (ICCT), Jaipur, India, 19–20 January 2023; pp. 1–7.

 *diagnostics*

Article

Validation of a Deep Learning Chest X-ray Interpretation Model: Integrating Large-Scale AI and Large Language Models for Comparative Analysis with ChatGPT

Kyu Hong Lee, Ro Woon Lee * and Ye Eun Kwon

Department of Radiology, College of Medicine, Inha University, Incheon 22212, Republic of Korea
* Correspondence: rowoon2@hanmail.net

Abstract: This study evaluates the diagnostic accuracy and clinical utility of two artificial intelligence (AI) techniques: Kakao Brain Artificial Neural Network for Chest X-ray Reading (KARA-CXR), an assistive technology developed using large-scale AI and large language models (LLMs), and ChatGPT, a well-known LLM. The study was conducted to validate the performance of the two technologies in chest X-ray reading and explore their potential applications in the medical imaging diagnosis domain. The study methodology consisted of randomly selecting 2000 chest X-ray images from a single institution's patient database, and two radiologists evaluated the readings provided by KARA-CXR and ChatGPT. The study used five qualitative factors to evaluate the readings generated by each model: accuracy, false findings, location inaccuracies, count inaccuracies, and hallucinations. Statistical analysis showed that KARA-CXR achieved significantly higher diagnostic accuracy compared to ChatGPT. In the 'Acceptable' accuracy category, KARA-CXR was rated at 70.50% and 68.00% by two observers, while ChatGPT achieved 40.50% and 47.00%. Interobserver agreement was moderate for both systems, with KARA at 0.74 and GPT4 at 0.73. For 'False Findings', KARA-CXR scored 68.00% and 68.50%, while ChatGPT scored 37.00% for both observers, with high interobserver agreements of 0.96 for KARA and 0.97 for GPT4. In 'Location Inaccuracy' and 'Hallucinations', KARA-CXR outperformed ChatGPT with significant margins. KARA-CXR demonstrated a non-hallucination rate of 75%, which is significantly higher than ChatGPT's 38%. The interobserver agreement was high for KARA (0.91) and moderate to high for GPT4 (0.85) in the hallucination category. In conclusion, this study demonstrates the potential of AI and large-scale language models in medical imaging and diagnostics. It also shows that in the chest X-ray domain, KARA-CXR has relatively higher accuracy than ChatGPT.

Keywords: ChatGPT; KARA-CXR; chest X-ray; LLM

check for
updates

Citation: Lee, K.H.; Lee, R.W.; Kwon, Y.E. Validation of a Deep Learning Chest X-ray Interpretation Model: Integrating Large-Scale AI and Large Language Models for Comparative Analysis with ChatGPT. *Diagnostics* **2024**, *14*, 90. https://doi.org/10.3390/diagnostics14010090

Academic Editor: Mugahed A. Al-antari

Received: 5 December 2023
Revised: 28 December 2023
Accepted: 29 December 2023
Published: 30 December 2023

1. Introduction

Artificial intelligence (AI) revolutionizes healthcare by improving clinical diagnosis, administration, and public health infrastructures. AI applications in healthcare include disease diagnosis, drug discovery, assisted surgeries, and patient care. AI can enhance healthcare outcomes, reduce costs, and optimize treatment planning [1]. However, challenges to be overcome include ensuring ethical boundaries, addressing bias in AI algorithms, and maintaining diversity, transparency, and accountability in algorithm development. AI is not meant to replace doctors and healthcare providers but to complement their skills through human—AI collaboration. The human-in-the-loop approach ensures safety and quality in healthcare services, where AI systems are guided and supervised by human expertise.

The rise in large language models (LLMs) in AI has garnered significant attention and investment from companies like Google, Amazon, Facebook, Tesla, and Apple. LLMs, such as OpenAI's GPT series and ChatGPT, have shown remarkable progress in tasks like text generation, language translation, and question answering. These models are trained on massive amounts of data and have the potential to display intelligence beyond their primary task of predicting the next word in a text.

LLMs have the potential to revolutionize healthcare by assisting medical professionals with administrative tasks, improving diagnostic accuracy, and engaging patients [2]. LLMs, such as GPT-4 and Bard, can be implemented in healthcare settings to facilitate clinical documentation, obtain insurance pre-authorization, summarize research papers, and answer patient questions [3]. They can generate personalized treatment recommendations, laboratory test suggestions, and medication prompts based on patient information [4]. It is essential to ensure LLMs' responsible and ethical use in medicine and healthcare, considering privacy, security, and the potential for perpetuating harmful, inaccurate, race-based content [5]. LLMs, like ChatGPT, can accelerate the creation of clinical practice guidelines by quickly searching and selecting evidence from numerous databases [6].

KakaoBrain AI for Radiology Assistant Chest X-ray (KARA-CXR) is a new medical technology that helps in radiological diagnosis. Developed by leveraging the cutting-edge capabilities of large-scale artificial intelligence and advanced language models, this cloud-based tool represents a significant leap in medical imaging analysis. The core functionality of KARA-CXR lies in its ability to generate detailed radiological reports that include findings and conclusions. This process is facilitated by its sophisticated AI, which has been trained on vast datasets of chest X-ray images. By interpreting these images, KARA-CXR can provide accurate and swift diagnostic insights essential in clinical decision-making.

Based on the GPT-4V architecture, ChatGPT has potential in the medical field, especially for interpreting chest X-ray images. This language model can analyze medical images, including chest X-ray data, to generate human-like reading reports. Although not yet available for clinical use, by providing a general interpretation of chest X-rays, ChatGPT has the potential to improve the diagnostic process, especially in settings with limited access to radiology expertise [7]. In this study, we analyze the diagnostic accuracy and utility of KARA-CXR and ChatGPT and discuss their potential for use in clinical settings.

2. Materials and Methods

2.1. Dataset

We randomly selected 2000 chest X-ray images (PA projection) from a single institution (Inha University Hospital, Incheon, Republic of Korea) from 2010 to 2022. The selected images were all of Asian individuals, with a male and female ratio of 46% and 54%, respectively. To ensure ease of reading, we excluded pediatric patients, poor-quality images (images not taken with digital equipment or taken with portable equipment) and selected only images of adult patients (aged 19 to 99 years, median age: 46.8 ± 2.5 SD). Furthermore, to ensure fairness in the assessment of reading difficulty, percentages were not separately established for each image' disease. Finally, the examination of the selected images and the decision to include them in the dataset was performed by a radiologist with 10 years of experience (Ro Woon, Lee). Furthermore, all included images were fully anonymized using Python (version 3.12) and then serially numbered for analysis. The files were exported as DICOM files.

2.2. Input Data

Anonymized DICOM files were uploaded to both KARA-CXR (Kakaobrain, Seoul, Republic of Korea) and ChatGPT (OpenAI, San Francisco, CA, USA). To anonymize medical images for analysis in ChatGPT and KARA-CXR, we removed all identifiable patient information to comply with the privacy and confidentiality standards set forth by the Health Insurance Portability and Accountability Act (HIPAA). Anonymization involved removing details such as patient names, dates of birth, medical record numbers, and other unique identifiers from the images.

Even after anonymization, we further enhanced privacy by turning off the "Chat History and Training" option in ChatGPT. This setting ensures that conversations and images shared during a session are not used for further training of the AI model or accessed in future sessions. This is a precautionary measure to ensure that residual or indirect information is not used in ways that could compromise patient confidentiality.

In KARA-CXR, a cloud-based analysis system, immediately deleted the input DICOM data after analysis for personal information protection. Unlike ChatGPT, KARA-CXR generates text-based readings shortly after uploading DICOM files without requiring separate prompts. KARA-CXR utilized a closed beta version prior to public release (Figure 1), and there are plans to make it publicly available via the website in December 2023.

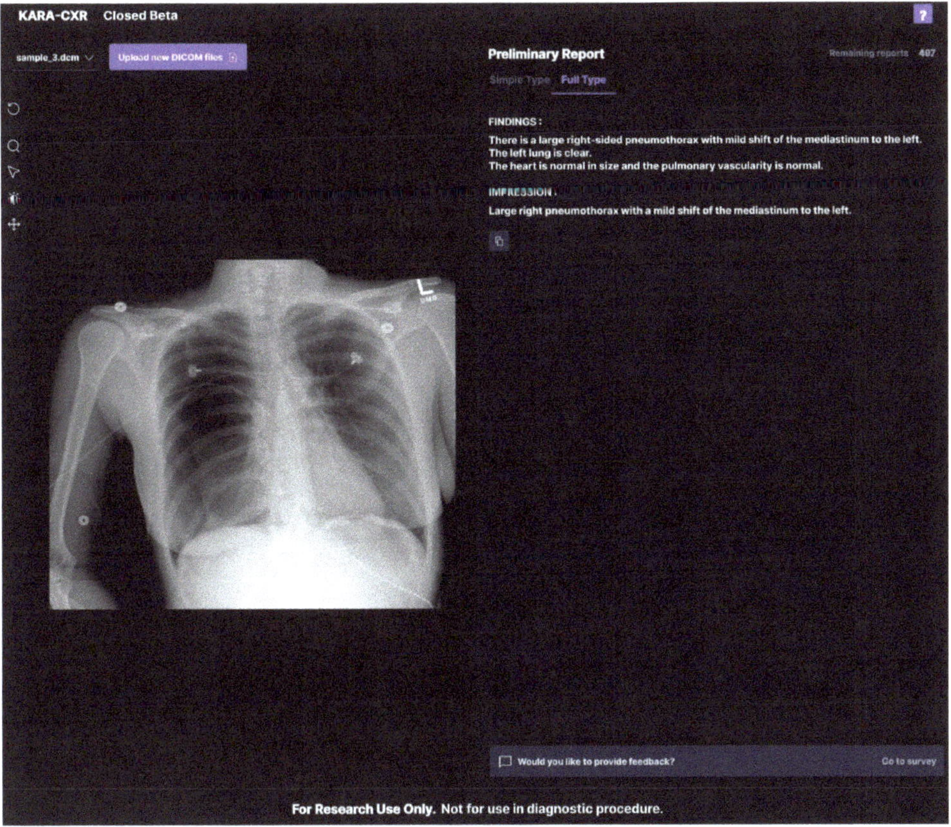

Figure 1. The appearance of KARA-CXR operating in a web browser. When a DICOM file is uploaded, findings and corresponding impressions are displayed on the right-side tab.

In the case of ChatGPT, to obtain the right results for our research, we first chose the prompt to be entered into ChatGPT. ChatGPT is designed with guidelines that prevent it from providing professional interpretations or diagnoses, especially in contexts requiring specialized expertise, such as medical imaging, including chest X-rays [8]. To overcome the limitations of this large language model, we employed a carefully crafted, non-directive bypass prompt: 'This is a chest PA image. Tell me more about what is going on?' This prompt was strategically chosen to navigate ChatGPT's usage policies and ethical constraints, allowing us to obtain a chest x-ray reading from ChatGPT. Furthermore, ChatGPT was used in its paid version, GPT-4V, and to protect personal information, the 'Chat history & training' option was disabled, ensuring data were not stored on OpenAI's servers.

The interpretation texts thus generated were qualitatively analyzed by two observers. A rough schematic of this process can be seen in Figure 2.

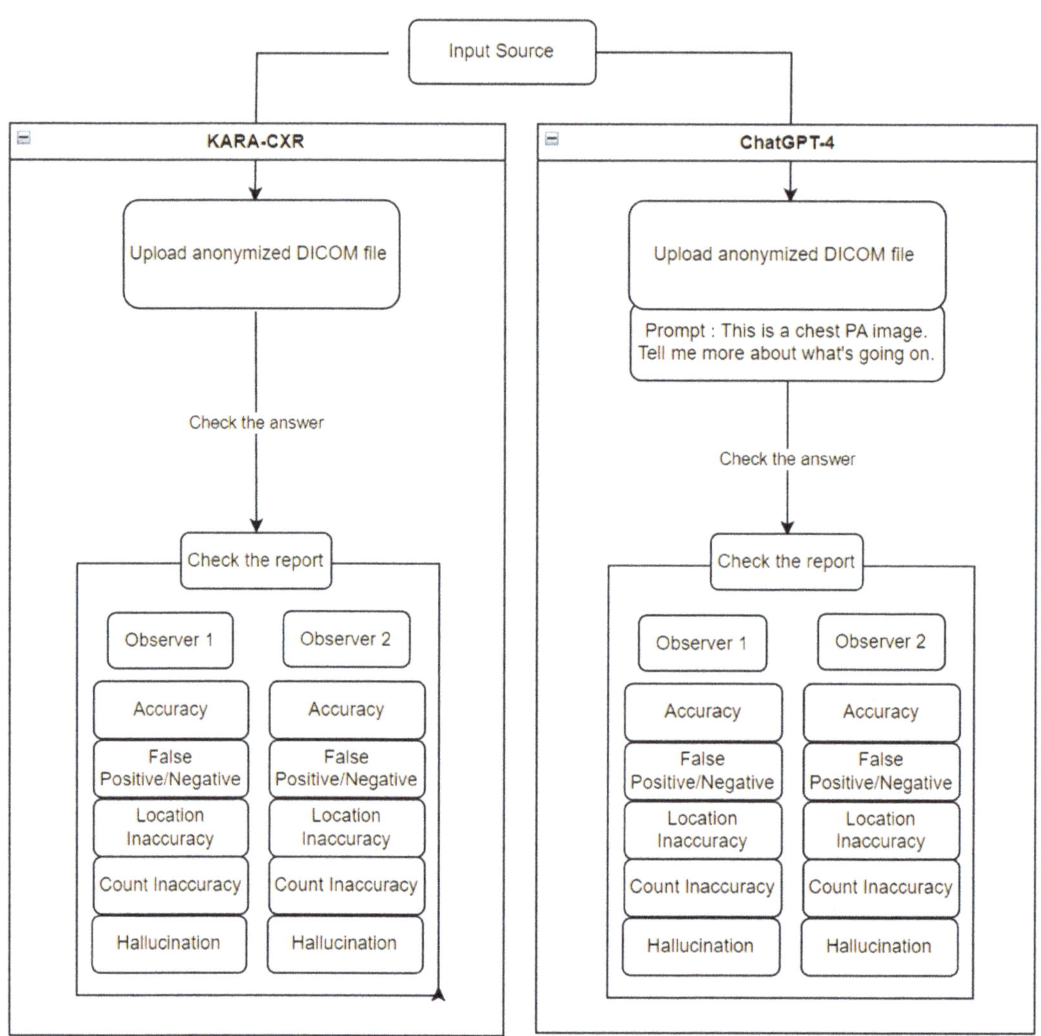

Figure 2. Schematic of data input and analysis process.

2.3. Analyzing Readings by LLMs

We selected five qualitative factors (accuracy, false findings, location inaccuracies, count inaccuracies, and hallucination) to evaluate the quality of the readings generated by KARA-CXR and ChatGPT. The detailed descriptions of the factors are shown in Table 1.

Table 1. The qualitative factors with which to evaluate the quality of the readings.

Assessment		Description
	Acceptable	The reading is accurate and clinically useful.
Accuracy	Questionable	There are errors in the reading, but it retains some clinical usability.
	Unacceptable	There are significant errors in the reading, rendering it clinically useless.

Table 1. *Cont.*

Assessment		Description
False Findings	None	There are no false findings.
	False Positive (FP)	The reading includes a false positive.
	False Negative (FN)	The reading includes a false negative.
	Both	The reading has both false positives and false negatives.
Location Inaccuracy	None	There is no location inaccuracy.
	Not significant	The location of lesions is inaccurately identified, but it does not significantly affect clinical judgment.
	Significant	The location of lesions is inaccurately identified, and it severely affects clinical judgment.
Count Inaccuracy	None	There is no count inaccuracy.
	Single	The count of lesions is inaccurate, but single error is noted.
	Multiple	The count of lesions is incorrect and multiple count errors of lesion are seen.
Hallucination	None	There are no hallucinations in the reading.
	Not significant	Hallucinations are present but do not significantly affect clinical judgment.
	Significant	Hallucinations are present and significantly affect clinical judgment.

For the five items mentioned in Table 1, two readers with ten years of experience in chest radiology reading evaluated the images in independent sessions. We evaluated the interpretation results of each model for chest X-ray images and recorded the evaluation results according to the case numbers of the anonymized images.

2.4. Statistical Analytics

We analyzed the percentages of details in each of the five assessment categories rated by each reader and obtained interobserver agreement between each reader. The statistical analysis of the data was performed in Python (version 3.12).

3. Results

In evaluating diagnostic accuracy, two observers assessed the performance of KARA and GPT4. Observer 1 found that KARA achieved 70.50% accuracy in the category deemed 'Acceptable', while GPT4 was reported at a notably lower value of 40.50% in the same category. Observer 2's assessments were slightly lower for KARA at 68.00% but higher for GPT4 at 47.00% in the 'Acceptable' category (Figure 3). The interobserver agreement rates, which reflect the consistency between observers, were relatively close, with KARA at 0.74 and GPT4 at 0.73, indicating moderate agreement.

In the category of 'False Findings' with no findings being the subcategory, both observers recorded similar results for KARA, with Observer 1 at 68.00% and Observer 2 at 68.50%. In comparison, GPT4 was observed at 37.00% by both observers (Figure 4). The interobserver agreement for KARA stood at a high rate of 0.96, and GPT4 also had a high agreement rate of 0.97. These high agreement rates suggest a consistent assessment of false findings between the two observers for KARA and GPT4.

When it came to 'Location Inaccuracy' with no inaccuracies noted, Observer 1 reported KARA at 76.00% and GPT4 at 46.50%, and Observer 2 reported KARA at 77.50% and GPT4 at 46.00% (Figure 5). The interobserver agreement for KARA was 0.93, indicating high consistency, whereas GPT4's agreement was lower at 0.83, signifying a moderate-to-high consistency between observers.

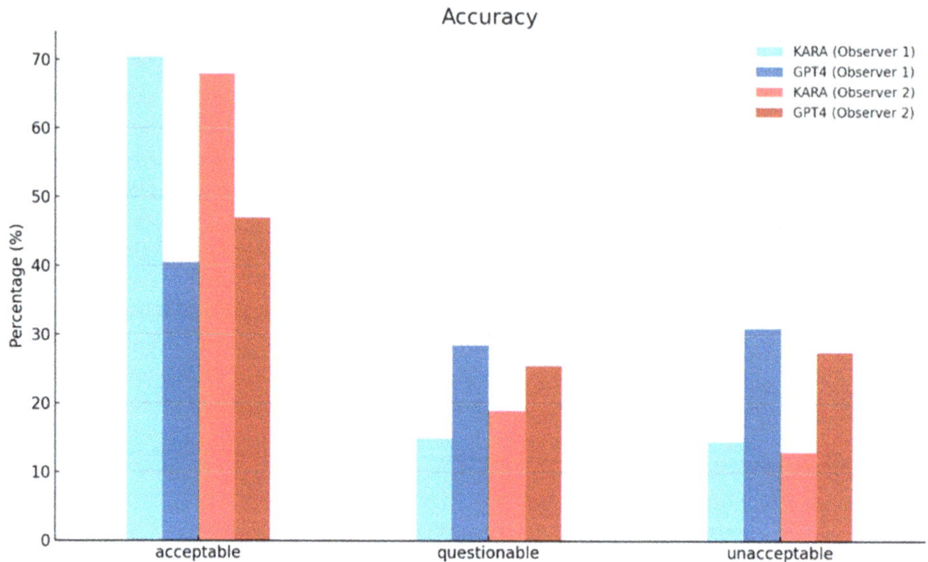

Figure 3. Diagnostic accuracy of KARA-CXR and GPT4.

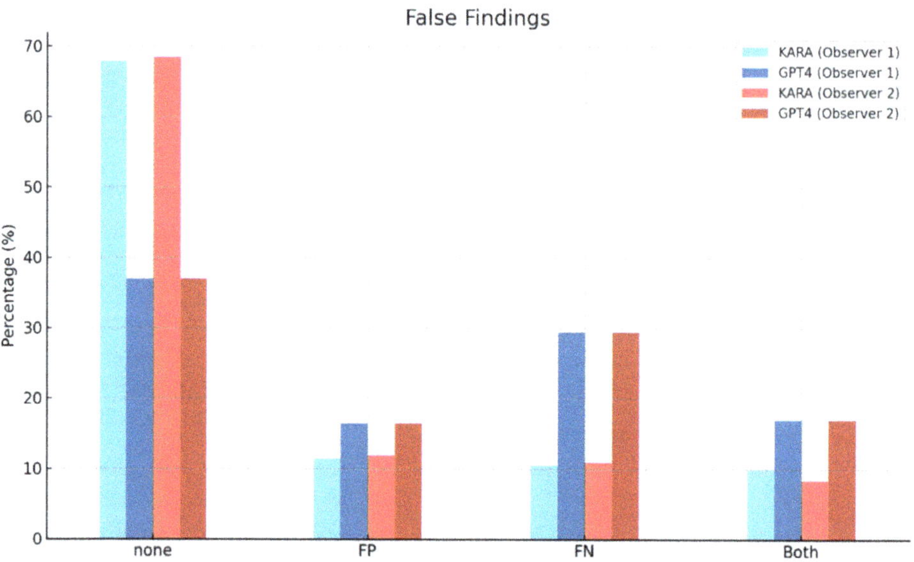

Figure 4. False findings of KARA-CXR and GPT4.

'Count Inaccuracy' with no inaccuracies was observed at a high rate by both observers for KARA (94.00%) and GPT4 (90.00%), reflecting a very high level of performance in this category (Figure 6). The interobserver agreement for KARA and GPT4 was at 0.99, indicating almost perfect consistency between the two observers.

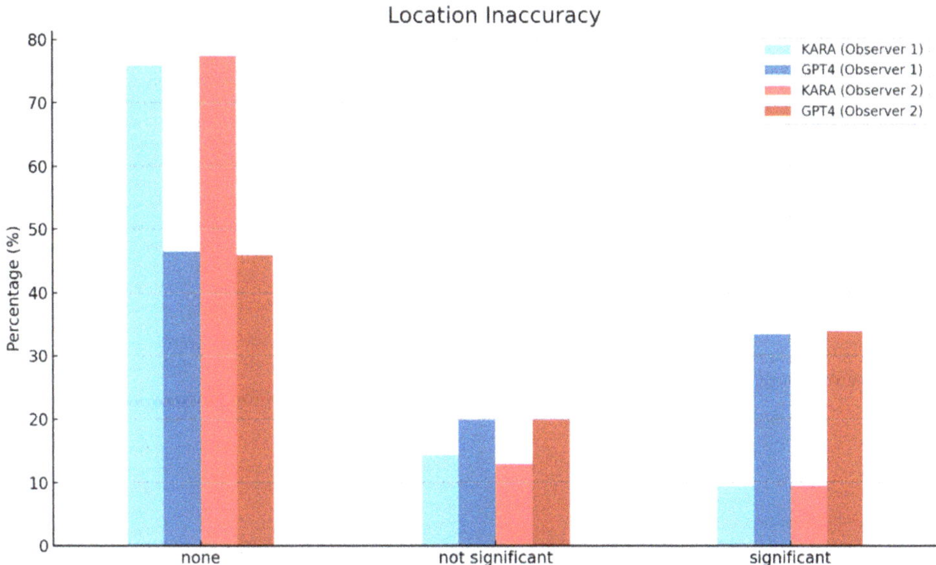

Figure 5. Location inaccuracy of KARA-CXR and GPT4.

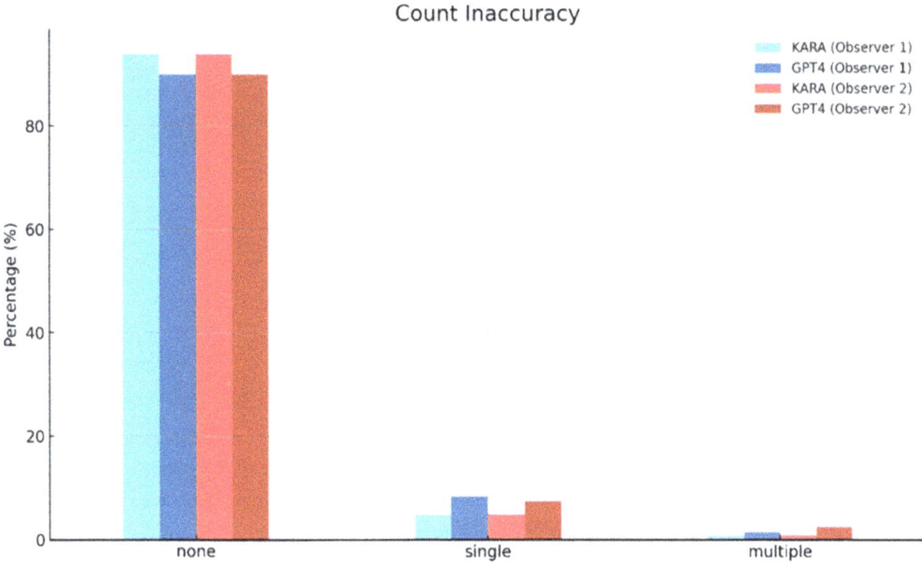

Figure 6. Count inaccuracy of KARA-CXR and GPT4.

Lastly, the 'Hallucination' category with no instances reported by Observer 1 showed KARA at 75.88% and GPT4 at 38.69%, while Observer 2 reported the same percentage for KARA and a slightly lower 38.19% for GPT4 (Figure 7). The interobserver agreement was 0.91 for KARA and 0.85 for GPT4, demonstrating high consistency for KARA assessments and moderate-to-high consistency for GPT4.

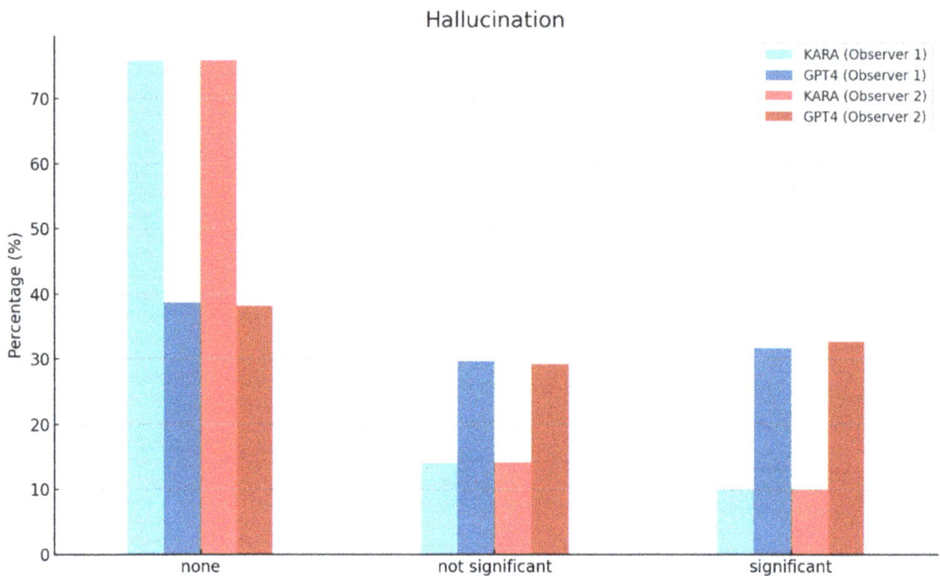

Figure 7. Hallucination of KARA-CXR and GPT4.

4. Discussion

AI is being increasingly used in the field of chest X-ray reading. It has various applications, including lung cancer risk estimation, detection, and diagnosis, reducing reading time, and serving as a second 'reader' during screening interpretation [9]. Doctors in a single hospital reported positive experiences and perceptions of using AI-based software for chest radiographs, finding it useful in the emergency room and for detecting pneumothorax [10]. A model for automatic diagnosis of different diseases based on chest radiographs using machine learning algorithms has been proposed [11]. In a multicenter study, AI was used as a chest X-ray screening tool and achieved good performance in detecting normal and abnormal chest X-rays, reducing turnaround time, and assisting radiologists in assessing pathology [12]. AI solutions for chest X-ray evaluation have been demonstrated to be practical, perform well, and provide benefits in clinical settings [13].

However, conventional labeling-based chest X-ray reading AI has limitations in terms of accuracy and efficiency. The manual labeling of large datasets is expensive and time-consuming. Automatic label extraction from radiology reports is challenging due to semantically similar words and missing annotated data [14]. In a multicenter evaluation, the AI algorithm for chest X-ray analysis showed lower sensitivity and specificity values during prospective validation compared to retrospective evaluation [15]. However, the AI model performed at the same level as or slightly worse than human radiologists in most regions of the ROC curve [15]. A method for standardized automated labeling based on similarity to a previously validated, explainable AI model-derived atlas has been proposed to overcome these limitations. Fine-tuning the original model using automatically labeled exams can preserve or improve performance, resulting in a highly accurate and more generalized model.

The effectiveness of deep-learning based computer-aided diagnosis has been demonstrated in disease detection [16]. However, one of the major challenges in training deep learning models for medical purposes is the need for extensive, high-quality clinical annotation, which is time-consuming and costly. Recently, CLIP [17] and ALIGN [18] have shown the ability to perform vision tasks without any supervision. However, vision-language pre-training (VLP) in the CXR domain still lacks sufficient image-text datasets because many public datasets consist of image-label pairs with different class compositions.

The rise in medical image reading with large language models has gained significant attention in recent research [19]. Language models have been explored to improve various tasks in medical imaging, such as image captioning, report generation, report classification, finding extraction, visual question answering, and interpretable diagnosis. Researchers have highlighted the potential benefits of accurate and efficient language models in medical imaging analysis, including improving clinical workflow efficiency, reducing diagnostic errors, and assisting healthcare professionals in providing timely and accurate diagnoses [20].

KARA-CXR is an innovative cloud-based medical technology that utilizes artificial intelligence and advanced language models to revolutionize radiological diagnostics. It operates over the web and offers a user-friendly interface for healthcare professionals. KARA-CXR generates detailed radiological reports with findings and conclusions by analyzing chest X-ray images uploaded in DICOM format. This is made possible by its sophisticated AI, which has been trained on vast datasets of chest X-ray images. The technology provides accurate and swift diagnostic insights, aiding radiologists in ensuring precise diagnoses and reducing report generation time. KARA-CXR is particularly valuable in high-volume or resource-limited settings where radiologist expertise may be scarce or overburdened.

In this study, ChatGPT based on GPT-4V architecture showed some potential in interpreting chest X-ray images but also revealed some limitations. ChatGPT can generate human-like diagnostic reports based on chest X-ray data through extensive reinforcement learning on the medical text and imaging data included during development. However, due to the limitations of reinforcement learning based on information openly available on the internet, we must recognize that the data generated by ChatGPT do not guarantee medical expertise. In conclusion, it is essential to note that ChatGPT is not a substitute for professional medical advice, diagnosis, or treatment [21].

In our study, detailed observations of reports indicate that KARA generally outperforms GPT4 across various categories of diagnostic accuracy, with consistently higher percentages and interobserver agreement rates. The data suggest a significant discrepancy between the two systems, with KARA displaying more reliable and accurate performance as per the observers' evaluations. Particularly in terms of hallucination, KARA-CXR demonstrated superior performance compared to ChatGPT. ChatGPT sometimes produced incorrect interpretation results, including hallucinations, even in cases with clinically significant and obvious abnormalities such as pneumothorax (Figure 8).

In our comparative analysis between KARA-CXR and ChatGPT, a striking advantage of KARA-CXR was observed in the hallucination. Notably, KARA-CXR demonstrated a significantly higher percentage in non-hallucinations with a non-hallucination rate of 75% as compared to that of only 38% for ChatGPT, as agreed upon by both observers. This substantial difference underscores the superior capability of KARA-CXR in providing reliable and accurate interpretations in chest X-ray diagnostics, a crucial aspect in the field of medical imaging where the precision of diagnosis can significantly impact patient outcomes. The propensity of ChatGPT to generate more hallucinations in medical contexts can be attributed to its foundational design and training methodology. As a large language model, ChatGPT is trained on a vast corpus of text from diverse sources, not specifically tailored for medical diagnostics. This generalist approach, while versatile, can lead to inaccuracies and hallucinations, especially in highly specialized fields like medical imaging [22]. Despite its potential, the accuracy and reliability of ChatGPT responses should be carefully assessed, and its limitations in understanding medical terminology and context should be addressed [23]. In contrast, KARA-CXR, designed explicitly for medical image analysis, benefits from a more focused training regime, enabling it to discern nuanced details in medical images more effectively and reducing the likelihood of generating erroneous interpretations.

 **You**

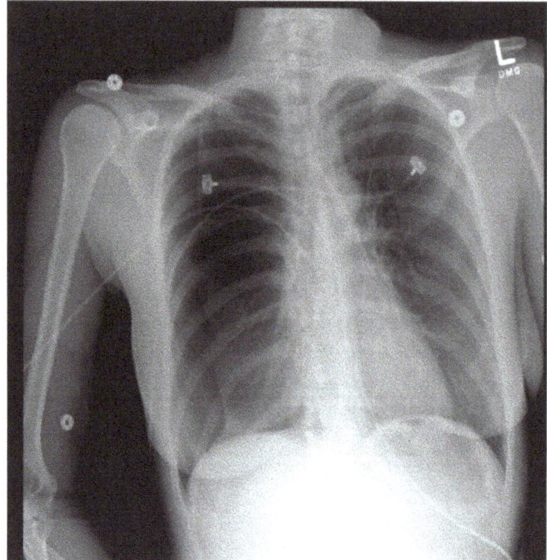

Tell me more about what's going on.

ChatGPT

The image appears to be a standard posteroanterior (PA) chest X-ray. Chest X-rays are used to diagnose and monitor a multitude of conditions involving the chest, heart, and lungs.

In this X-ray:

1. **Lungs**: The lungs should appear mostly black due to the air content, with the pulmonary vessels visible as lighter, branching structures. There are no obvious abnormal opacities, but a detailed analysis would be required to rule out any subtle findings.
2. **Heart**: The heart is situated in the center of the thorax; its size and contour can give information about cardiac health and morphology.
3. **Diaphragm**: The diaphragm appears as a smooth, curved line above the abdominal area, and its position can indicate proper lung inflation and function.
4. **Bones**: The ribs, clavicle, and part of the humerus are visible. The bone structures should be evaluated for any signs of fractures or lesions.
5. **Soft Tissues**: The soft tissues around the chest, including the skin and muscles, can sometimes show signs of swelling or other issues.
6. **Medical Devices**: There appear to be markers or devices with alphanumeric codes, which are typically used for identification and orientation during the X-ray (the 'L' indicates this is the left side of the patient).

Figure 8. A case where ChatGPT misidentified pneumothorax as normal.

In our exploration of ChatGPT's application to medical imaging, particularly in chest X-ray interpretation, a notable limitation emerged, meriting explicit mention. ChatGPT, in its current design, is programmed to refuse direct requests for the professional interpretation of medical images, such as X-rays [8]. This usage policy and ethical boundary, built into ChatGPT to avoid the non-professional practice of medicine, significantly impacts its clinical application in this context. In the initial process of our study, we observed that direct prompts requesting chest X-ray interpretation were consistently declined by ChatGPT, aligning with its programming to avoid assuming the role of a radiologist or other medical professional. This limitation is critical to understand for any future research utilizing ChatGPT or similar language models in medical image interpretation. Despite the impressive capabilities of AI in healthcare, such as KARA-CXR and ChatGPT, hallucinations can cause serious problems in real-world clinical applications of AI. Such hallucinations may be of minimal consequence in casual conversation or other contexts but can pose significant risks when applied to the healthcare sector, where accuracy and reliability are of paramount importance. Misinformation in the medical domain can lead to severe health consequences on patient care and outcomes. The accuracy and reliability of information provided by language models can be a matter of life or death. They pose real-life risks, as they could potentially affect healthcare decisions, diagnosis, and treatment plans. Hence, the development of methods to evaluate and mitigate such hallucinations is not just of academic interest but of practical importance.

While promising, the integration of SaMD (software as medical device), including KARA-CXR and ChatGPT, into medical diagnostics faces several challenges that must be addressed in future research. A primary concern is the diversity of data used to train these AI models. Often, AI systems are trained on datasets that may only adequately represent some population groups, leading to potential biases and inaccuracies in diagnostics, particularly for underrepresented demographics. Moreover, these AI systems' "black box" nature poses a significant challenge. The internal mechanisms of how they analyze and interpret chest X-ray images are only partially transparent, making it difficult for healthcare professionals to understand and thus trust the conclusions drawn by these technologies.

Another notable limitation is the integration of these AI tools into clinical practice. Healthcare professionals may be hesitant to depend on AI for critical diagnostic tasks due to concerns about the accuracy and reliability of these systems, as well as potential legal and ethical implications. Building trust in AI technologies is essential for their successful adoption in medical settings [24]. In addition to these concerns, it is essential to keep in mind that even if an AI-powered diagnostic solution is highly accurate, the final judgment should still be made by a medical professional—a doctor.

To overcome these challenges, future research should focus on enhancing the diversity of training datasets, including a broader range of demographic data, to ensure that AI models can deliver accurate diagnostics across different populations [25]. It is also crucial to improve the transparency and explainability of AI algorithms, developing methods to demystify the decision making process and increase acceptability and trustworthiness among medical practitioners [26]. Although this paper evaluates the diagnostic accuracy of two potential SaMDs, ChatGPT and KARA-CXR, one of the limitations is that there needs to be a clear rationale or recommendation for evaluating or approving such software, legally or within the academic community. The limitation in approving software for medical use (SaMD) stems from the need for a clear definition of SaMD, which makes it difficult to create standards and regulations for its development and implementation [27]. Without clear boundaries, there are risks to patient safety because not all components potentially impacting SaMD are covered by regulations [27]. This lack of clarity also affects innovation and design in the field of SaMD, as new technology applications that support healthcare monitoring and service delivery may need to be more effectively regulated [28]. We believe that gradually, along with software development, we will need to establish factors and regulations that will define the clinical accuracy and safety of these SaMDs. Extensive clinical validation studies are necessary to establish the reliability and accuracy of AI-based diagnostic tools, adhering to high ethical standards and regulatory compliance [29]. These studies should also address patient privacy, data security, and the potential ramifications of misdiagnoses [29]. By focusing on these areas, the potential of AI in medical diagnostics can be more fully realized, leading to enhanced patient care and more efficient healthcare delivery.

The limitations of this study include that this research was conducted as a single-institution study, which presents certain limitations. One of the primary constraints was the limited number of images that could be analyzed due to the restricted number of researchers involved in the study. This limitation could potentially impact the generalizability of our findings to broader image populations and diverse clinical settings. Another significant limitation was the lack of a reference standard for the chest X-ray interpretations. Although we analyzed the interobserver agreement between the readers, the absence of a definitive standard means that even interpretations by experienced readers cannot be considered definitive answers. This aspect could affect the reliability and validity of the diagnostic conclusions drawn in our study. Additionally, ethical considerations programmed into ChatGPT led to the refusal of direct requests for chest image interpretation, necessitating the use of indirect prompts to obtain diagnostic interpretations. This workaround might have influenced the quality and accuracy of the results derived from ChatGPT. We acknowledge that the possibility of obtaining more accurate results from ChatGPT cannot be entirely ruled out if direct requests for chest X-ray interpretation were permissible.

5. Conclusions

This study underscores the potential of AI in improving medical diagnostic processes, with specific emphasis on chest X-ray interpretation. While KARA demonstrates superior precision in image analysis, ChatGPT excels in contextual data interpretation. The key take-away is the complementary nature of these technologies. A hybrid approach, integrating KARA's imaging expertise with ChatGPT's comprehensive analysis, could lead to more accurate and efficient diagnostic processes, ultimately improving patient care.

The future of AI in healthcare is about more than replacing human expertise but augmenting it. Combining AI systems like KARA and ChatGPT with human oversight could offer a robust diagnostic tool, maximizing the strengths of both artificial intelligence and human judgment. As AI continues to evolve, its integration into healthcare systems must be approached thoughtfully, ensuring that it supports and enhances the work of medical professionals for the betterment of patient outcomes.

Author Contributions: Conceptualization, R.W.L.; Resources, K.H.L.; Writing—original draft, K.H.L. and R.W.L.; Writing—review & editing, K.H.L., R.W.L. and Y.E.K.; Supervision, R.W.L.; Project administration, K.H.L., R.W.L. and Y.E.K. All authors have read and agreed to the published version of the manuscript.

Funding: This research was funded by KakaoBrain, grant number [2023-22-001].

Institutional Review Board Statement: This study was approved by the institutional review board of Inha University Hospital (IRB No. 2023-10-049, approval date: 15 October 2023), which waived the requirement for informed patient consent.

Informed Consent Statement: The requirement for written informed consent was waived by the Institutional Review Board.

Data Availability Statement: The data presented in this study are available on request from the corresponding author. The data are not publicly available due to need approval from the affiliated institution's DRB (Data review board) is required for disclosure or export.

Conflicts of Interest: Ro Woon Lee is a co-researcher in developing Kakaobrain KARA-CXR and has received research funding from Kakaobrain. The other authors have no potential conflict of interest to disclose.

References

1. Sezgin, E. Artificial intelligence in healthcare: Complementing, not replacing, doctors and healthcare providers. *Digit. Health* **2023**, *9*, 20552076231186520. [CrossRef] [PubMed]
2. Meskó, B.; Topol, E.J. The imperative for regulatory oversight of large language models (or generative AI) in healthcare. *Npj Digit. Med.* **2023**, *6*, 120. [CrossRef] [PubMed]
3. Yang, H.; Li, J.; Liu, S.; Du, L.; Liu, X.; Huang, Y.; Shi, Q.; Liu, J. Exploring the Potential of Large Language Models in Personalized Diabetes Treatment Strategies. *medRxiv* **2023**. [CrossRef]

4.	Omiye, J.A.; Lester, J.; Spichak, S.; Rotemberg, V.; Daneshjou, R. Beyond the hype: Large language models propagate race-based medicine. *medRxiv* **2023**. [CrossRef]

5.	Tustumi, F.; Andreollo, N.A.; Aguilar-Nascimento, J.E.D. Future of the language models in healthcare: The role of chatbot. *Arq. Bras. De Cir. Dig.* **2023**, *36*, e1727. [CrossRef] [PubMed]

6.	Zhu, L.; Mou, W.; Chen, R. Can the ChatGPT and other Large Language Models with internet-connected database solve the questions and concerns of patient with prostate cancer? *J. Transl. Med.* **2023**, *21*, 269. [CrossRef] [PubMed]

7.	Beaulieu-Jones, B.R.; Shah, S.; Berrigan, M.T.; Marwaha, J.S.; Lai, S.L.; Brat, G.A. Evaluating Capabilities of Large Language Models: Performance of GPT4 on American Board of Surgery Qualifying Exam Question Banks. *medRxiv* **2023**. [CrossRef]

8.	OpenAI. Usage Policies. 2023. Available online: https://openai.com/policies/usage-policies (accessed on 28 December 2023).

9.	Vedantham, S.; Shazeeb, M.S.; Chiang, A.; Vijayaraghavan, G.R. Artificial Intelligence in Breast X-ray Imaging. *Semin. Ultrasound CT MRI* **2022**, *44*, 2–7. [CrossRef]

10.	Shin, H.J.; Lee, S.; Kim, S.; Son, N.H.; Kim, E.K. Hospital-wide survey of clinical experience with artificial intelligence applied to daily chest radiographs. *PLoS ONE* **2023**, *18*, e0282123. [CrossRef]

11.	Tembhare, N.P.; Tembhare, P.U.; Chauhan, C.U. Chest X-ray Analysis using Deep Learning. *Int. J. Sci. Technol. Eng.* **2023**, *11*, 1441–1447. [CrossRef]

12.	Govindarajan, A.; Govindarajan, A.; Tanamala, S.; Chattoraj, S.; Reddy, B.; Agrawal, R.; Iyer, D.; Srivastava, A.; Kumar, P.; Putha, P. Role of an Automated Deep Learning Algorithm for Reliable Screening of Abnormality in Chest Radiographs: A Prospective Multicenter Quality Improvement Study. *Diagnostics* **2022**, *12*, 2724. [CrossRef] [PubMed]

13.	Ridder, K.; Preuhs, A.; Mertins, A.; Joerger, C. Routine Usage of AI-based Chest X-ray Reading Support in a Multi-site Medical Supply Center. *arXiv* **2022**, arXiv:2210.10779.

14.	Vasilev, Y.; Vladzymyrskyy, A.; Omelyanskaya, O.; Blokhin, I.; Kirpichev, Y.; Arzamasov, K. AI-Based C.X.R. First Reading: Current Limitations to Ensure Practical Value. *Diagnostics* **2023**, *13*, 1430. [CrossRef] [PubMed]

15.	Kim, D.; Chung, J.; Choi, J.; Succi, M.D.; Conklin, J.; Longo, M.G.F.; Ackman, J.B.; Little, B.P.; Petranovic, M.; Kalra, M.K.; et al. Accurate auto-labeling of chest X-ray images based on quantitative similarity to an explainable AI model. *Nat. Commun.* **2022**, *13*, 1867. [CrossRef] [PubMed]

16.	Qin, C.; Yao, D.; Shi, Y.; Song, Z. Computer-aided detection in chest radiography based on artificial intelligence: A survey. *Biomed. Eng. Online* **2018**, *17*, 113. [CrossRef]

17.	Radford, A.; Kim, J.W.; Hallacy, C.; Ramesh, A.; Goh, G.; Agarwal, S.; Sastry, G.; Askell, A.; Mishkin, P.; Clark, J.; et al. Learning transferable visual models from natural language supervision. *arXiv* **2021**, arXiv:2103.00020.

18.	Jia, C.; Yang, Y.; Xia, Y.; Chen, Y.; Parekh, Z.; Pham, H.; Le, Q.V.; Sung, Y.; Li, Z.; Duerig, T. Scaling up visual and vision-language representation learning with noisy text supervision. *arXiv* **2021**, arXiv:2102.05918.

19.	Srivastav, S.; Chandrakar, R.; Gupta, S.; Babhulkar, V.; Agrawal, S.; Jaiswal, A.; Prasad, R.; Wanjari, M.B.; Agarwal, S.; Wanjari, M. ChatGPT in Radiology: The Advantages and Limitations of Artificial Intelligence for Medical Imaging Diagnosis. *Cureus* **2023**, *15*, e41435. [CrossRef]

20.	Hu, M.; Pan, S.; Li, Y.; Yang, X. Advancing Medical Imaging with Language Models: A Journey from N-grams to ChatGPT. *arXiv* **2023**, arXiv:2304.04920.

21.	Biswas, S.; Logan, N.S.; Davies, L.N.; Sheppard, A.L.; Wolffsohn, J.S. Assessing the utility of ChatGPT as an artificial intelligence-based large language model for information to answer questions on myopia. *Ophthalmic Physiol. Opt.* **2023**, *43*, 1562–1570. [CrossRef]

22.	Zhang, J.; Sun, K.; Jagadeesh, A.; Ghahfarokhi, M.; Gupta, D.; Gupta, A.; Gupta, V.; Guo, Y. The Potential and Pitfalls of using a Large Language Model such as ChatGPT or GPT-4 as a Clinical Assistant. *arXiv* **2023**, arXiv:2307.08152.

23.	DeGrave, A.J.; Cai, Z.R.; Janizek, J.D.; Daneshjou, R.; Lee, S.I. Dissection of medical AI reasoning processes via physician and generative-AI collaboration. *medRxiv* **2023**. [CrossRef]

24.	Jha, D.; Rauniyar, A.; Srivastava, A.; Hagos, D.H.; Tomar, N.K.; Sharma, V.; Keles, E.; Zhang, Z.; Demir, U.; Topcu, A.; et al. Ensuring Trustworthy Medical Artificial Intelligence through Ethical and Philosophical Principles. *arXiv* **2023**, arXiv:2304.11530.

25.	Polat Erdeniz, S.; Kramer, D.; Schrempf, M.; Rainer, P.P.; Felfernig, A.; Tran, T.N.; Burgstaller, T.; Lubos, S. Machine Learning Based Risk Prediction for Major Adverse Cardiovascular Events for ELGA-Authorized Clinics1. In *dHealth*; Studies in health technology and informatics; IOS Press: Amsterdam, The Netherlands, 2023. [CrossRef]

26.	Chaddad, A.; Lu, Q.; Li, J.; Katib, Y.; Kateb, R.; Tanougast, C.; Bouridane, A.; Abdulkadir, A. Explainable, Domain-Adaptive, and Federated Artificial Intelligence in Medicine. *IEEE/CAA J. Autom. Sin.* **2022**, *10*, 859–876. [CrossRef]

27.	Lal, A.; Dang, J.; Nabzdyk, C.; Gajic, O.; Herasevich, V. Regulatory oversight and ethical concerns surrounding software as medical device (SaMD) and digital twin technology in healthcare. *Ann. Transl. Med.* **2022**, *10*, 950. [CrossRef]

28.	Hewitt, A.W. Dr AI will see you now. *Clin. Exp. Ophthalmol.* **2023**, *51*, 409–410. [CrossRef]

29.	Fowler, G.E.; Blencowe, N.S.; Hardacre, C.; Callaway, M.P.; Smart, N.J.; Macefield, R. Artificial intelligence as a diagnostic aid in cross-sectional radiological imaging of surgical pathology in the abdominopelvic cavity: A systematic review. *BMJ Open* **2023**, *13*, e064739. [CrossRef]

Systematic Review

Computational Intelligence-Based Stuttering Detection: A Systematic Review

Raghad Alnashwan [1], Noura Alhakbani [1], Abeer Al-Nafjan [2,*], Abdulaziz Almudhi [3] and Waleed Al-Nuwaiser [2]

[1] Information Technology Department, College of Computer and Information Sciences, King Saud University, Riyadh 11543, Saudi Arabia; 444203361@student.ksu.edu.sa (R.A.); nhakbani@ksu.edu.sa (N.A.)

[2] Computer Science Department, College of Computer and Information Sciences, Imam Mohammad Ibn Saud Islamic University (IMSIU), Riyadh 11432, Saudi Arabia; wmalnuwaiser@imamu.edu.sa

[3] Department of Medical Rehabilitation Sciences, College of Applied Medical Sciences, King Khalid University, Abha 62529, Saudi Arabia; almudhi@kku.edu.sa

* Correspondence: annafjan@imamu.edu.sa; Tel.: +966-112597570

Abstract: Stuttering is a widespread speech disorder affecting people globally, and it impacts effective communication and quality of life. Recent advancements in artificial intelligence (AI) and computational intelligence have introduced new possibilities for augmenting stuttering detection and treatment procedures. In this systematic review, the latest AI advancements and computational intelligence techniques in the context of stuttering are explored. By examining the existing literature, we investigated the application of AI in accurately determining and classifying stuttering manifestations. Furthermore, we explored how computational intelligence can contribute to developing innovative assessment tools and intervention strategies for persons who stutter (PWS). We reviewed and analyzed 14 refereed journal articles that were indexed on the *Web of Science* from 2019 onward. The potential of AI and computational intelligence in revolutionizing stuttering assessment and treatment, which can enable personalized and effective approaches, is also highlighted in this review. By elucidating these advancements, we aim to encourage further research and development in this crucial area, enhancing in due course the lives of PWS.

Keywords: stuttering detection; systematic review; rehabilitation; machine learning

Citation: Alnashwan, R.; Alhakbani, N.; Al-Nafjan, A.; Almudhi, A.; Al-Nuwaiser, W. Computational Intelligence-Based Stuttering Detection: A Systematic Review. *Diagnostics* **2023**, *13*, 3537. https://doi.org/10.3390/diagnostics13233537

Academic Editor: Mugahed A. Al-antari

Received: 8 October 2023
Revised: 10 November 2023
Accepted: 21 November 2023
Published: 27 November 2023

1. Introduction

Stuttering, a prevalent speech disorder that affects millions all over the globe [1], lacks comprehensive research in terms of accurately determining and categorizing its manifestations. Even though speech is a fundamental medium to convey ideas and emotions, not all individuals can flawlessly verbally communicate. The efficacy of speech is dependent on its fluency, which denotes the natural flow between phonemes that constitute a message [2]. Dysfluencies, which include stuttering, disrupt this flow and represent a complexity that impacts over 80 million people worldwide, that is, approximately 1% of the world's population [3].

Stuttering is characterized by the repetition of sounds, syllables, or words; the prolongation of sounds; and the interruption of speech through blocks. PWS often have a clear understanding of their intended speech but struggle with its fluid expression. These disruptions in speech can manifest with accompanying struggle behaviors, including secondary behaviors such as rapid eye blinks and quivering lip movements. The impacts of stuttering extend beyond its surface manifestations; it impairs effective communication, thus affecting interpersonal relationships and the overall quality of life of people suffering from it [4]. People with varying degrees of stuttering severity may find difficulties in both social interactions and professional settings, with heightened severity correlated with potential emotional struggles [5].

The traditional approach to evaluating stuttering is to manually tally the instances of different stuttering types and express them as a ratio that is relative to the total words in a speech segment. Nevertheless, owing to its time-intensive and subjective nature, this method is not without limitations that lead to inconsistencies and potential errors when different evaluators are involved [2]. Manually detecting stuttering exhibits several challenges. First, distinguishing stuttering from other speech disfluencies can be difficult, considering that subtle instances may resemble hesitations or pauses. Furthermore, consistent detection of stuttering becomes a complex task because the severity and frequency of stuttering can vary widely among people and across different contexts. Moreover, factors such as the speaker's age, gender, and language as well as the speaking task and the context in which the speech is produced can further complicate the identification of stuttering [6].

Considering the increasing need for improved detection and management of stuttering, there is a noticeable trend in adopting innovative technologies, particularly artificial intelligence (AI) [7]. The application of AI in identifying and classifying stuttering indicates an essential development in the study of speech-related issues. AI has a special ability to understand complex speech patterns that might not be easy for humans to notice, and this capability can help in the early detection of stuttering. This potential has sparked a wave of novel research efforts, each influenced by the prospect of revolutionizing the understanding and treatment of stuttering. The way AI and stuttering research work together can modify how speech assessment and management are carried out and enhance the quality of life of PWS. This exciting progress demonstrates that AI can greatly assist in managing stuttering and can even alter how speech therapy is performed, which makes it more personalized and effective.

This study highlights a perspective on the utilization of AI technologies for determining and classifying stuttering. While AI holds promise for the assessment of stuttering, this area has received limited attention, likely due to the complexity of the disorder, the need for extensive and diverse datasets, and the challenges of developing robust and accurate AI models. This study analyzes the existing research to extract recent efforts and methods in the field. The primary objective is to categorize and summarize the relevant literature concerning to stuttering identification, offering insights and organizing these articles for future research focused on the use of AI in stuttering identification. This approach aims to facilitate advancements in the field by highlighting the recent developments and methodologies employed in automated stuttering identification.

In the field of ASD research, it is essential to acknowledge the prior systematic reviews that have explored machine learning approaches for stuttering identification. Two notable systematic reviews have been published in recent years, namely Sheikh et al. (2022) [8] and Barrett et al. (2022) [9], who conducted a comprehensive review that encompassed various aspects of stuttering identification, including stuttered speech characteristics, datasets, and automatic stuttering-identification techniques. On the other hand, Barrett et al. focused specifically on machine learning techniques for detecting developmental stuttering. Their systematic review concentrated on studies utilizing supervised learning models trained on speech data from individuals who stutter. They emphasized the importance of accuracy reporting, sample sizes, and specific inclusion criteria in their analysis.

This study aims to answer the following questions: (i) What are the recent advancements in AI and computational intelligence for stuttering detection and treatment, and how can they contribute to improving assessment and intervention strategies? (ii) What are the challenges and future directions in computational intelligence-based stuttering detection, and how can they be addressed to enhance accuracy and effectiveness?

Our review aims to differentiate itself by providing a distinct contribution to the field. We focus on the latest developments in AI and computational intelligence within the timeframe of 2019–2023, specifically addressing the challenges and advancements in stuttering detection. Our review offers a comprehensive analysis of the datasets used, the specific types of stuttering investigated, the techniques employed for feature extraction, and the choice of classifiers. Furthermore, we aim to highlight potential pathways for improving accuracy and effectiveness in computational intelligence-based stuttering detection.

The rest of this paper is structured as follows: Section 2 explains the method utilized for this organized review. Section 3 covers the results and discussion. Section 4 provides insights for future research. Finally, Section 5 presents the conclusions.

2. Research Methodology

Research articles that are focused on automating the identification of stuttering are spread throughout different conference proceedings and journals, encompassing distinct areas for enhancement. These areas include refining methods for improving stuttering-identification accuracy, assessing various forms of stuttering, refining severity evaluation, and enhancing accessible datasets. This section describes the approach employed to locate relevant articles, along with the criteria for article selection and the procedures for filtering. Despite being a relatively novel and emerging field, research into stuttering detection using AI has received attention from researchers from various fields, eventually becoming a significant area of academic investigation. The outcomes of studies in this domain have been published in scholarly journals and conferences and indexed in the *Web of Science (WoS)* database [10].

The *Web of Science (WoS)* database is widely recognized for its comprehensive coverage of academic literature and its commitment to delivering high-quality research. The *WoS Core Collection* database provides us with robust access to prominent citation databases, including *Science Direct* (Elsevier), *IEEE/IEE Library*, *ACM Digital Library*, *Springer Link Online Libraries*, and *Taylor & Francis*. Utilizing the WoS for research provides numerous advantages owing to its comprehensive coverage of scholarly literature. This robust platform provides access to an extensive range of high-quality journals, conference proceedings, and research articles from different disciplines. The platform's precise indexing and citation tracking help researchers determine key studies and trends. Moreover, its rigorous evaluation and inclusion of reputable sources enhance the reliability and credibility of accessed materials, elevating the overall quality of research efforts. Articles and review articles published between 2019 and 2023 were precisely searched within the *WoS Core Collection* database (Figure 1).

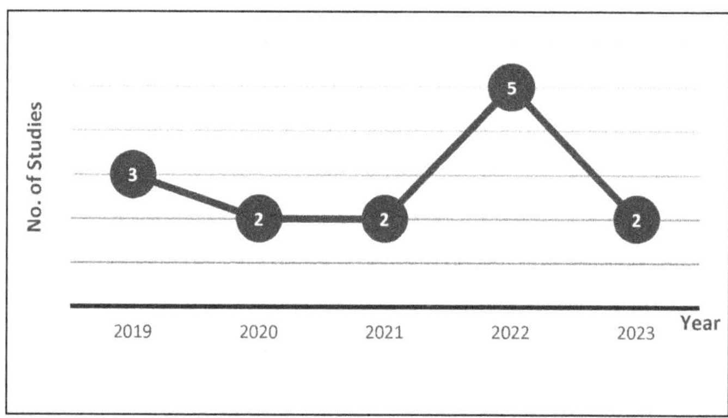

Figure 1. Number of published articles per year.

By using the search field, to guarantee the inclusion of all studies pertinent to the identification of stuttering via several AI technologies, we entered the following terms: (stuttering detection using machine learning) or (stuttering detection with the use of AI) or (stuttering detection) or (stuttering detection or stuttering recognition) or (stuttering classification) or (automatic stutter detection). Figure 2 illustrates the search process.

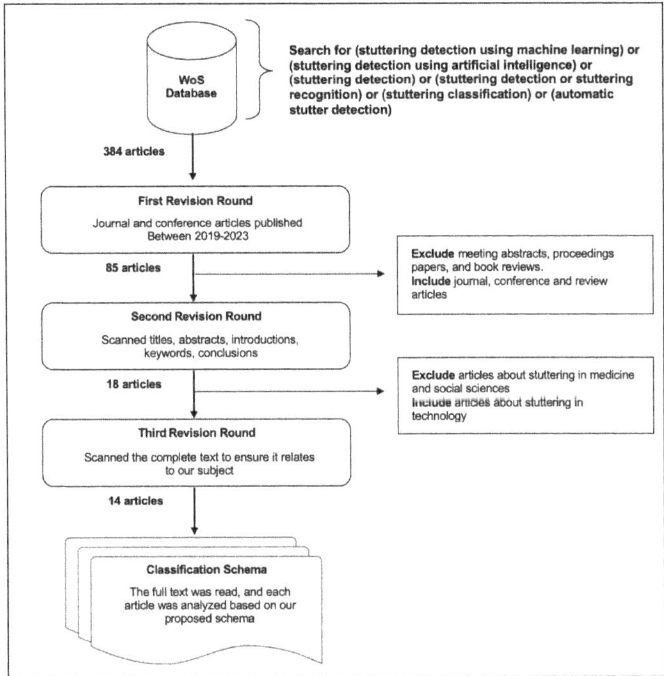

Figure 2. Article filtering process.

Initially, the search yielded a total of 384 research papers. From this pool, we selected journal and conference articles for inclusion and excluded meeting abstracts, proceedings papers, and book reviews. This selection also focused on articles published between 2019 and 2023. Following this initial stage, we finally obtained a refined collection of 85 search results. Afterward, a secondary screening procedure was performed, where we conducted an in-depth manual assessment to identify whether each article was relevant to the subject matter. Consequently, any articles that were considered unrelated were excluded from consideration.

In the second phase of revision, we reviewed the titles, abstracts, introductions, keywords, and conclusions of the articles. Then, we refined the collection by specifically choosing papers that centered around the application of machine learning and AI in stuttering detection. This yielded a set of 18 articles. In the third phase of revision, we meticulously examined the complete texts of the remaining 18 articles. After careful consideration, four articles were found to be unrelated to our designated theme, as they focused on the medical field. Consequently, these articles were excluded from our analysis, resulting in a final selection of 14 articles.

These 14 articles served as the basis for a further in-depth analysis and comparison of important aspects in automatic stuttering-detection (ASD) research. Our analysis was conducted based on specific criteria, taking into account the need for a comprehensive evaluation. These criteria included the dataset utilized in each study, the specific type of stuttering investigated (such as prolongation, block, or repetition), the techniques employed for feature extraction, the choice of classifier used, and the achieved performance accuracy.

By considering these five dimensions—dataset, classified stuttering type, feature-extraction approach, classifier selection, and performance evaluation—we aim to provide a structured and comprehensive overview of the diverse studies within the field of ASD. This framework will facilitate a deeper understanding and effective comparison of the different research approaches in this domain.

3. Results

3.1. Datasets

The success of stuttering detection using deep learning and AI models crucially depends on the quality and diversity of the data applied for training these systems. This critical dependency underscores the need for encompassing datasets that capture a wide spectrum of stuttering patterns, speech variations, and linguistic contexts.

In the literature, researchers have applied various datasets, including *University College London's Archive of Stuttered Speech* (*UCLASS*) [11], *SEP-28k* [12], *FluencyBank* [13], and *LibriStutter* [14], and have also used resources like *VoxCeleb* [15] (see Table 1). Furthermore, some researchers such as [16–18] have even made their own customized datasets to cater to their specific research needs. This combined endeavor emphasizes the importance of carefully curated and extensive data in advancing the field of stuttering-detection technology via deep learning and AI methods. Table 1 presents the benchmark datasets along with their respective descriptions.

Table 1. Benchmark datasets.

Dataset	Classes	Description
UCLASS (2009) [11]	Interjection, sound repetition, part-word repetition, word repetition, phrase repetition, prolongation, and no stutter	The *University College London's Archive of Stuttered Speech* (*UCLASS*) is a widely used dataset in stuttering research. It includes monologs, conversations, and readings, totaling 457 audio recordings. Although small, UCLASS is offered in two releases by UCL's Department of Psychology and Language Sciences. Notably, *UCLASS3* release 1 contains 138 monolog samples, namely 120 and 18 from male and female participants, respectively, from 81 individuals who stutter, aged 5–47 years. Conversely, release 2 contains a total of 318 monologs, reading, and conversation samples from 160 speakers suffering from stuttering, aged 5–20 years, with samples from 279 male and 39 female participants. Transcriptions, including orthographic versions, are available for some recordings, making them suitable for stutter labeling.
VoxCeleb (2017) [15]	The dataset does not have classes in the traditional sense, as it is more focused on identifying and verifying individual speakers	It is developed by the VGG, Department of Engineering Science, University of Oxford, UK. It is a large-scale dataset designed for speaker-recognition and verification tasks. It contains a vast collection of speech segments extracted from celebrity interviews, talk shows, and online videos. This dataset covers a diverse set of speakers and is widely employed in research that is related to speaker recognition, speaker diarization, and voice biometrics.
SEP-28k (2021) [12]	Prolongations, repetitions, blocks, interjections, and instances of fluent speech	Comprising a total of 28,177 samples, the *SEP-28k* dataset stands as the first publicly available annotated dataset to include stuttering labels. These labels encompass various disfluencies, such as prolongations, repetitions, blocks, interjections, and instances of fluent speech without disfluencies. Alongside these, the dataset covers nondisfluent labels such as natural pauses, unintelligible speech, uncertain segments, periods of no speech, poor audio quality, and even musical content.

Table 1. *Cont.*

Dataset	Classes	Description
FluencyBank (2021) [13]	Individuals who stutter (IWS) and individuals who do not stutter (IWN)	The *FluencyBank* dataset is a collection of audio recordings of people who stutter. It was created by researchers from the United States and Canada and contains over 1000 h of recordings from 300 speakers. The dataset is divided into two parts, namely research and teaching. The research data are password-protected, and the teaching data are open-access. The teaching data include audio recordings of 10 speakers who stutter, transcripts, and annotations of stuttering disfluencies. The dataset is valuable for researchers and clinicians studying stuttering.
LibriStutter (2021) [14]	Sound, word, and phrase repetitions; prolongations; and interjections.	The *LibriStutter* dataset is a corpus of audio recordings of speech with synthesized stutters. It was created by the Speech and Language Processing group at Queen's University in Canada. The dataset contains 100 h of audio recordings of 10 speakers, each of whom stutters differently. The stutters were synthesized via a technique known as the hidden Markov model. It is a valuable resource for researchers who are developing automatic speech-recognition (ASR) systems for people who stutter. The dataset can also be used to train models for detecting and classifying different types of stutters.

In Table 1, the dataset most commonly applied is *UCLASS*, which has been utilized in studies [14,19–24], closely followed by the *SEP-28k* dataset, which is featured in various works [6,23,25–27]. Notably, the latter dataset has gained popularity in recent research endeavors. *FluencyBank* contributes to investigations in previous studies [6,23,24], whereas the utilization of the *LibriStutter/LibriSpeech* dataset is relatively less frequent, as seen in previous studies [6,25]. Conversely, the *VoxCeleb* dataset plays a more minor role, appearing only once in [25].

Furthermore, several studies [16–18] chose to create their own tailored datasets to address their specific research objectives. For instant, in [16], the dataset comprised 20 individuals aged between 15 and 35, all of whom had been diagnosed with stuttering by qualified speech language pathologists. This group consisted of 17 males and 3 females. They were all directed to read a specific passage, and their speech was recorded in a room at the All India Institute of Speech and Hearing (AIISH) in Mysuru using the PRAAT tool, which has a sampling rate of 44 KHz. The participants displayed characteristics of stuttering, including repeating sounds or syllables, prolongations, and blocks.

Pravin et al. [17] created a dataset that consists of recordings of natural speech from children who came from bilingual families (speaking both Tamil and English). The children included in the dataset were between the ages of 4 and 7. Furthermore, the dataset included recordings of the pronunciation of phonemes as well as mono-syllabic and multi-syllabic words in both English and Tamil.

Asci et al. [18] recruited 53 individuals with stuttering (24 females and 29 males aged 7–30) alongside 71 age- and sex-matched controls (29 females and 44 males aged 7–30). All participants were native Italian speakers, non-smokers, and had no cognitive or mood impairments, hearing loss, respiratory disorders, or other conditions that could affect their vocal cords. None were taking central nervous-system-affecting drugs at the time of the study, and demographic and anthropometric data were collected during the enrollment visit.

Some studies [14,16–22,26,27] used just one dataset, whereas others [6,23–25] opted for multiple datasets. In one particular study [6], the researchers went a step further by enhancing their data with the *MUSAN* dataset for added diversity. In terms of stuttering, implementing data augmentation can be difficult, considering that many common techniques such as time stretch and high rate of speech fundamentally alter the structure of disfluent speech samples. The proposed approach, however, uses techniques such as speech shadowing, threshold masking, and delayed auditory feedback, which closely mimic real-world conditions without significantly altering the underlying stuttering characteristics of the speech sample. This diversity in dataset choice and augmentation methods showcases the evolving nature of research approaches in this field.

3.2. Classified Stuttering Type

Diverse forms of stuttering have been the subject of investigation across various research studies. These encompass repetition, prolongation, block, interjection, sound repetitions, part-word repetitions, word repetitions, phrase repetitions, syllable repetition, and revision.

Table 2 illustrates each type and its definition. Among these categories, prolongation emerged as the most frequently explored, being referenced in 13 out of the 14 studies, followed by interjection, which was cited in 9 out of the 14 studies. Contrarily, part-word repetition and syllable repetition were the least discussed, having only a single mention each.

Table 2. Classification of stuttering type.

Type	Definition	Example
Repetition	Repeating a sound, syllable, or word multiple times.	"I-I-I want to go to the park."
Prolongation	Extending or elongating sounds or syllables within words.	"Sssssend me that email, please."
Block	Temporary interruption or cessation of speech flow.	"I can't... go to the... park tonight."
Interjection	Spontaneous and abrupt interruption in speech with short exclamations.	"Um, I don't know the answer."
Sound repetitions	Repeating individual sounds within a word.	"Th-th-that movie was great."
Part-word repetitions	Repetition of part of a word, usually a syllable or sound.	"Can-c-c-come over later?"
Word repetitions	Repeating entire words within a sentence.	"I like pizza, pizza, pizza."
Phrase repetitions	Repeating phrases or groups of words.	"He said, "he said it too."
Syllable repetition	Repeating a syllable within a word.	"But-b-but I want to go."
Revision	Rewording or revising a sentence during speech to avoid stuttering.	"I'll take the, um, the bus."

The predominant stuttering categories, namely repetition, prolongation, and block, were discussed in previous works [6,16,18,22,24]. Other variations of stuttering, which can be considered subcategories of the primary classes, were addressed in previous studies [14,19,20], including sound repetitions, word repetitions, and phrase repetitions. Interjection was highlighted in various studies [6,14,19,20,23–25], whereas sound repetition and word repetition were jointly explored in different studies [21,23,26,27]. Additional examinations of prolongation, block, and interjection occurred in studies [26,27]. Revisions and prolongations were linked in previous works [14,20], and prolongation was connected with syllable repetition in one study [21]. Repetition and prolongation were jointly examined in one study [25], and singularly, prolongation was explored in another [23].

3.3. Feature-Extraction Approach

Feature extraction stands as a pivotal step within speech-recognition systems, serving to convert raw audio signals into informative data for subsequent processing. It is a foundational element in the translation of spoken language into digital information, which facilitates human–technology communication. Previous studies have targeted various speech features, such as the Mel frequency cepstral coefficient (MFCC) [28]. Table 3 presents each type and its description.

Table 3. Computational methods for the feature-extraction phase.

Method	No. of Studies	Ref.
Mel frequency cepstral coefficient (MFCC)	6	Sheikh et al., 2023 [6] Manjula et al., 2019 [16] Sheikh et al., 2021 [22] Jouaiti and Dautenhahn, 2022 [23] Sheikh et al., 2022 [25] Filipowicz and Kostek, 2023 [27]
Weighted MFCC (WMFCC)	1	Gupta et al., 2020 [21]
Spectrograms	3	Kourkounakis et al., 2020 [20] Al-Banna et al., 2022 [24] Prabhu and Seliya, 2022 [26]
Phonation features	1	Pravin and Palanivelan, 2021 [17]
Ngram	1	Alharbi et al., 2020 [19]
Character-based features	1	Alharbi et al., 2020 [19]
Utterance-based features	1	Alharbi et al., 2020 [19]
Acoustic analysis of voice recordings	1	Asci et al., 2023 [18]
Word distance features	1	Alharbi et al., 2020 [19]
Phoneme features	2	Sheikh et al., 2023 [6] Sheikh et al., 2022 [25]
Squeeze-and-excitation (SE) residual networks	1	Kourkounakis et al., 2021 [14]
Bidirectional long short-term memory (BLSTM) layers	1	Kourkounakis et al., 2021 [14]
Speaker embeddings from the ECAPA-TDNN model	1	Sheikh et al., 2022 [25]
Contextual embeddings from the Wav2Vec2.0 model	1	Sheikh et al., 2022 [25]
Pitch-determining feature	1	Filipowicz and Kostek, 2023 [27]
Two-dimensional speech representations	1	Filipowicz and Kostek, 2023 [27]

In research, the method of feature extraction has undergone diverse exploration, with various techniques being employed to extract valuable insights from speech data. Among these methods, MFCC emerges as the most prevalent choice, with mentions in 6 out of the total 14 studies. The primary feature employed in automatic speech-recognition systems is the Mel frequency cepstral coefficient (MFCC). MFCC is obtained by applying the discrete cosine transform to the logarithm of the power spectrum, which is computed on a Mel scale frequency. It offers a more effective representation of speech, capitalizing on human auditory perception, and is widely applied in the majority of speech-recognition research. Its popularity can be attributed to its effectiveness in translating audio signals into a format that facilitates further analysis [29]. Table 3 also summarizes the computational methods to extract features from speech signals.

Nevertheless, in one particular study [21], a departure from the conventional MFCC approach is observed. Instead, the researchers opted for the utilization of the weighted MFCC (WMFCC). This distinctive choice stems from WMFCC's unique ability to capture dynamic information inherent in speech samples, consequently bolstering the accuracy in detecting stuttering events. Furthermore, this alternative method offers the added advantage of reducing the computational overhead during the subsequent classification process, making it an intriguing avenue of exploration.

Spectrograms, a graphical representation of audio signals over time, have gained attention in multiple studies, notably in several studies [20,24,26]. These studies leverage spectrograms as a feature-extraction tool, emphasizing their utility in speech analysis.

Exploring more specialized domains, one study [17] delved into phonation features such as pitch, jitter, shimmer, amplitude perturbation quotient, pitch-period perturbation quotient, logarithmic energy, and the duration of voiceless speech. This nuanced approach offers a comprehensive understanding of the acoustic characteristics of speech.

Beyond the aforementioned methods, various other feature-extraction techniques have also been explored, including Ngram, character-based features, and utterance-based features. The combination of squeeze-and-excitation (SE) residual networks and bidirectional long short-term memory (BLSTM) layers, as witnessed in one study [14], illustrates the innovative strides taken to extract spectral features from input speech data, pushing the boundaries of feature extraction.

In Asci et al. [18], acoustic analysis of voice recordings was employed to further augment the array of feature-extraction methods applied. Intriguingly, Alharbi et al. [19] focused on word distance features, whereas Sheikh et al. [5] and Jouaiti and Dautenhahn [16] delved into the utilization of phoneme features. On a different note, Sheikh et al. [25] took a unique approach by extracting speaker embeddings from the ECAPA-time-delay neural network (TDNN) model and contextual embeddings from the Wav2Vec2.0 model, further enriching the feature-extraction landscape.

Lastly, Filipowicz and Kostek [27] introduced a pitch-determining feature into the signal processing toolkit, also exploring various 2D speech representations and their impact on classification results. This multifaceted exploration of feature-extraction techniques within the research realm highlights the dynamic and evolving nature of this crucial aspect of speech analysis.

3.4. Classifier Selection

In the world of ASD, different AI models have been employed in research with varying levels of accuracy and performance. These AI models have been investigated to see how well they can identify and understand stuttering, which has led to a range of results. Table 4 summarizes the computational methods for classifying speech features.

Table 4. Computational methods for classifying speech features.

Method	Ref.
Artificial neural network (ANN)	Manjula et al., 2019 [16] Sheikh et al., 2022 [25]
K-nearest neighbor (KNN)	Sheikh et al., 2022 [25] Filipowicz and Kostek, 2023 [27]
Gaussian back-end	Sheikh et al., 2022 [25]
Support vector machine (SVM)	Asci et al., 2023 [18] Filipowicz and Kostek, 2023 [27]
Bidirectional long short-term memory (BLSTM)	Pravin and Palanivelan, 2021 [17] Asci et al., 2023 [18] Alharbi et al., 2020 [19] Gupta et al., 2020 [21]
Convolutional neural networks (CNNs)	Kourkounakis et al., 2020 [20] Prabhu et al., 2022 [26]
Two-dimensional atrous convolutional network	Al-Banna et al., 2022 [24]
Conditional random fields (CRF)	Alharbi et al., 2020 [19]
ResNet18	Filipowicz and Kostek, 2023 [27]
ResNetBiLstm	Filipowicz and Kostek, 2023 [27]

Table 4. *Cont.*

Method	Ref.
Wav2Vec2	Filipowicz and Kostek, 2023 [27]
Deep LSTM autoencoder (DLAE)	Pravin and Palanivelan, 2021 [17]
FluentNet	Kourkounakis et al., 2021 [14]
StutterNet	Sheikh et al., 2023 [6] Sheikh et al., 2021 [22]

In our exploration of the existing literature, it was apparent that most studies have leaned toward applying deep learning models. By contrast, only 3 out of the 14 studies exclusively utilized machine learning. Additionally, several studies chose to combine both machine learning and deep learning models, whereas others opted for the creation of innovative architectural approaches.

Machine learning has emerged as a powerful tool in the domain of stuttering detection and classification. In this context, several studies have leveraged various traditional machine learning models, such as artificial neural network (ANN), K-nearest neighbor (KNN), and support vector machine (SVM), to develop faster diagnostic approaches.

Manjula et al. [8] employed ANN, which was fine-tuned using the adaptive fish swarm optimization (AFSO) algorithm. This ANN was purposefully trained to distinguish between various types of speech disfluencies, including repetitions, prolongations, and blocks, commonly observed in disfluent speech. The integration of the AFSO algorithm was instrumental in enhancing the network's architectural design, thereby optimizing its performance in the disfluency classification task. In Sheikh et al. [25], a range of classifiers, including KNN, Gaussian back-end, and neural network classifiers, were applied for stuttering detection. Utilizing Wav2Vec2.0 contextual embedding-based stuttering-detection methods, the study achieved a notable improvement over baseline methods, showcasing superior performance across all disfluent categories. Asci et al. [18] utilized a support vector machine (SVM) classifier to extract acoustic features from audio recordings, achieving high accuracy in classifying individuals with stuttering. The study also identified age-related changes in acoustic features associated with stuttering, holding potential applications in clinical assessment and telehealth practice. In the ever-changing field of stuttering detection and classification, the use of deep learning techniques has led to significant advancements. The following studies highlighted how deep neural networks, especially the BLSTM and convolutional neural networks (CNNs), have played a transformative role in tackling the complexities of stuttering analysis.

In Kourkounakis et al. [20], a deep learning model that combines CNNs for extracting features from spectrograms with BLSTM layers for capturing temporal dependencies was introduced. This system outperformed existing methods in terms of detecting sound repetitions and revisions, boasting high accuracy and low miss rates across all stutter types. Moreover, Gupta et al. [21] employed the BLSTM model, achieving an impressive overall classification accuracy of 96.67% in detecting various types of stuttered events. This achievement was attributed to the utilization of WMFCC for feature extraction and BLSTM for classification, which outperformed conventional methods and displayed heightened accuracy in recognizing speech disfluencies. Furthermore, the utility of BLSTM emerged again in Jouaiti and Dautenhahn [23], where a deep neural network incorporating BLSTM was introduced for stuttering detection and dysfluency classification. The network's architecture comprised multiple layers, including BLSTM layers, dense layers, batch normalization, and dropout layers, along with an embedding layer for processing phoneme-estimation data. Impressively, this network matched or exceeded state-of-the-art results for both stuttering detection and dysfluency classification.

Additionally, Al-Banna et al. [24] introduced a novel detection model comprising a 2D atrous convolutional network designed to learn spectral and temporal features from log Mel spectrogram data. This network architecture featured multiple layers, including convolutional layers with varying dilation rates, batch normalization, dropout layers, and softmax activation for predicting stuttering classes. When compared with other stuttering-detection methods, the proposed model exhibited superior performance, especially in the detection of prolongations class and fluent speech class. Meanwhile, in Prabhu and Seliya [26], a CNN-based classifier was designed for stutter detection, distinguishing itself from previous models relying on long short-term memory (LSTM)-based structures. This CNN-based model demonstrated high accuracy and precision, albeit with varying recall, making it exceptionally adept at achieving high F1 scores and surpassing other models across different datasets.

Several studies adopted a comprehensive approach by integrating both machine learning and deep learning models. These studies aim to identify the most effective approach for detecting and analyzing stuttering. In Alharbi et al. [19], a combination of conditional random fields (CRF) and BLSTM classifiers was utilized to detect and transcribe stuttering events in children's speech. The study's findings revealed that BLSTM outperformed CRFngram when evaluated using human-generated reference transcripts. Notably, the CRFaux variant, which incorporated additional features, achieved superior results compared to both CRFngram and BLSTM. Nevertheless, it is worth noting that when these classifiers were evaluated with ASR (automatic speech recognition) transcripts, all of them experienced a decrease in performance because of ASR errors and data mismatches.

In a parallel study, Sheikh et al. [25] used a range of classifiers, including KNN, Gaussian back-end, and neural network classifiers, to detect stuttering detection. Utilizing Wav2Vec2.0 contextual embedding-based stuttering-detection methods, the study obtained a noteworthy improvement over baseline methods, showcasing superior performance across all disfluent categories. Moreover, in Filipowicz and Kostek [27], various classifiers, including KNN, SVM, deep neural networks (ResNet18 and ResNetBiLstm), and Wav2Vec2, were evaluated for the classification of speech disorders. Notably, ResNet18 displayed superior performance over the other algorithms tested in the research.

Recently, in the field of ASD, various innovative approaches have surfaced to address the complex aspects of this speech condition. These studies embody significant advancements, each presenting fresh methods and structures for improving the comprehension of stuttering. Pravin and Palanivelan [17] presented a novel approach in the form of a deep long short-term memory (LSTM) autoencoder (DLAE) using long short-term memory (LSTM) cells, which are specialized recurrent neural network units designed for sequential data. The DLAE model was evaluated against various baseline models, including shallow LSTM autoencoder, deep autoencoder, and stacked denoising autoencoder, showing superior accuracy in predicting the severity class of phonological deviations. Meanwhile, Kourkounakis et al. [14] introduced FluentNet, a cutting-edge end-to-end deep neural network architecture designed exclusively for automated stuttering speech detection. FluentNet's architecture comprises components such as SE-ResNet blocks, BLSTM networks, and an attention mechanism, achieving state-of-the-art results for stutter detection and classification across different stuttering types in both the *UCLASS* and *LibriStutter* datasets.

Furthermore, Sheikh et al. [22] proposed the StutterNet architecture, based on a time-delay neural network (TDNN) and specifically designed to detect and classify various types of stuttering. This architecture treats stuttering detection as a multiclass classification problem, featuring components such as an input layer, time-delay layers, statistical pooling, fully connected layers, and a softmax layer. Experimental results demonstrated the StutterNet model's promising recognition performance across various stuttering types, even surpassing the performance of the ResNet + BiLSTM method in some cases, particularly in detecting fluent speech and core behaviors. Notably, the StutterNet model was also adopted by Sheikh et al. [6].

3.5. Performance Evaluation

Within this section, an overview of the best accuracy across all studies is presented. In some of the studies, accuracy numbers were not explicitly provided, such as in [16], where the authors stated that the proposed AOANN effectively predicts the occurrences of repetitions, prolongations, and blocks with accuracy.

In contrast, the majority of the studies presented specific numerical data to illustrate their results. For instance, according to Al Harbi et al. [19], the BLSTM classifiers outperformed the CRF classifiers by a margin of 33.6%. However, incorporating auxiliary features for the CRFaux classifier led to performance enhancements of 45% compared to the CRF baseline (CRFngram) and 18% compared to the BLSTM outcomes.

Kourkounakis et al. [20] reported that their proposed model achieved a 26.97% lower miss rate on the *UCLASS* dataset compared to the previous state of the art. It also slightly outperforms the unidirectional LSTM baseline across all stutter types. Similarly, the DLAE model proposed by [17] outperformed the baseline models, achieving an AUC of 1.00 and a perfect test accuracy of 100%. This capability enables accurate discrimination between "mild" and "severe" cases of phonation deviation, ensuring precise assessment of speech disorders and avoiding any conflicting diagnoses.

The method of Gupta et al. [21] achieved the best accuracy of 96.67%, outperforming the LSTM model. Promising recognition accuracies were also observed for fluent speech (97.33%), prolongation (98.67%), syllable repetition (97.5%), word repetition (97.19%), and phrase repetition (97.67%). Furthermore, the FluentNet model proposed by Kourkounakis et al. [14] achieved an average miss rate and accuracy of 9.35% and 91.75% on the *UCLASS* dataset. The StutterNet model [22] outperformed the state-of-the-art method utilizing a residual neural network and BiLSTM, with a considerable gain of 4.69% in overall average accuracy and 3% in MCC. Also, the methodology proposed by Sheikh et al. [6] achieved a 4.48% improvement in F1 over the single-context-based MB StutterNet. Furthermore, data augmentation in the cross-corpora scenario improved the overall SD performance by 13.23% in F1 compared to clean training.

Moreover, the method proposed by Sheikh et al. [25] showed a 16.74% overall accuracy improvement over the baseline. Combining two embeddings and multiple layers of Wav2Vec2.0 further enhanced SD performance by up to 1% and 2.64%, respectively. During the training phase using *SEP-28K + FluencyBank + UCLASS* datasets, Jouaiti and Dautenhahn [23] achieved the following F1 scores: 82.9% for word repetition, 83.9% for sound repetition, 82.7% for interjection, and 83.8% for prolongation. In another training scenario using *FluencyBank* and *UCLASS*, the obtained F1 scores were 81.1% for word repetition, 87.1% for sound repetition, 86.6% for interjection, and 81.5% for prolongation.

The model proposed by Al-Banna et al. [24] surpassed the state-of-the-art models in detecting prolongations, with F1 scores of 52% and 44% on the *UCLASS* and *FluencyBank* datasets. It also achieved gains of 5% and 3% in classifying fluent speech on the *UCLASS* and *FluencyBank* datasets. In the study proposed by Prabhu and Seliya [26], interjection had the best performance with F1 score: 97.8%. Furthermore, in [18], machine learning accurately differentiated individuals who stutter from controls with an 88% accuracy. Age-related effects on stuttering were demonstrated with a 92% accuracy when classifying children and younger adults with stuttering. Additionally, in [27], ResNet18 was able to classify speech disorders at the F1 measure of 93% for the general class.

4. Discussion

In this section, the importance of understanding the challenges and identifying future directions in computational intelligence-based stuttering detection is examined further. By exploring these aspects, we gain valuable insights into the current limitations of existing approaches and pave the way for advancements in the field. Figure 3 provides insights into the challenges and future directions of computational intelligence-based stuttering detection.

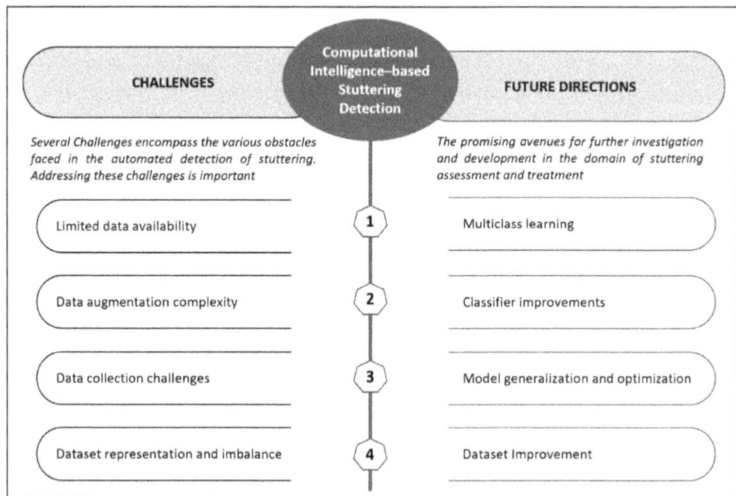

Figure 3. Challenges and future directions of computational intelligence-based stuttering detection.

4.1. Challenges

In this section, an overview of the challenges that automatic stuttering-identification systems encounter is provided, and possible solutions that could be explored in the field of stuttering research are suggested. Although there have been notable developments in the automated detection of stuttering, several issues must still be addressed to ensure a robust and effective identification of stuttering.

A significant obstacle that should be addressed is the limited availability of data for research in stuttering identification. A notable challenge is the scarcity of natural speech datasets that include disfluent speech. The limitations posed by the availability of a limited dataset have been a recurring concern in several studies [19,20,23,25]. This constraint can significantly impact the outcomes of their proposed methods, which often leads to results that do not meet expectations. The deficiency in extensive and diverse datasets has emerged as a pivotal factor that influences the overall performance and robustness of the methods explored in these studies. Hence, addressing the challenge of dataset scarcity is a fundamental step toward enhancing the accuracy and reliability of research findings in this domain.

Medical data collection is generally a costly and resource-intensive endeavor, and stuttering research is no exception in this regard [25]. Moreover, the complexity of the stuttering domain is compounded by the need for a diverse set of speakers and sentences for comprehensive analysis. One of the obstacles that contribute to the scarcity of available datasets is the challenge of data collection itself. This challenge arises since it involves organizing meetings and recording sessions with PWS while they engage in spontaneous speech. This approach is necessary to capture authentic instances of stuttering speech considering that requesting PWS to read from a predetermined list can often result in a reduction in the frequency of stuttering occurrences [30].

Sheikh et al. [6] introduced a possible solution to the challenge posed by the limited availability of datasets in the field. They proposed that data augmentation could prove beneficial in the context of stuttering research. However, notably, the application of data augmentation in the realm of stuttering is not a straightforward process. This complexity arises because several conventional data augmentation techniques, such as time stretch and the fundamentally high rate of speech, can significantly alter the underlying structure of stuttering speech samples.

To make data augmentation more effective for stuttering research, specialized data augmentation techniques tailored specifically to the unique characteristics of stuttering speech must be developed. Such domain-specific data augmentation methods would enable researchers to enhance their datasets while preserving the vital features of disfluent speech, ultimately contributing to more accurate and meaningful results in this field of study.

Another significant concern that affects the outcomes of proposed procedures in various studies is the class imbalance within the available datasets, as highlighted by Jouaiti and Dautenhahn [23] and Filipowicz and Kostek [27]. This problem arises from several factors, one of which is the limited availability of datasets. Notably, certain datasets may exhibit an underrepresentation of specific types of core stuttering behaviors, such as the case of "prolongations" in the *FluencyBank* dataset. Stuttering, which is a highly diverse speech disorder, can manifest in diverse ways among individuals. Some individuals may predominantly display repetitions, whereas others may experience more prolonged speech sounds or blocks. This inherent variability in stuttering presentations contributes to the disparities observed in stuttering datasets.

Speech data collection from PWS, especially in authentic conversational settings, introduces its own set of challenges. Some core stuttering behaviors may occur less frequently or may be less perceptible because PWS may be more likely to stutter when they are feeling anxious or stressed [30]. Moreover, PWS may be more likely to use avoidance strategies (e.g., pausing and substituting a word) in authentic conversational settings, such as avoiding words or phrases that they know are likely to trigger stuttering, rendering it more challenging to capture the events during data collection endeavors [31]. Furthermore, stuttering datasets often suffer from limited size because of the relatively low prevalence of stuttering in the general population. This limited sample size increases the likelihood of imbalances that arise purely by chance. Fundamentally, the issue of class imbalance in stuttering datasets stems from multifaceted factors, which include the diverse nature of stuttering, data collection challenges, and the inherent constraints associated with dataset size. Recognizing and addressing these factors are necessary steps in striving for more balanced and representative datasets in stuttering research.

4.2. Future Directions

Based on the findings of our systematic review, we have identified several potential future directions for research in the field. These directions include the exploration of multiclass learning techniques, improvements in classifier algorithms, advancements in model generalization and optimization methods, as well as enhancements in dataset quality and diversity. These areas present promising avenues for further investigation and development in the domain of stuttering assessment and treatment.

4.2.1. Multiclass Learning

Numerous researchers have sought to enhance their systems by incorporating the concept of multiclass learning. This means that instead of restricting their proposed models to identifying only one type of stuttering at a time, these models can recognize multiple stuttering types concurrently. PWS can exhibit various forms of stuttering within a single sentence, which occur simultaneously in their speech.

Kourkounakis et al. [20] aimed to build upon existing models and embarked on research regarding multiclass learning for different stuttering types. Considering that multiple stuttering types can manifest simultaneously in a sentence (e.g., "I went to uh to to uh to"), this approach has the potential to yield a more robust classification of stuttering.

Furthermore, Sheikh et al. [22] outlined future work in which they would examine thoroughly the realm of multiple disfluencies. They intended to explore advanced variations of TDNN for stuttering detection in real-world settings. This advancement aims to address the complexity of stuttering, where different disfluency types can co-occur, contributing to a more comprehensive understanding of stuttering patterns in spontaneous speech.

4.2.2. Classifier Improvements

This research strongly highlights the importance of enhancing the classifiers as a central focus for future work. Improving the classifiers is a pivotal aspect of our future research agenda, which reflects its importance in achieving more accurate and effective ASR classification.

Alharbi et al. [19] proposed enhancements to the ASR stage, intending to reduce the word error rate. They also suggested exploring alternative methods for detecting prolongation events, which proved challenging using the current approach.

Pravin and Palanivelan [17] aimed to employ deep learning for the solitary classification of disfluencies, contributing to a more descriptive severity assessment for subjects and enhanced self-assessment. Nevertheless, notably, an increase in the number of training epochs resulted in longer model run times, which requires consideration for future improvements. Moreover, Gupta et al. [21] suggested the exploration of different feature-extraction and classification techniques to improve stuttering detection. Kourkounakis et al. [14] proposed experimenting with FluentNet's architecture, potentially implementing different attention mechanisms, including transformers, to investigate their impact on results.

Sheikh et al. [6] proposed exploring the combination of various types of neural networks in stuttering detection to pinpoint precisely where stuttering occurs in speech frames. Furthermore, investigating different context variations, depths, and convolutional kernel numbers in stutter detection poses a promising area of focus. The study also identified blocks as particularly difficult to detect, prompting further analysis and ablation studies on speakers with hard-to-determine disfluencies.

Future work involves exploring self-supervised models that utilize unlabeled audio data, building upon the research of Sheikh et al. [25], which aimed at fine-tuning the Wav2Vec2.0 model to identify and locate stuttering in speech frames. Al-Banna et al. [24] suggested the incorporation of atrous spatial pyramid pooling and local and global attention mechanisms to enhance detection scores.

Prabhu and Seliya [26] identified potential enhancements to the model as possibly including the incorporation of LSTM layers to capture temporal relationships in data and improve performance. The possibility of utilizing different machine models, such as LSTM models, has also been considered for better data interpretation with the *SEP-28k* dataset. Finally, Filipowicz and Kostek [27] proposed extending the training duration for each model and potentially expanding the ResNet18 model with additional convolutional layers, which are dependent on available resources.

4.2.3. Model Generalization and Optimization

Several researchers have stressed the importance of a crucial area for future research: enhancing their models' ability to generalize. This means making their models better at working effectively in different situations, not just the specific ones for which they were originally designed. The reason behind this recommendation is slightly straightforward: Researchers want their models to be versatile and adaptable. They understand that a model's success should not be confined to specific datasets or conditions. By focusing on improving generalization, researchers aim to strengthen their methods, which enables them to handle a wide range of real-world challenges and variations effectively.

Kourkounakis et al. [14], for example, proposed the idea of conducting more studies that bridge synthetic datasets such as *LibriStutter* with real-world stutter datasets. They suggested exploring domain adaptation techniques, which include the use of adversarial networks, to enhance the transfer of learning and create more adaptable and broadly applicable solutions. Furthermore, researchers could investigate optimizing various parameters concurrently, such as context, filter bank size, and layer dimensions within the proposed system.

Additionally, Sheikh et al. [25] suggested that it would be interesting to study how well the proposed method can generalize across multiple datasets. This exploration would shed light on the method's adaptability and effectiveness in various data scenarios. Finally,

Al-Banna et al. [24] recommended investigating ways to improve model generalization and robustness by analyzing different datasets and using domain-adaptation techniques. These recommendations collectively highlight researchers' dedication to advancing methods that are not only effective in specific situations but also excel in various real-world contexts.

4.2.4. Dataset Improvement

In terms of dataset enhancement, Prabhu and Seliya [26] highlighted the potential for improving the *SEP-28k* dataset. Notably, the *SEP-28k* dataset is recognized for its robustness, comprising a substantial volume of stuttering data that serve as a valuable resource for model development. Nevertheless, this dataset still offers room for refinement and optimization.

A key proposal involves revising the labeling scheme currently applied to each 3-second audio clip. By creating more specific labels with distinct start and end points, the dataset could pave the way for the development of more effective classifiers. Such an adjustment would render disfluent and fluent events entirely independent of each other, potentially facilitating the detection of these events within arbitrary speech contexts.

5. Conclusions

Stuttering is a complex speech disorder that necessitates accurate detection for effective assessment and treatment. This paper has discussed the challenges, advancements, and future directions in computational intelligence-based stuttering detection.

The challenges of limited datasets and dataset imbalance were determined, with proposed solutions including specialized data-augmentation techniques and balanced dataset creation. Advancements in stuttering detection using computational intelligence techniques have shown promising results from employing various algorithms and feature-extraction methods.

Future research directions include multiclass learning approaches, classifier enhancements, model generalization, and optimization. When these challenges and the exploration of future directions are addressed, we can enhance the accuracy and reliability of stuttering-detection systems, benefiting individuals who stutter and improving their quality of life.

Author Contributions: R.A., literature review, analysis and interpretation of the data, and drafting the manuscript; A.A.-N., design and supervision of the analysis, review of the manuscript, and contribution to the discussion; N.A., A.A. and W.A.-N., review and editing of the manuscript. All authors have read and agreed to the published version of the manuscript.

Funding: This work was supported and funded by the Deanship of Scientific Research at Imam Mohammad Ibn Saud Islamic University (IMSIU) (grant number IMSIU-RG23151).

Acknowledgments: The authors would like to thank the Deanship of Scientific Research at Imam Mohammad Ibn Saud Islamic University (IMSIU) for funding and supporting this research.

Conflicts of Interest: The authors declare no conflict of interest.

References

1. Etchell, A.C.; Civier, O.; Ballard, K.J.; Sowman, P.F. A Systematic Literature Review of Neuroimaging Research on Developmental Stuttering between 1995 and 2016. *J. Fluen. Disord.* **2018**, *55*, 6–45. [CrossRef] [PubMed]
2. Guitar, B. *Stuttering: An Integrated Approach to Its Nature and Treatment*; Lippincott Williams & Wilkins: Philadelphia, PA, USA, 2013; ISBN 978-1-4963-4612-4.
3. FAQ. Available online: https://www.stutteringhelp.org/faq (accessed on 8 August 2023).
4. What Is Stuttering? Diagnosis & Treatment | NIDCD. Available online: https://www.nidcd.nih.gov/health/stuttering (accessed on 8 August 2023).
5. Craig, A.; Blumgart, E.; Tran, Y. The Impact of Stuttering on the Quality of Life in Adults Who Stutter. *J. Fluen. Disord.* **2009**, *34*, 61–71. [CrossRef]
6. Sheikh, S.A.; Sahidullah, M.; Hirsch, F.; Ouni, S. Advancing Stuttering Detection via Data Augmentation, Class-Balanced Loss and Multi-Contextual Deep Learning. *IEEE J. Biomed. Health Inform.* **2023**, *27*, 2553–2564. [CrossRef]
7. Korinek, A.; Schindler, M.; Stiglitz, J. *Technological Progress, Artificial Intelligence, and Inclusive Growth*; IMF Working Paper no. 2021/166; International Monetary Fund: Washington, DC, USA, 2021.

8. Sheikh, S.A.; Sahidullah, M.; Hirsch, F.; Ouni, S. D-Machine Learning for Stuttering Identification: Review, Challenges and Future Directions. *Neurocomputing* **2022**, *514*, 385–402. [CrossRef]
9. Barrett, L.; Hu, J.; Howell, P. Systematic Review of Machine Learning Approaches for Detecting Developmental Stuttering. *IEEE/ACM Trans. Audio Speech Lang. Process.* **2022**, *30*, 1160–1172. [CrossRef]
10. Document Search—Web of Science Core Collection. Available online: https://www-webofscience-com.sdl.idm.oclc.org/wos/woscc/basic-search (accessed on 8 August 2023).
11. Howell, P.; Davis, S.; Bartrip, J. The UCLASS Archive of Stuttered Speech. *J. Speech Lang. Hear. Res.* **2009**, *52*, 556–569. [CrossRef]
12. Lea, C.; Mitra, V.; Joshi, A.; Kajarekar, S.; Bigham, J.P. SEP-28k: A Dataset for Stuttering Event Detection from Podcasts with People Who Stutter. In Proceedings of the ICASSP 2021—2021 IEEE International Conference on Acoustics, Speech and Signal Processing (ICASSP), Toronto, ON, Canada, 6 June 2021; pp. 6798–6802.
13. FluencyBank. Available online: https://fluency.talkbank.org/ (accessed on 13 September 2023).
14. Kourkounakis, T.; Hajavi, A.; Etemad, A. FluentNet: End-to-End Detection of Stuttered Speech Disfluencies With Deep Learning. *IEEE/ACM Trans. Audio Speech Lang. Process.* **2021**, *29*, 2986–2999. [CrossRef]
15. Nagrani, A.; Chung, J.S.; Zisserman, A. VoxCeleb: A Large-Scale Speaker Identification Dataset. *arXiv* **2017**, arXiv:1706.08612.
16. Manjula, G.; Shivakumar, M.; Geetha, Y.V. Adaptive Optimization Based Neural Network for Classification of Stuttered Speech. In Proceedings of the 3rd International Conference on Cryptography, Security and Privacy, Kuala Lumpur, Malaysia, 19 January 2019; pp. 93–98.
17. Pravin, S.C.; Palanivelan, M. Regularized Deep LSTM Autoencoder for Phonological Deviation Assessment. *Int. J. Patt. Recogn. Artif. Intell.* **2021**, *35*, 2152002. [CrossRef]
18. Asci, F.; Marsili, L.; Suppa, A.; Saggio, G.; Michetti, E.; Di Leo, P.; Patera, M.; Longo, L.; Ruoppolo, G.; Del Gado, F.; et al. Acoustic Analysis in Stuttering: A Machine-Learning Study. *Front. Neurol.* **2023**, *14*, 1169707. [CrossRef] [PubMed]
19. Alharbi, S.; Hasan, M.; Simons, A.J.H.; Brumfitt, S.; Green, P. Sequence Labeling to Detect Stuttering Events in Read Speech. *Comput. Speech Lang.* **2020**, *62*, 101052. [CrossRef]
20. Kourkounakis, T.; Hajavi, A.; Etemad, A. *Detecting Multiple Speech Disfluencies Using a Deep Residual Network with Bidirectional Long Short-Term Memory*; IEEE: Barcelona, Spain, 2020; p. 6093.
21. Gupta, S.; Shukla, R.S.; Shukla, R.K.; Verma, R. Deep Learning Bidirectional LSTM Based Detection of Prolongation and Repetition in Stuttered Speech Using Weighted MFCC. *Int. J. Adv. Comput. Sci. Appl.* **2020**, *11*, 345–356. [CrossRef]
22. Sheikh, S.A.; Sahidullah, M.; Hirsch, F.; Ouni, S. StutterNet: Stuttering Detection Using Time Delay Neural Network. In *2021 29th European Signal Processing Conference (EUSIPCO)*; IEEE: Dublin, Ireland, 2021; pp. 426–430.
23. Jouaiti, M.; Dautenhahn, K. Dysfluency Classification in Stuttered Speech Using Deep Learning for Real-Time Applications. In Proceedings of the ICASSP 2022—2022 IEEE International Conference on Acoustics, Speech and Signal Processing (ICASSP), Singapore, 23–27 May 2022; pp. 6482–6486.
24. Al-Banna, A.-K.; Edirisinghe, E.; Fang, H. Stuttering Detection Using Atrous Convolutional Neural Networks. In Proceedings of the 2022 13th International Conference on Information and Communication Systems (ICICS), Irbid, Jordan, 21–23 June 2022; pp. 252–256.
25. Sheikh, S.A.; Sahidullah, M.; Hirsch, F.; Ouni, S. Introducing ECAPA-TDNN and Wav2Vec2.0 Embeddings to Stuttering Detection. *arXiv* **2022**, arXiv:2204.01564. [CrossRef]
26. Prabhu, Y.; Seliya, N. A CNN-Based Automated Stuttering Identification System. In Proceedings of the 2022 21st IEEE International Conference on Machine Learning and Applications (ICMLA), Nassau, Bahamas, 12–14 December 2022; pp. 1601–1605.
27. Filipowicz, P.; Kostek, B. D-Rediscovering Automatic Detection of Stuttering and Its Subclasses through Machine Learning—The Impact of Changing Deep Model Architecture and Amount of Data in the Training Set. *Appl. Sci.* **2023**, *13*, 6192. [CrossRef]
28. Automatic Speaker Recognition Using MFCC and Artificial Neural Network. Available online: https://www.researchgate.net/publication/338006282_Automatic_Speaker_Recognition_using_MFCC_and_Artificial_Neural_Network?enrichId=rgreq-4822a0f9838aa087ae99ea77c2ec27ce-XXX&enrichSource=Y292ZXJQYWdlOzMzODAwNjI4MjtBUzo4NTA1MDQxNDU5NzczNTVAMTU3OTc4NzM5ODY0MA==&el=1_x_3&_esc=publicationCoverPdf (accessed on 7 November 2023).
29. Ancilin, J.; Milton, A. Improved Speech Emotion Recognition with Mel Frequency Magnitude Coefficient. *Appl. Acoust.* **2021**, *179*, 108046. [CrossRef]
30. Constantino, C.D.; Leslie, P.; Quesal, R.W.; Yaruss, J.S. A Preliminary Investigation of Daily Variability of Stuttering in Adults. *J. Commun. Disord.* **2016**, *60*, 39–50. [CrossRef] [PubMed]
31. Jackson, E.S.; Yaruss, J.S.; Quesal, R.W.; Terranova, V.; Whalen, D.H. Responses of Adults Who Stutter to the Anticipation of Stuttering. *J. Fluen. Disord.* **2015**, *45*, 38–51. [CrossRef] [PubMed]

Article

Supervised Machine Learning Methods for Seasonal Influenza Diagnosis

Edna Marquez [1,*], Eira Valeria Barrón-Palma [1], Katya Rodríguez [2], Jesus Savage [3] and Ana Laura Sanchez-Sandoval [1]

[1] Genomic Medicine Department, General Hospital of México "Dr. Eduardo Liceaga", Mexico City 06726, Mexico; valeirabarron@gmail.com (E.V.B.-P.)

[2] Institute for Research in Applied Mathematics and Systems, National Autonomous University of Mexico, Mexico City 04510, Mexico; katya.rodriguez@iimas.unam.mx

[3] Signal Processing Department, Engineering School, National Autonomous University of Mexico, Mexico City 04510, Mexico; robotssavage@gmail.com

* Correspondence: cednam@gmail.com

Abstract: Influenza has been a stationary disease in Mexico since 2009, and this causes a high cost for the national public health system, including its detection using RT-qPCR tests, treatments, and absenteeism in the workplace. Despite influenza's relevance, the main clinical features to detect the disease defined by international institutions like the World Health Organization (WHO) and the United States Centers for Disease Control and Prevention (CDC) do not follow the same pattern in all populations. The aim of this work is to find a machine learning method to facilitate decision making in the clinical differentiation between positive and negative influenza patients, based on their symptoms and demographic features. The research sample consisted of 15480 records, including clinical and demographic data of patients with a positive/negative RT-qPCR influenza tests, from 2010 to 2020 in the public healthcare institutions of Mexico City. The performance of the methods for classifying influenza cases were evaluated with indices like accuracy, specificity, sensitivity, precision, the f1-measure and the area under the curve (AUC). Results indicate that random forest and bagging classifiers were the best supervised methods; they showed promise in supporting clinical diagnosis, especially in places where performing molecular tests might be challenging or not feasible.

Keywords: machine learning; decision support system; medical diagnosis; influenza; artificial intelligence

1. Introduction

Influenza is a respiratory disease that can increase the incidence of pneumonia and cause a high number of hospitalizations [1]. In March 2009, Mexico, the United States and Canada were the focus of international attention when the influenza A H1N1 virus burst onto the epidemiological scene [2]. In June of that same year, the World Health Organization (WHO) declared an influenza pandemic of moderate severity. Since 2009, respiratory diseases due to influenza have recurred in numerous nations during the colder months annually, thus acquiring the category of seasonal influenza. There are four types of influenza viruses: A, B, C and D. Influenza A and B viruses cause seasonal epidemics of disease, and have been responsible for thousands of deaths worldwide, despite the annual vaccination campaigns [3]. In Mexico, between 2020 and 2021, the incidence of influenza decreased substantially due to the Coronavirus disease (COVID-19) caused by the Severe Acute Respiratory Syndrome Coronavirus 2 (SARS-CoV-2); however, influenza cases increased again in 2022 [4]. Before the COVID-19 pandemic, an estimated 291,000 to 646,000 respiratory deaths occurred worldwide each year due to seasonal influenza [5,6]. In Mexico, information regarding influenza infections has been registered since 2010 in the Influenza Epidemiological Surveillance System to identify its behavior and be able to predict how the next influenza season will develop [7]. The data in this system are obtained from symptomatic patients treated at healthcare centers and who had undergone a quantitative reverse transcription polymerase chain reaction (RT-qPCR) test to detect the presence of influenza viral RNA.

The Centers for Disease Control and Prevention (CDC) specify the common symptoms experienced by influenza patients, such as fever or feeling feverish/chills, cough, sore throat, runny or stuffy nose, muscle or body aches, headaches, and fatigue (tiredness) [8]. The presence of these symptoms is not a guarantee of having been infected by the influenza virus; moreover, they vary among the population. Distinguishing the causal agent of the illness between the influenza virus and other viral or bacterial agents proves to be challenging through clinical evaluation alone. Therefore, other tests should be applied to confirm the diagnosis of influenza, RT-qPCR being the most successful test for the molecular diagnosis [9]. However, in developing countries, this test is not routinely performed due to high costs and the limited availability of testing facilities. Not all hospitals and clinics have the necessary equipment and supplies to perform these tests, leading to potentially lengthy turnaround times [10].

We propose to use alternative methods to facilitate the diagnosis of influenza, like methods based on Artificial Intelligence (AI), as they could serve as tools to assist in medical attention for diagnosis prior to RT-qPCR tests or can be applied in locations with difficult access to molecular analysis. In Figure 1, the workflow applied in this work to find the best Machine Learning method to diagnose influenza is shown.

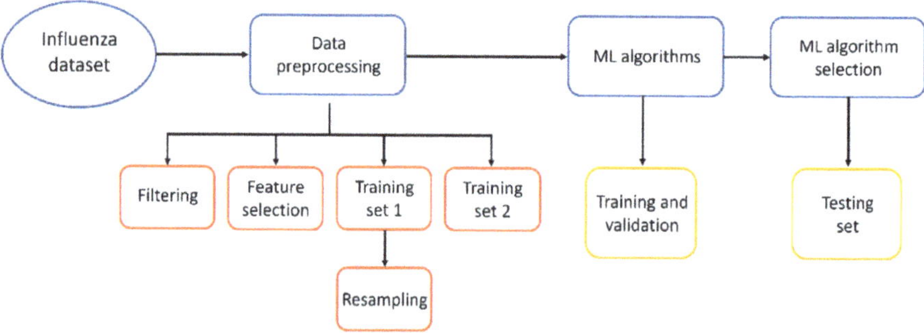

Figure 1. Workflow to find the best ML method to diagnose influenza.

Machine Learning (ML) is a component of AI that encompasses a set of techniques which enable the implementation of adaptive algorithms to make predictions and self-organize input data based on common characteristics. When ML is trained with a correct data set and the algorithm is standardized to accurately respond to all inputs, it is called a supervised ML [11].

Since the 1970s, interest in applying AI-ML in the health sector has grown [12]. In the medical field, a significant amount of patient data is managed, including sociodemographic and epidemiological information, results from physical examinations, diagnostic support test outcomes, and procedures performed, among others [10,13–47]. Because ML can effectively handle a large number of attributes (features), and due to its ability to identify and leverage the interactions among these numerous attributes, it becomes a particularly compelling tool in this domain [20]. ML algorithms have found extensive application across numerous medical specialties, serving purposes in prevention, diagnosis, treatment, and survival analysis alike.

In the case of influenza, ML has been applied to achieve several objectives, one of which is predicting the incidence of cases in the upcoming influenza seasons [38–40], including predicting the most prevalent types of influenza viruses for the season [41–43]. In the diagnostic stage, studies have demonstrated how metabolomic data from patients can be used to infer whether they are positive or negative for influenza [44]. There is even a report in which open access data were employed to develop a classifier for influenza diagnosis; however, not all included patients had a RT-qPCR result to confirm the diagnosis and validate the classifier's functionality [10]. ML has also been applied to forecast the efficacy of influenza vaccines [45–47].

2. Materials and Methods

2.1. Data Set

In this study, a clinical data set comprising 19,160 patient records from Mexico City was used. The database was made to track influenza's seasonal behavior and make prognosis for the next season. Data excluded patients' names, home addresses and hospital registration numbers. The data set was exported to the authorized researcher for this retrospective study, which was reviewed and approved by the institutional ethical committee (D1/19/501-T/03/096).

We applied three exclusion criteria: (1) age < 7 years, (2) patients with a negative influenza test but positive for another respiratory virus, and (3) no RT-qPCR result record. After these criteria, the study included 15,480 records of patients aged between 7 and 119 years old. Figure 2 shows the data distribution. The age ranges were 7–19 (41.5%), 20–39 (20.1%), 40–59 (21.2%), and age $\geq$ 60 (17.2%); according to the RT-qPCR test, 11,268 (72.8%) were negative and 4212 (27.2%) were positive for influenza virus, and the distribution by sex was 7710 (49.8%) men, and 7770 (50.2%) women. The data set consisted of 24 attributes encompassing clinical and demographic information collected from patients upon arrival at healthcare institutions for clinical examination and before the sample taking (nasopharyngeal and/or oropharyngeal exudate) for a RT-qPCR test.

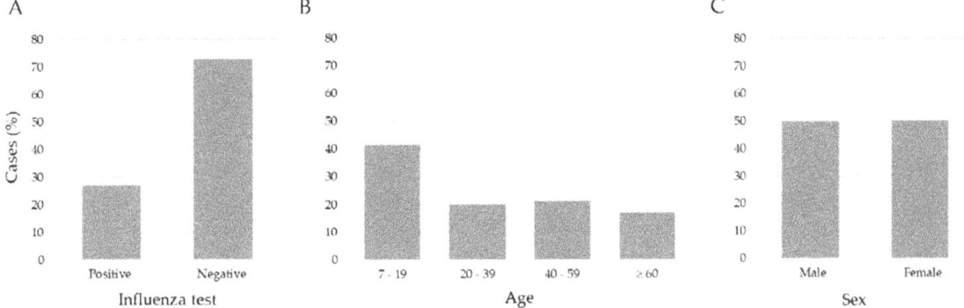

Figure 2. Distribution of influenza cases, ages and sex in the sample of 15,480 patients. (**A**) The number of negative (72.8) and positive (27.2) cases of influenza, (**B**) the ages separated into age ranges; (**C**) the % of feminine (50.2) and masculine (49.8) samples were very similar.

The set of symptoms and demographic features selected was the vector. The data were subjected to manual labeling by a clinician who assessed each patient and assigned values of 'yes' or 'no' to denote the presence or absence of symptoms, and 'unknown' otherwise.

2.2. Data Preprocessing

The symptoms and demographic data were labeled by binary numbers to indicate the presence or absence of a feature (1, 0, respectively); an unknown case was labeled as 0. Age range was mapped into the range {0...1}, to normalize the data.

In this study, two-step approaches were used to select the main features for training and testing the supervised ML methods. In the first step, Spearman's correlation was used to determine the correlation coefficient between features, as these are categorical variables. We selected them with a weak correlation ($r < 0.75$). In the second approach, chi-squared was computed to analyze the association between independent variables and influenza. The features selected had a strong association ($p < 0.001$).

To examine ML models and evaluate their performance, the data set was randomly split into 80% for the training set and 20% for the testing set. The models were evaluated with 10-fold cross-validation to select the best one. Python 3 functions were applied to create the k-Fold distribution and stratification.

The original data set was unbalanced, with the majority of the cases being negative for influenza (72.8%) and a minority of cases with a positive influenza test (27.2%). In this study, the target classes of the training set were balanced 50:50, and the skew was eliminated to obtain a most appropriate performance of the ML methods [48]. The minority oversampling technique used was Random Oversampling (ROS), which increases the size of the data set by randomly resampling the original minority class without creating new samples or changing the sample variability [49]. All samples from the majority class (negative for influenza) were used, and through ROS, data were added to the minority class (positive for influenza), obtaining an equal number of samples in both classes.

2.3. Machine Learning Algorithms

ML methods are an automatic and objective way to classify the samples into two classes, positive or negative for influenza, using records with inputs and outputs for the process, features and their classification [50–52]. In this way, we tried to find patterns in the known data in order to apply them to the new unclassified data.

We had the pair (X,Y) in all cases, $X = \{x_1, x_2, \ldots, x_n\}$ and Y = [positive | negative] for influenza, according to the RT-qPCR test.

The set of signs, symptoms and demographic features selected are represented by the vector X, and the data are binary numbers that indicate the absence or presence of the feature. The age was normalized to a range $\{0\ldots1\}$.

The aim was to find a model F of ML to represent the approximation between inputs and outputs.

$$F: X \rightarrow Y \tag{1}$$

For this study, 10 popular supervised ML methods through python.sklearn libraries for binary classification were used [53,54]: Adda Boost, Decision Tree, Bagging classification, Gradient Boosting Classifier (GBC), Random Forest (RF), K-nearest neighbors (Neighbors KNN), Naive Bayes (NB), Support Vector Machine (SVM), Logistic Regression (LR) and Discriminant Analysis. Taking advantage of the implementation of supervised ML models available in the python 3.7 programming language, the tests were carried out with several algorithms to evaluate their performance and be able to select the most accurate for this case study, and not only with the classical ML algorithms such as SVM, decision tree, KNN and NB.

Adaptive Boosting, named the AdaBoost algorithm, is a ML Meta-algorithm that can be used with many other ML algorithms to enhance its performance [55]. AdaBoost is sometimes denominated as the strongest out-of-the-box classifier for the so-called weak learners [56].

Decision trees used to predict categorical variables are called classification trees and decision trees [49]. The decision tree classifier is a flowchart-like tree structure; each internal node represents a test on an attribute, each branch represents an outcome of the test, and the class label is represented by each leaf nodes (or terminal nodes). Decision trees can be transformed into classification rules [57]. This ML algorithm is used to create the ensemble ML methods.

Random forest is a set of decision tree classifiers; in this ML model, each decision tree depends on a random vector of the training data set. They vote independently for the most popular class, and their classification is ensembled to give the final output using the classes given by each model [58]. The random forest algorithm is a special type of ensemble method. A random forest consists of many small classification decision or regression trees. Each tree, individually, is a weak learner; however, all the decision trees together can build a strong learner. It is random because (a) when building trees, a random sampling of training data sets is followed; and (b) when splitting nodes, a random subset of features is considered [59].

Bagging is a classifier which generates different subsets of the training data set by selecting data points randomly and with replacement. It can select the same instance multiple times. It is also called bootstrap aggregation and was created before the random forest. Given that a small change in the data can bring diverse effects in the model, the structure of the tree can completely change each tree to randomly sample the data set with a replacement, results on different trees [60].

The random forest algorithm is considered an extension of the bagging method. The difference is the number of features used in the decision tree construction: in bagging, all attributes are used for every decision tree, whereas in random forest, the decision trees have a random sample of attributes. Both ML methods are based on the decision tree method. The decision tree method works only with one tree to represent all samples and can be overfitting. Bagging and random forest work with several trees to represent different types of samples for each one.

Gradient boosting is a class of ensemble algorithms for machine learning that is used for regression or classification prediction modeling problems. It combines several sequential classifiers [61]. At an iteration, trees are added to the ensemble to fit correctly to the prediction errors made by prior models (boosting) and model fittings, using any arbitrary differentiable loss function and gradient descent optimization algorithm. The techniques is known as gradient boosting (Gboost) [62].

In the case of the KNN classifier, the main goal is to predict the closest value using distance as a basis. The Euclidean distance is a widely used technique [48]. The classification of the input data is based mainly on the selection of the majority class among its nearest neighbors [63].

The Naive Bayes classifier is considered a powerful probabilistic algorithm, based on Bayes theorem: the word "naïve" indicates interdependencies between characteristics. The version Bernoulli Naive Bayes (BNB) is used for Boolean variables as predictors [64,65].

Support Vector Machine (SVM) is a popular machine learning tool, which offers solutions to problems in classification and regression [66]. SVM separates classes and finds the hyperplane that best separates the data into different classes with a maximal margin between the classes. Initially using a linear decision boundary called a hyperplane to classify the data, Vapnik introduced a way of building a nonlinear classifier by using kernel functions [67]; it is placed at a location that maximizes the distance between the hyperplane and instances [68].

On the other hand, Discriminant Analysis aims to classify objects, by a set of independent variables, into one of two or more mutually exclusive and exhaustive categories. Discriminant analysis can be used only for classification (i.e., with a categorical target variable), not for regression. The target variable may have two or more categorical data by the use of multivariate information from the samples studied [69]. There are two methods: linear and quadratic discrimination. The first is the most widely used where classes are linearly separated. When a multi-classes analysis is needed, the two-groups method is used repeatedly in the analysis of pairs of data and the separation is linearity. Quadratic discrimination is used with nonlinearly separable classes.

2.4. Validation

The model performances were evaluated with a k-fold cross-validation method, which is an objective way to find the most robust ML algorithm, and we used the contingency table with the classification results [70].

In k-fold cross validation, the entire data set was divided into k equal parts, with k-1 subsets used for training while the remaining set was for testing. Each algorithm was trained and tested k times and the model output for each sample configuration obtained using cross validation was averaged to provide the global performance output of the model. The partition of the set with the folds in the subsets was arbitrary and with equal numbers of positive and negative cases in the training.

The confusion matrix helps to compare the classification result, and it has 4 values: true positives (TP), false positives (FP), true negatives (TN), and false negatives (FN). The columns in the matrix are the results obtained, while in the rows are the expected and results.

$$\begin{pmatrix} TP & FP \\ FN & TF \end{pmatrix}$$

We compared the performance of the methods in cross validation with the results using the following metrics:

$$\text{Accuracy, Acc} = \frac{TP + TF}{TN + FP + FN + TP} \tag{2}$$

This measure is for the samples correctly classified.

$$\text{Precision, Prec} = \frac{TP}{TP + FP} \tag{3}$$

measures the positive samples correctly classified vs. only positive samples.

$$\text{Recall, Rec} = \frac{TP}{TP + FN} \tag{4}$$

calculates the positive samples correctly classified vs. the samples expected to be positive. This is also called sensitivity.

$$\text{Specificity, Spec} = \frac{TN}{TN + FP} \tag{5}$$

measures the fraction of negative samples classified as negative.

$$\text{F1-score, F1} = \frac{2TP}{2TP + FP + FN} \tag{6}$$

It is an average between recall and precision.

Additionally, all the models were evaluated with the area under the curve (AUC) score in ROC analysis. This measure uses the ROC curve that shows the ability to make the difference between 2 classes with a graphical method using recall and specificity, and the AUC summarizes the performance of the classifiers in the training.

3. Results

At the beginning, the database had 24 attributes. According to Spearman's correlation, the feature arthralgia was highly correlated with myalgia ($r = 0.87$); hence, only myalgia was selected in the analysis, as it had fewer missing values. In the chi-squared test (p-value in Table 1), factors like sex, diarrhea and vaccination presented a low association with influenza ($p > 0.01$); therefore, these factors were also dismissed. Finally, 20 features were selected, many more than the 8 main symptoms of influenza indicated by the CDC: fever or feeling feverish/chills, cough, sore throat, runny or stuffy nose, muscle or body aches, headaches, fatigue (tiredness). Age data were used like one factor normalized in the range {0...1}.

Table 1. Attributes from influenza database of Mexico City.

Attributes	All Patients $n = 15,480$ n (%)	Positive $n = 4212$ n (%)	Negative $n = 11,268$ n (%)	p-Value
Demographic information				
Sex—Feminine	7770 (50.2)	2162 (51.3)	5608 (49.8)	0.087
Sex—Masculine	7710 (49.8)	2050 (48.7)	5660 (50.2)	
Hospitalized	10516 (67.9)	2449 (58.1)	8067 (71.6)	<0.001
Contact influenza-patients	2012 (13.0)	715 (17.0)	1297 (11.5)	<0.001
Vaccinated for influenza	2096 (13.5)	534 (12.7)	1562 (13.9)	0.059
Age 7–19 years	6417 (41.5)	1511 (35.9)	4906 (43.5)	
Age 20–39 years	3111 (20.1)	967 (23.0)	2144 (19.0)	<0.001
Age 40–59 years	3283 (21.2)	1056 (25.0)	2227 (19.8)	
Age $\geq$ 60 years	2669 (17.2)	678 (16.1)	1991 (17.7)	
Symptoms				
Fever	13,112 (84.7)	3853 (84.2)	9259 (82.2)	<0.001
Cough	13,953 (90.1)	3918 (85.7)	10,035 (89.1)	<0.001
Chest pain	3750 (24.2)	1160 (25.4)	2590 (23.0)	<0.001
Dyspnea	8642 (55.8)	2079 (45.5)	6563 (58.2)	<0.001
Irritability	4688 (30.3)	1159 (25.3)	3529 (31.3)	<0.001
Diarrhea	1833 (11.8)	492 (10.8)	1341 (11.9)	0.727
Shaking chills	5738 (37.1)	2003 (43.8)	3735 (33.1)	<0.001
Headache	8692 (56.1)	2896 (63.3)	5796 (51.4)	<0.001
Myalgia	6255 (40.4)	2279 (49.8)	3976 (35.3)	<0.001
Arthralgia	5539 (35.8)	2014 (44.0)	3525 (31.3)	<0.001
Malaise	9826 (63.5)	2947 (64.4)	6879 (61.0)	<0.001
Rhinorrhea	9277 (59.9)	2817 (61.6)	6460 (57.3)	<0.001
Polypnea	4602 (29.7)	1073 (23.5)	3529 (31.3)	<0.001
Vomiting	1958 (12.6)	606 (13.2)	1352 (12.0)	<0.001
Abdominal pain	2114 (13.7)	683 (14.9)	1431 (12.7)	<0.001
Sore throat	5321 (34.4)	1850 (40.4)	3471 (30.8)	<0.001
Conjunctivitis	3074 (19.9)	1104 (24.1)	1970 (17.5)	<0.001
Cyanosis	1703 (11.0)	395 (8.6)	1308 (11.6)	<0.001

The p-value corresponds to chi-squared. For the study, the attributes sex, vaccinated for influenza, and diarrhea were excluded with p-value > 0.001, and the others were selected.

The number of samples to test with ML was 15,840. This database was unbalanced with 11,268 (72.8%) negative and 4212 (27.2%) positive for influenza, according to the RT-qPCR tests. In this work, Random Over Sampling (ROS) was used to increase the number of positive records in the training set to improve the method performance with an equal number of samples in positive and negative classes.

The features sex, vaccination and diarrhea were not significant in the chi-squared test, with the dependent variable influenza: these features showed small variation between the positive and the negative classes.

Finally, our balanced training set had 7918 of positive and the same number of negative rows and 20 columns of features for training and testing 10 supervised ML algorithms, validated with k-fold cross validation (k = 10). In Table 2 are the results of ML methods with cross-validation. RF had the best evaluation (AUC = 0.94, Acc = 0.86, Rec = 0.91 and Spec = 0.88 were the best); in second place was the bagging classifier, which works similarly to RF. With the resampling technique, the number of the minor class (positive for influenza) increased, and the scores reflected the equilibrium. Figure 3 shows the ROC curves of the four best ML methods in the 10-fold cross-validation.

Table 2. Results of supervised machine learning algorithms.

Algorithm	AUC	Acc	Rec	Prec	Spec	F1
Random Forest	0.94	0.86	0.91	0.82	0.88	0.86
Bagging	0.93	0.85	0.90	0.82	0.87	0.85
Decision Tree	0.85	0.70	0.71	0.73	0.73	0.72
Kneighbors (7)	0.73	0.63	0.67	0.63	0.60	0.63
Gradient Boosting	0.69	0.62	0.69	0.61	0.56	0.62
SVM rbf	0.67	0.62	0.65	0.61	0.59	0.62
Quadratic Discriminant	0.66	0.62	0.70	0.60	0.54	0.62
Ada Boost	0.66	0.62	0.62	0.61	0.61	0.62
Linear Discriminant *	0.65	0.61	0.62	0.61	0.61	0.61
Linear SVM *	0.65	0.61	0.62	0.61	0.61	0.61
Logistic Regression	0.65	0.61	0.62	0.61	0.61	0.61
BernoulliNB	0.65	0.61	0.59	0.61	0.62	0.61

* Discriminant Analysis and SVM were used twice with different parameters. AUC, area under the curve; Acc, accuracy; Rec, recall; Prec, precision; Spec, specificity; D1, F1-score.

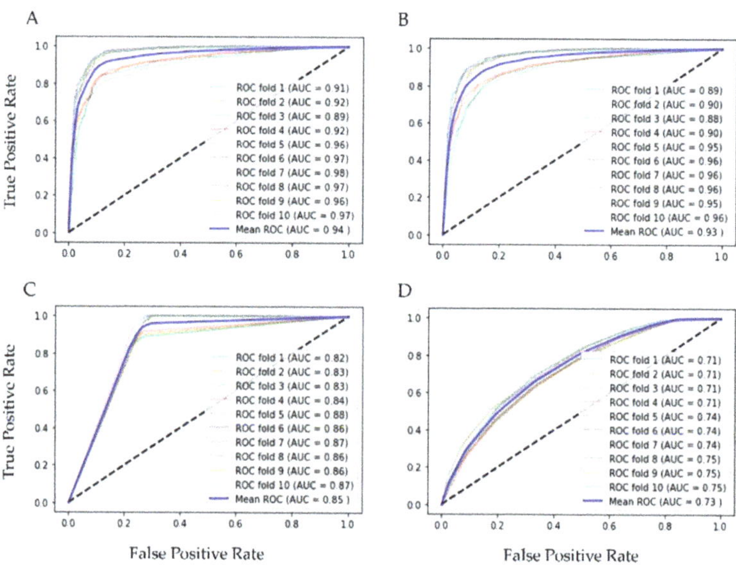

Figure 3. ROC graphics of the best ML algorithms with training set. (**A**) Random Forest, AUC = 0.94 ± 0.004; (**B**) Bagging, AUC = 0.93 ± 0.004; (**C**) Decision Tree, AUC = 0.85 ± 0.006; and (**D**) Kneighbors (7), AUC = 0.73 ± 0.014.

Random Forest, Bagging, and Decision Tree are supervised learning algorithms that can effectively handle categorical or binary features like the ones we have. The applied resampling seemed to have benefited the sensitivity results as it increased the number of positive samples in the training set. Bagging and Random Forest are ensemble techniques, based on Decision Tree, which ranked third, and along with Kneighbors, are algorithms capable of capturing non-linear relationships in the data.

The performance of the top four ML methods is shown in Figure 4, and their respective ROC plots are shown in Figure 5, both showing the results obtained with the test set. The RF and bagging methods demonstrated the highest scores when applied to the independent samples in the test set. However, it is important to note a substantial disparity between the sensitivity and specificity results in the test set. random forest achieved a sensitivity (rec) of 0.30 and a specificity (spec) of 0.90, while bagging had a sensitivity of 0.29 and a specificity of 0.88. Additionally, the significant differences observed between the results during the training and testing phases highlight certain limitations in this study. One potential factor contributing to these limitations is the sensitivity of machine learning models to class imbalance. Even after implementing random oversampling (ROS) on the training set to address the issue of imbalance, especially considering the considerably higher proportion of negative samples compared to positive samples, the desired variability was not effectively introduced into the training data. Another contributing factor might be attributed to the nature of the data set itself. In this study, we utilized binary data to represent the presence or absence of specific characteristics, with the exception of age, which was represented as a continuous variable. In contrast, other studies in the medical field that have achieved superior results have not only incorporated binary variables but have also integrated continuous variables obtained from laboratory tests and biometric measurements to assess patients' conditions [19,20,32,37]. An illustrative case from [21] involved the transformation of continuous data into binary values, but this approach also yielded unsatisfactory results. It is plausible that, in our case, the lack of relevant information and the use of subjective values to evaluate the health status of patients may have led to weaker associations between these characteristics and the occurrence of influenza.

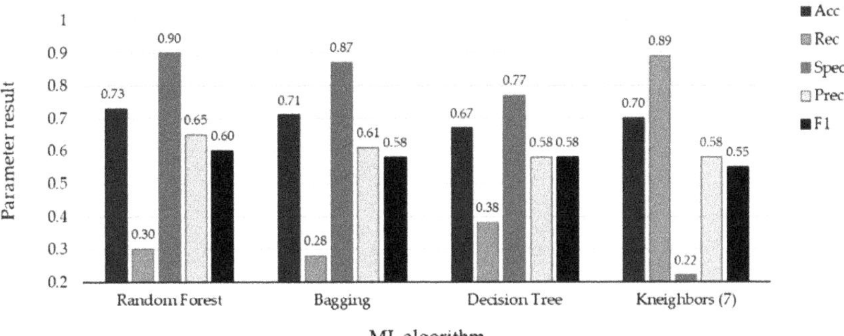

Figure 4. Results with test set using the four ML algorithms. The top scores are associated with specificity for the four algorithms: RF (spec = 0.90), Bagging (spec = 0.87), DT (spec = 0.77), and KNN (spec = 0.89). Conversely, the remaining metrics indicate the misclassification of positive influenza cases.

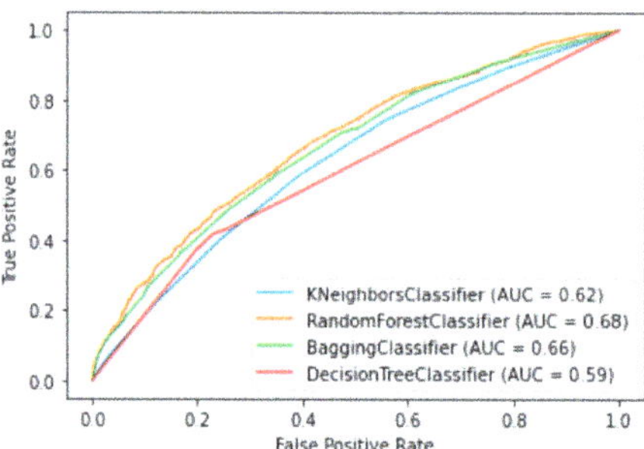

Figure 5. ROC graphics of the four best ML algorithms. Random Forest, Bagging, Decision Tree and Kneighbors (7).

4. Discussion

The use of artificial intelligence techniques through ML has increased its possibility as an alternative or powered tool in the diagnosis of infectious diseases [71–73]. With this idea, in this work we searched for an ML method as an alternative to the PCR test to perform diagnostics of influenza. Here, 19 category features and age were used, collected when the patient arrived with symptoms of influenza. In order to eliminate the unbalanced data between the positive and negative influenza cases of 15840 samples and to reduce the skew, we applied a resampling technique, random over sampling to positives. With resampling techniques, the Ensemble ML methods like Random Forest and Bagging could be favored—RF with AUC = 0.94, Acc = 0.86, Spec = 0.88 and, sensibility = 0.91, and Bagging with AUC = 0.93, Acc = 0.85, Spec = 0.87 and sensibility = 0.90—in the cross-validation. In other works [74,75] RF and Bagging were also combined with resampling techniques and showed good performance.

Nevertheless, it is important to note that these ML methods may serve as valuable screening tools to assist medical practitioners in distinguishing between positive and negative influenza cases, yielding promising results that could aid in decision making. This is particularly relevant for scenarios where the RT-qPCR test results are expected to be negative, potentially leading to reduced costs associated with testing.

Our research findings were based exclusively on data obtained from Mexican patients. This approach was chosen due to the unique health conditions prevalent in Mexico, which may differ significantly from those in other countries. It is important to consider that COVID-19 has changed the patterns of respiratory diseases [76,77]. Even though vaccines are applied every year, many people around the world are infected with influenza, causing a large number of deaths [4,78,79].

The potential advantage highlighted in our study is the use of an alternative decision-support tool, particularly relevant to regions where healthcare providers or patients, armed solely with basic questioning information, can assess the necessity for treatment and the conduction of PCR tests for the influenza disease.

4.1. Limitations of Work

Our results show problems in the prediction of positive influenza cases, maybe because the data set is imbalanced, and the binary features lose representativeness of the patient's health status. ML techniques hold potential in diverse applications; however, it is crucial to acknowledge that, in this study, these methods play a limited role as detection tools. They should not be perceived as a complete substitute for clinical diagnosis. RT-qPCR tests retain their indispensable status for precise influenza results, and therefore, machine learning models should be considered as complementary rather than as a complete replacement for conventional diagnostic approaches. Our findings indicate challenges in predicting positive influenza cases, possibly due to data imbalance and the diminished representativeness of binary features concerning a patients' health status.

It is possible that the low prediction values could be improved with our data set through several avenues. One approach could involve grouping individuals according to age, as symptoms may exhibit more pronounced patterns within specific age groups. Exploring alternative machine learning models is also a worthwhile consideration in our quest for improved predictions. Additionally, expanding the data set by including more positive cases from different Mexican regions could enhance the models' performance.

4.2. Future Work

This research has the potential for ongoing improvement and broader application. Comprehensive ablation studies can provide deeper insights into the algorithm's capabilities, allowing for a clearer grasp of its strengths and weaknesses. These studies encompass various facets, including feature selection, the incorporation of additional continuous data to enhance patient health assessment, the adoption of class balancing techniques, and the use of advanced machine learning models like convolutional neural networks to handle larger data sets and continuous data. Furthermore, it is essential to explore the creation of comprehensive models that effectively differentiate between COVID-19 and influenza cases.

5. Conclusions

In this study, machine learning models showcased a notably higher specificity compared to sensitivity, suggesting their potential utility in the identification of negative cases. This capability could help minimize the number of unnecessary molecular tests for individuals presenting with symptoms resembling influenza. This aspect is particularly pertinent in Mexico, where, for epidemiological reasons, during the influenza season around 10% of the population with symptoms resembling influenza are randomly selected for RT-qPCR testing, with approximately 70% of those cases turning out to be negative. By incorporating a tool akin to the one outlined in this study, clinicians can make more informed decisions about which patients require PCR testing, ultimately enhancing data quality for national-level decision making. Furthermore, given the limited availability of RT-qPCR testing facilities in certain areas, this tool can serve as valuable support for healthcare practitioners, aiding them in determining the necessity of conducting tests. This approach has the potential to reduce costs for patients and ease the burden on the healthcare sector.

Author Contributions: Conceptualization and methodology, E.M. and E.V.B.-P.; validation, E.V.B.-P. and J.S.; investigation, E.M. and K.R.; data curation, E.M.; writing—original draft preparation, E.M. and K.R.; writing—review and editing, E.V.B.-P. and A.L.S.-S.; supervision, J.S. All authors have read and agreed to the published version of the manuscript.

Funding: This research received no external funding.

Institutional Review Board Statement: The work follows the principles of the Declaration of Helsinki and the Research Ethics Committee of the General Hospital of Mexico "Dr. Eduardo Liceaga" institutional ethical committee reviewed and approved the study design (D1/19/501-T/03/096).

Informed Consent Statement: Informed consent was not required by patients because this is a retrospective study.

Data Availability Statement: The data presented in this study are available on request from the corresponding author.

Conflicts of Interest: The authors declare no conflict of interest.

References

1.   Centro Nacional de Programas de Control y Preventivos de Enfermedades. Manual de Atención a la Salud Ante Emergencias. Available online: https://epidemiologia.salud.gob.mx/gobmx/salud/documentos/manuales/ (accessed on 1 February 2022).
2.   LaRussa, P. Pandemic novel 2009 H1N1 influenza: What have we learned? *Semin. Respir. Crit. Care Med.* **2011**, *32*, 393–399. [CrossRef] [PubMed]
3.   Gordon, A.; Reingold, A. The Burden of Influenza: A Complex Problem. *Curr. Epidemiol. Rep.* **2018**, *5*, 1–9. [CrossRef] [PubMed]

4. Epidemiology General Vigilance of Mexico, Informe Semanal de Vigilancia Epidemilógica. Available online: https://www.gob. mx/cms/uploads/attachment/file/737555/INFLUENZA_OVR_SE26_2022.pdf (accessed on 8 July 2022).
5. Krammer, F.; Smith, G.J.D.; Fouchier, R.A.M.; Peiris, M.; Kedzierska, K.; Doherty, P.C.; Palese, P.; Shaw, M.L.; Treanor, J.; Webster, R.G.; et al. Influenza. *Nat. Rev. Dis. Primers* **2018**, *4*, 3. [CrossRef]
6. Chow, E.J.; Doyle, J.D.; Uyeki, T.M. Influenza virus-related critical illness: Prevention, diagnosis, treatment. *Crit. Care* **2019**, *23*, 214. [CrossRef]
7. Ruiz-Matus, C.; Kuri-Morales, P.; Narro-Robles, J. Comportamiento de las temporadas de influenza en México de 2010 a 2016, análisis y prospectiva. *Gac. Med. Mex.* **2017**, *153*, 205–213.
8. Centers for Disease Control and Prevention. Flu Symptoms and Complications. Available online: https://www.cdc.gov/flu/ symptoms/symptoms.htm (accessed on 13 July 2022).
9. World Health Organization. Influenza (Seasonal). Available online: https://www.who.int/news-room/fact-sheets/detail/ influenza-(seasonal) (accessed on 13 July 2022).
10. López-Pineda, A.; Ye, Y.; Visweswaran, S.; Cooper, G.F.; Wagner, M.M.; Tsui, F.R. Comparison of machine learning classifiers for influenza detection from emergency department free-text reports. *J. Biomed. Inform.* **2015**, *58*, 60–69. [CrossRef]
11. Bonaccorso, G. *Machine Learning Algorithms*, 1st. ed.; Pack Publishing: Birmingham, UK, 2017; pp. 8–16.
12. Pandya, S.; Thakur, A.; Saxena, S.; Jassal, N.; Patel, C.; Modi, K.; Shah, P.; Joshi, R.; Gonge, S.; Kadam, K.; et al. A Study of the Recent Trends of Immunology: Key Challenges, Domains, Applications, Datasets, and Future Directions. *Sensors* **2021**, *21*, 7786. [CrossRef]
13. Vijayan, V.V.; Anjali, C. Prediction and diagnosis of diabetes mellitus—A machine learning approach. In Proceedings of the IEEE Recent Advances in Intelligent Computational Systems (RAICS), Trivandrum, India, 20 April 2021; pp. 122–127. [CrossRef]
14. Pecht, M.G.; Kang, M. Machine Learning: Diagnostics and Prognostics. In *Prognostics and Health Management of Electronics: Fundamentals, Machine Learning, and the Internet of Things*; John Wiley and Sons Ltd.: Hoboken, NJ, USA, 2019; pp. 163–191. [CrossRef]
15. Shigueoka, L.S.; de Vasconcellos, J.P.C.; Schimiti, R.B.; Reis, A.S.C.; de Oliveira, G.O.; Gomi, E.S.; Vianna, J.A.R.; Lisboa, R.D.d.R.; Medeiros, F.A.; Costa, V.P. Automated algorithms combining structure and function outperform general ophthalmologists in diagnosing glaucoma. *PLoS ONE* **2018**, *13*, e0207784. [CrossRef]
16. Ullah, R.; Khan, S.; Ali, H.; Chaudhary, I.I.; Bilal, M.; Ahmad, I. A comparative study of machine learning classifiers for risk prediction of asthma disease. *Photodiagn. Photodyn. Ther.* **2019**, *28*, 292–296. [CrossRef]
17. Akazawa, M.; Hashimoto, K. Artificial Intelligence in Ovarian Cancer Diagnosis. *Anticancer Res.* **2020**, *40*, 4795–4800. [CrossRef]
18. Lu, W.; Tong, Y.; Yu, Y.; Xing, Y.; Chen, C.; Shen, Y. Applications of Artificial Intelligence in Ophthalmology: General Overview. *J. Ophthalmol.* **2018**, *2018*, 5278196. [CrossRef] [PubMed]
19. Silva, F.R.; Vidotti, V.G.; Cremasco, F.; Días, M.; Gomi, E.S.; Costa, V.P. Sensitivity and specificity of machine learning classifiers for glaucoma diagnosis using Spectral Domain OCT and standard automated perimetry. *Arq. Bras. Oftalmol.* **2013**, *76*, 170–174. [CrossRef] [PubMed]
20. Guncar, G.; Kukar, M.; Notar, M.; Brvar, M.; Cernelc, P. An application of machine learning to haematological diagnosis. *Sci. Rep.* **2018**, *8*, 411. [CrossRef] [PubMed]
21. Smith, J.P.; Milligan, K.; McCarthy, K.D.; Mchembere, W.; Okeyo, E.; Musau, S.K.; Okumu, A.; Song, R.; Click, E.S.; Cain, K.P. Machine learning to predict bacteriologic confirmation of *Mycobacterium. tuberculosis.* in infants and very young children. *PLoS Digit. Health* **2023**, *2*, 249. [CrossRef]
22. Peng, B.; Gong, H.; Tian, H.; Zhuang, Q.; Li, J.; Cheng, K.; Ming, Y. The study of the association between immune monitoring and pneumonia in kidney transplant recipients through machine learning models. *J. Transl. Med.* **2020**, *18*, 370. [CrossRef]
23. Saybani, M.R.; Shamshirband, S.; Hormozi, S.G.; Wah, T.Y.; Aghabozorgi, S.; Pourhoseingholi, M.A.; Olariu, T. Diagnosing tuberculosis with a novel support vector machine-based artificial immune recognition system. *Iran. Red Crescent Med. J.* **2015**, *17*, e24557. [CrossRef]
24. Melendez, J.; Sánchez, C.I.; Philipsen, R.H.H.M.; Maduskar, P.; Dawson, R.; Theron, G.; Dheda, K.; van Ginneken, B. An automated tuberculosis screening strategy combining X-ray-based computer-aided detection and clinical information. *Sci. Rep.* **2016**, *6*, 25265. [CrossRef]
25. Er, O.; Temurtas, F.; Tanrikulu, A.Ç. Tuberculosis disease diagnosis using artificial neural networks. *J. Med. Syst.* **2010**, *34*, 299–302. [CrossRef]
26. e Souza, J.B.D.O.; Sanchez, M.; de Seixas, J.M.; Maidantchik, C.; Galliez, R.; Moreira, A.D.S.R.; da Costa, P.A.; Oliveira, M.M.; Harries, A.D.; Kritski, A.L. Screening for active pulmonary tuberculosis: Development and applicability of artificial neural network models. *Tuberculosis* **2018**, *111*, 94–101. [CrossRef]
27. Revuelta-Zamorano, P.; Sánchez, A.; Rojo-Álvarez, J.L.; Álvarez-Rodríguez, J.; Ramos-López, J.; Soguero-Ruiz, C. Prediction of healthcare associated infections in an intensive care unit using machine learning and big data tools. In Proceedings of the XIV Mediterranean Conference on Medical and Biological Engineering and Computing, Paphos, Cyprus, 31 March–2 April 2016; Springer International Publishing: Cham, Switzerland, 2016; Volume 57, p. 840e5. [CrossRef]
28. Hernandez, B.; Herrero, P.; Rawson, T.M.; Moore, L.S.P.; Evans, B.; Toumazou, C.; Holmes, A.H.; Georgiou, P. Supervised learning for infection risk inference using pathology data. *BMC Med. Inform. Decis. Mak.* **2017**, *17*, 168. [CrossRef]

29. Van Steenkiste, T.; Ruyssinck, J.; De Baets, L.; Decruyenaere, J.; De Turck, F.; Ongenae, F.; Dhaene, T. Accurate prediction of blood culture outcome in the intensive care unit using long short-term memory neural networks. *Artif. Intell. Med.* **2018**, *97*, 38–43. [CrossRef] [PubMed]

30. Ke, C.; Jin, Y.; Evans, H.; Lober, B.; Qian, X.; Liu, J.; Huang, S. Prognostics of surgical site infections using dynamic health data. *J. Biomed. Inform.* **2017**, *65*, 22–33. [CrossRef] [PubMed]

31. Horng, S.; Sontag, D.A.; Halpern, Y.; Jernite, Y.; Shapiro, N.I.; Nathanson, L.A. Creating an automated trigger for sepsis clinical decision support at emergency department triage using machine learning. *PLoS ONE* **2017**, *12*, e0174708. [CrossRef]

32. Taylor, R.A.; Moore, C.L.; Cheung, K.; Brandt, C. Predicting urinary tract infections in the emergency department with machine learning. *PLoS ONE* **2018**, *13*, e0194085. [CrossRef]

33. Revett, K.; Gorunescu, F.; Ene, M. A machine learning approach to differentiating bacterial from viral meningitis. In Proceedings of the IEEE John Vincent Atanasoff 2006 International Symposium on Modern Computing, Sofia, Bulgaria, 3–6 October 2006; pp. 155–162. [CrossRef]

34. D'Angelo, G.; Pilla, R.; Tascini, C.; Rampone, S. A proposal for distinguishing between bacterial and viral meningitis using genetic programming and decision trees. *Soft Comput.* **2019**, *23*, 11775–11791. [CrossRef]

35. Shamshirband, S.; Hessam, S.; Javidnia, H.; Amiribesheli, M.; Vahdat, S.; Petković, D.; Gani, A.; Kiah, L.M. Tuberculosis disease diagnosis using artificial immune recognition system. *Int. J. Med. Sci.* **2014**, *11*, 508–514. [CrossRef]

36. Jayatilake, S.M.; Ganegoda, G.U. Involvement of Machine Learning Tools in Healthcare Decision Making. *J. Healthc. Eng.* **2021**, *2021*, 6679512. [CrossRef]

37. Altini, N.; Brunetti, A.; Mazzoleni, S.; Moncelli, F.; Zagaria, I.; Prencipe, B.; Lorusso, E.; Buonamico, E.; Carpagnano, G.E.; Bavaro, D.F.; et al. Predictive Machine Learning Models and Survival Analysis for COVID-19 Prognosis Based on Hematochemical Parameters. *Sensors* **2021**, *21*, 8503. [CrossRef]

38. Borkenhagen, L.K.; Allen, M.W.; Runstadler, J.A. Influenza virus genotype to phenotype predictions through machine learning: A systematic review. *Emerg. Microbes Infect.* **2021**, *10*, 1896–1907. [CrossRef]

39. Cheng, H.-Y.; Wu, Y.-C.; Lin, M.-H.; Liu, Y.-L.; Tsai, Y.-Y.; Wu, J.-H.; Pan, K.-H.; Ke, C.-J.; Chen, C.-M.; Liu, D.-P.; et al. Applying Machine Learning Models with An Ensemble Approach for Accurate Real-Time Influenza Forecasting in Taiwan: Development and Validation Study. *J. Med. Internet Res.* **2020**, *22*, e15394. [CrossRef]

40. Reich, N.G.; McGowan, C.J.; Yamana, T.K.; Tushar, A.; Ray, E.L.; Osthus, D.; Kandula, S.; Brooks, L.C.; Crawford-Crudell, W.; Gibson, G.C.; et al. Accuracy of real-time multi-model ensemble forecasts for seasonal influenza in the U.S. *PLoS Comput. Biol.* **2019**, *15*, e1007486. [CrossRef] [PubMed]

41. Hayati, M.; Biller, P.; Colijn, C. Predicting the short-term success of human influenza virus variants with machine learning. *Proc. Biol. Sci.* **2020**, *287*, 20200319. [CrossRef] [PubMed]

42. Kwon, E.; Cho, M.; Kim, H.; Son, S. A Study on Host Tropism Determinants of Influenza Virus Using Machine Learning. *Curr. Bioinform.* **2020**, *15*, 121–134. [CrossRef]

43. Al Dalbhi, S.; Alshahrani, H.A.; Almadi, A.; Busaleh, H.; Alotaibi, M.; Almutairi, W.; Almukhrq, Z. Prevalence and mortality due to acute kidney injuries in patients with influenza A (H1N1) viral infection: A systemic narrative review. *Int. J. Health Sci.* **2019**, *13*, 56–62.

44. Hogan, C.A.; Rajpurkar, P.; Sowrirajan, H.; Phillips, N.A.; Le, A.T.; Wu, M.; Garamani, N.; Sahoo, M.K.; Wood, M.L.; Huang, C.; et al. Nasopharyngeal metabolomics and machine learning approach for the diagnosis of influenza. *EBioMedicine* **2021**, *71*, 103546. [CrossRef]

45. Fukuta, H.; Goto, T.; Wakami, K.; Kamiya, T.; Ohte, N. The effect of influenza vaccination on mortality and hospitalization in patients with heart failure: A systematic review and meta-analysis. *Heart Fail. Rev.* **2019**, *24*, 109–114. [CrossRef]

46. Tomic, A.; Tomic, I.; Rosenberg-Hasson, Y.; Dekker, C.L.; Maecker, H.T.; Davis, M.M. SIMON, an Automated Machine Learning System, Reveals Immune Signatures of Influenza Vaccine Responses. *J. Immunol.* **2019**, *203*, 749–759. [CrossRef]

47. Wolk, D.M.; Lanyado, A.; Tice, A.M.; Shermohammed, M.; Kinar, Y.; Goren, A.; Chabris, C.F.; Meyer, M.N.; Shoshan, A.; Abedi, V. Prediction of Influenza Complications: Development and Validation of a Machine Learning Prediction Model to Improve and Expand the Identification of Vaccine-Hesitant Patients at Risk of Severe Influenza Complications. *J. Clin. Med.* **2022**, *11*, 4342. [CrossRef]

48. López, V.; Fernández, A.; García, S.; Herrera, F. An insight into classification with imbalanced data: Empirical results and current trends on using data intrinsic characteristics. *Inf. Sci.* **2013**, *250*, 113–141. [CrossRef]

49. Chawla, N.V. Data Mining for Imbalanced Datasets: An Overview. In *Data Mining and Knowledge Discovery Handbook*; Springer: Boston, MA, USA, 2005.

50. Kamal, K.H.; Ritesh, K.J.; Kamlesh, L.; Ruchi, D. *Machine Learning: Master Supervised and Unsupervised Learning Algorithms with Real Examples*; BPB Publications: Noida, India, 2021.

51. Alpaydin, E. *Machine Learning: The New AI*; MIT Press Essential Knowledge Series; MIT Press: London, UK, 2016.

52. Singh, A.; Thakur, N.; Sharma, A. A review of supervised machine learning algorithms. In Proceedings of the 3rd International Conference on Computing for Sustainable Global Development (INDIACom), New Delhi, India, 16–18 March 2016; pp. 1310–1315.

53. Raschka, S. *Python Machine Learning*, 2nd ed.; Packt Publishing: Birmingham, UK, 2017.

54. Müller, A.C.; Guido, S. *Introduction to Machine Learning with Python: A Guide for Data Scientists*, 1st ed.; O'Reilly Media, Inc.: Sebastopol, CA, USA, 2016.

55. Mathanker, S.K.; Weckler, P.R.; Bowser, T.J.; Wang, N.; Maness, N.O. AdaBoost classifiers for pecan defect classification. *Comput. Electron. Agric.* **2011**, *77*, 60–68. [CrossRef]
56. Freund, Y.; Schapire, R.E. A decision-theoretic generalization of on-line learning and an application to boosting. *J. Comput. Syst. Sci.* **1997**, *55*, 119–139. [CrossRef]
57. Decision-Trees. Available online: https://www.ibm.com/topics/decision-trees (accessed on 5 October 2022).
58. Breiman, L. Random Forests. *Mach. Learn.* **2001**, *5*, 5–32. [CrossRef]
59. Satpathy, R.; Choudhury, T.; Satpathy, S.; Mohanty, S.; Zhang, X. Introduction to supervised learning. In *Data Analytics in Bioinformatics: A Machine Learning Perspective*, 1st ed.; John Wiley & Sons: Hoboken, NJ, USA, 2021; pp. 18–20.
60. Breiman, L. Bagging Predictors. *Mach. Learn.* **1996**, *24*, 123–140. [CrossRef]
61. Marsland, S. Boosting. In *Machine Learning: An Algorithmic Perspective*, 2nd ed.; Taylor and Francis Group: Boca Raton, FL, USA; CRC Press: Oxford, NY, USA, 2015; pp. 268–273.
62. Natekin, A.; Knoll, A. Gradient boosting machines, a tutorial. *Front. Neurorobot.* **2013**, *7*, 21. [CrossRef] [PubMed]
63. Taunk, K.; Verma, S.; Swetapadma, A. A Brief Review of Nearest Neighbor Algorithm for Learning and Classification. In Proceedings of the International Conference on Intelligent Computing and Control Systems (ICCS), Madurai, India, 15–17 May 2019. [CrossRef]
64. Bafjaish, S. Comparative Analysis of Naive Bayesian Techniques in Health-Related for Classification Task. *J. Soft Comput. Data Min.* **2020**, *1*, 1–10. Available online: https://penerbit.uthm.edu.my/ojs/index.php/jscdm/article/view/7144 (accessed on 24 October 2023).
65. Ranjitha, K.V. Classification and optimization scheme for text data using machine learning Naïve Bayes classifier. In Proceedings of the 2018 IEEE World Symposium on Communication Engineering (WSCE), Singapore, 28–30 December 2018; pp. 33–36.
66. Sulaiman, M.A. Evaluating Data Mining Classification Methods Performance in Internet of Things Applications. *J. Soft Comput. Data Min.* **2020**, *1*, 11–25. Available online: https://penerbit.uthm.edu.my/ojs/index.php/jscdm/article/view/7127 (accessed on 13 August 2021).
67. Ghosh, S.; Dasgupta, A.; Swetapadma, A. A Study on Support Vector Machine based Linear and Non-Linear Pattern Classification. In Proceedings of the 2019 International Conference on Intelligent Sustainable Systems (ICISS), Palladam, India, 21–22 February 2019; pp. 24–28.
68. Cortes, C.; Vapnik, V. Support-vector networks. *Mach. Learn.* **1995**, *20*, 273–297. [CrossRef]
69. Alayande, S.; Adekunle, B. An overview and application of discriminant analysis in data analysis. *IOSR J. Math.* **2015**, *11*, 12–15.
70. Varoquaux, G.; Colliot, O. Evaluating machine learning models and their diagnostic value. In *Machine Learning for Brain Disorders*; Springer: New York, NY, USA, USA, 2022; pp. 1–30.
71. Chiu, H.-Y.R.; Hwang, C.-K.; Chen, S.-Y.; Shih, F.-Y.; Han, H.-C.; King, C.-C.; Gilbert, J.R.; Fang, C.-C.; Oyang, Y.-J. Machine learning for emerging infectious disease field responses. *Sci. Rep.* **2022**, *12*, 328. [CrossRef]
72. Mondal, M.R.; Bharati, S.; Podder, P. Diagnosis of COVID-19 Using Machine Learning and Deep Learning: A Review. *Curr. Med. Imaging* **2021**, *17*, 1403–1418. [CrossRef]
73. Mayrose, H.; Bairy, G.M.; Sampathila, N.; Belurkar, S.; Saravu, K. Machine Learning-Based Detection of Dengue from Blood Smear Images Utilizing Platelet and Lymphocyte Characteristics. *Diagnostics* **2023**, *13*, 220. [CrossRef] [PubMed]
74. Kamalov, F.; Elnagar, A.; Leung, H. Ensemble Learning with Resampling for Imbalanced Data. In *ICIC 2021: Intelligent Computing Theories and Application*; Lecture Notes in Computer, Science; Huang, D.S., Jo, K.H., Li, J., Gribova, V., Hussain, A., Eds.; Springer: Cham, Switzerland, 2021; Volume 12837. [CrossRef]
75. Geetha, R.; Sivasubramanian, S.; Kaliappan, M.; Vimal, S.; Annamalai, S. Cervical Cancer Identification with Synthetic Minority Oversampling Technique and PCA Analysis using Random Forest Classifier. *J. Med. Syst.* **2019**, *43*, 286. [CrossRef] [PubMed]
76. Chow, E.J.; Uyeki, T.M.; Chu, H.Y. The effects of the COVID-19 pandemic on community respiratory virus activity. *Nat. Rev. Microbiol.* **2023**, *21*, 195–210. [CrossRef]
77. Aloui, K.; Hamza, C.; Mefteh, K.; Hancn, S. Epidemiologic changes of Respiratory syncytial virus in the COVID-19 Era. *Med. Mal. Infect. Form.* **2022**, *1*, 109. [CrossRef]
78. Kandeel, A.; Fahim, M.; Deghedy, O.; Roshdy, W.H.; Khalifa, M.K.; El Shesheny, R.; Kandeil, A.; Naguib, A.; Afifi, S.; Mohsen, A.; et al. Resurgence of influenza and respiratory syncytial virus in Egypt following two years of decline during the COVID-19 pandemic: Outpatient clinic survey of infants and children. *BMC Public Health* **2022**, *23*, 1067. [CrossRef]
79. Barraza, M.F.O.; Fasce, R.A.; Nogareda, F.; Marcenac, P.; Mallegas, N.V.; Alister, P.B.; Loayza, S.; Chard, A.N.; Arriola, C.S.; Couto, P.; et al. Influenza Incidence and Vaccine Effectiveness During the Southern Hemisphere Influenza Season—Chile. *MMWR Morb. Mortal. Wkly.* **2022**, *71*, 1353–1358. [CrossRef]

 *diagnostics*

Article

Parkinson's Disease Detection Using Filter Feature Selection and a Genetic Algorithm with Ensemble Learning

Abdullah Marish Ali [1], **Farsana Salim** [2] and **Faisal Saeed** [2,*]

[1] Department of Computer Science, Faculty of Computing and Information Technology, King Abdulaziz University, Jeddah 21589, Saudi Arabia; ammali@kau.edu.sa
[2] DAAI Research Group, College of Computing and Digital Technology, Birmingham City University, Birmingham B4 7XG, UK; farsana.salim@bcu.ac.uk
* Correspondence: faisal.saeed@bcu.ac.uk

Abstract: Parkinson's disease (PD) is a neurodegenerative disorder marked by motor and non-motor symptoms that have a severe impact on the quality of life of the affected individuals. This study explores the effect of filter feature selection, followed by ensemble learning methods and genetic selection, on the detection of PD patients from attributes extracted from voice clips from both PD patients and healthy patients. Two distinct datasets were employed in this study. Filter feature selection was carried out by eliminating quasi-constant features. Several classification models were then tested on the filtered data. Decision tree, random forest, and XGBoost classifiers produced remarkable results, especially on Dataset 1, where 100% accuracy was achieved by decision tree and random forest. Ensemble learning methods (voting, stacking, and bagging) were then applied to the best-performing models to see whether the results could be enhanced further. Additionally, genetic selection was applied to the filtered data and evaluated using several classification models for their accuracy and precision. It was found that in most cases, the predictions for PD patients showed more precision than those for healthy individuals. The overall performance was also better on Dataset 1 than on Dataset 2, which had a greater number of features.

Keywords: Parkinson's disease (PD); filter feature selection; ensemble learning; genetic selection

 check for
updates

Citation: Ali, A.M.; Salim, F.; Saeed, F. Parkinson's Disease Detection Using Filter Feature Selection and a Genetic Algorithm with Ensemble Learning. *Diagnostics* **2023**, *13*, 2816. https://doi.org/10.3390/diagnostics13172816

Academic Editors: Christian la Fougere and Mugahed A. Al-antari

Received: 28 June 2023
Revised: 17 August 2023
Accepted: 25 August 2023
Published: 31 August 2023

1. Introduction

Parkinson's disease (PD) is a neurodegenerative disorder that affects millions of individuals worldwide. It is characterized by motor symptoms such as tremors, rigidity, bradykinesia (slowness of movement), and postural instability. PD not only impairs the quality of life for patients but also poses significant challenges for accurate and timely diagnosis. The presence of voice deficits, which are frequently defined by alterations in speech patterns, cadence, and tone, emerges as an important element of Parkinson's disease symptomatology. The study by Tjaden, Lam, and Wilding [1] revealed that speakers with PD displayed expanded peripheral and non-peripheral vowel space areas during articulate speech, accompanied by a reduction in speech rate and an increased vocal intensity. Furthermore, the study by Tsanas et al. [2] highlighted the feasibility of utilizing straightforward, self-administered, and non-intrusive speech tests as a potential strategy for regular, remote, and precise monitoring of PD symptom progression with the employment of the Unified Parkinson's Disease Rating Scale (UPDRS). These studies showcase the potential of voice-related changes to act as valuable indicators for the early detection of Parkinson's disease, despite receiving less recognition than motor symptoms.

Recent advances in machine learning techniques, as well as the availability of large-scale datasets, have opened new avenues for the automated identification of PD utilizing various forms of data, including voice recordings. Furthermore, machine learning-based PD detection systems have the potential to be non-invasive, low-cost, and easily scalable.

A voice recording can be collected easily through commonly available devices such as smartphones, making it a convenient and accessible tool for screening and monitoring PD.

A series of studies have delved into the domain of Parkinson's disease (PD) classification, harnessing voice data as a diagnostic indicator. However, one notable gap is the limited size and diversity of the datasets employed in many prior studies. This limitation raises concerns about the generalizability and reliability of the resulting classification models. This study makes a significant contribution to the field by decisively addressing this issue through the utilization of two distinct datasets. Another gap has been the lack of comprehensive feature selection methods employed in PD classification studies. While some efforts have been made to apply feature selection techniques, this study takes a step forward by introducing a novel combination of filter feature selection methods with ensemble learning and genetic selection. This fusion holds the promise of uncovering more relevant and discriminative features inherent in the voice data, potentially leading to a substantial enhancement in the accuracy of PD classification. Furthermore, the limited exploration of model ensemble techniques in prior studies has presented a significant gap, which this research effectively addresses. While several investigations have focused primarily on individual classification algorithms, the untapped potential of leveraging the strengths of various algorithms through ensemble methods has been underutilized. Ensemble learning methods have the inherent advantage of integrating the diverse strengths of different algorithms, thereby enhancing the overall predictive power and accuracy of the classification process. By exploring this avenue, this research provides a vital contribution to the field by demonstrating the potential of ensemble techniques to significantly elevate the performance and efficacy of PD classification models.

In this study, a combination of filter feature selection methods with ensemble learning and genetic selection was used to detect PD from voice clips. The filtered data was fed into different classification models, which were then evaluated based on their accuracy and precision. By evaluating the models on these diverse datasets with varying characteristics and complexities, the generalizability and scalability of the approach may be assessed. The outcomes of this study may enhance our understanding and augment the efficacy of early PD detection, ultimately leading to improved patient care and prognosis.

2. Related Work

Several studies have investigated the use of machine learning and statistical modelling techniques to extract discriminative features from voice recordings and to develop classification models for PD detection. These have been summarized in Table 1.

Table 1. Summary of studies that utilized machine learning for PD detection.

Study	Dataset	Method	Results
Sheikhi and Kheirabadi, 2022 [3]	Voice UCI PD dataset	Combination of the Random Forest (RF) and Rotation Forest algorithms for classifying prediction outcomes as severe or non-severe	Accuracy for: total UPDRS—76.09% motor UPDRS—79.49%
Mohammed et al., 2021 [4]	Voice UCI PD dataset	Feature selection and classification	Accuracy—96.6%
Velmurugan and Dhinakaran, 2022 [5]	UCI machine learning repository	Combination of linear regression and Adaboost ensemble methods with Random Forest (RF) and extreme gradient boosting (XGBoost)	Accuracy—90.13%

Table 1. *Cont.*

Study	Dataset	Method	Results
Sharma et al., 2021 [6]	Dataset collected by Max Little of Oxford University for voice disorders by collaborating with the National Centre for Voice and Speech	Feature selection using the Rao algorithm and classification using the k-Nearest Neighbors (KNN) classifier	Accuracy—99.25%
Sabeena et al., 2022 [7]	Voice dataset from the UCI machine learning repository	Feature selection using optimization-based ensembles and different classification algorithms	Accuracy ranging from 83.66% to 98.77%.
Ul Haq et al., 2020 [8]	Voice dataset	Feature selection using different feature selection methods and different classification using Support Vector Machine (SVM)	Accuracy ranging from 98.20% to 99.50%
Sarankumar et al., 2022 [9]	Voice dataset	Feature selection using the firming bacteria foraging algorithm and classification using the Deep Brooke inception net algorithm	Accuracy—99.88%
Pahuja and Nagabhushan, 2021 [10]	Voice dataset from the UCI repository	Classification using Artificial Neural Network (ANN), Support Vector Machine (SVM) and k-Nearest Neighbors (KNN) algorithms	Accuracy—95.89%, 88.21%, and 72.31%, respectively
Yücelbaş, 2021 [11]	Voice dataset	Feature selection using the Information gain algorithm-based KNN hybrid model (IGKNN)	Accuracy—98%
Pramanik et al., 2021 [12]	Acoustic features from the UCI machine learning repository	Feature selection using Correlated Feature Selection (CFS), Fisher Score Feature Selection (FSFS), and Mutual Information-based Feature Selection (MIFS) techniques, and classification using Naïve Bayes classifier	Accuracy—78.97%
Salmanpour, Shamsaei, Saberi, et al., 2021 [13]	Combination of non-imaging, imaging, and radiomic features from DAT-SPECT images	Sixteen algorithms for feature reduction, eight algorithms for clustering, and 16 classifiers	Subdivided the PD into three subtypes, namely mild, intermediate, and severe
Nahar et al., 2021 [14]	Acoustic features from the UCI machine learning repository	Feature selection using Boruta, Recursive Feature Elimination (RFE), and Random Forest (RF), and classification using Gradient Boosting, Extreme Gradient Boosting, Bagging, and Extra Tree Classifier	Accuracy—82.35% from applying the RFE feature selection methods and Bagging classifier

In a recent study by Sheikhi and Kheirabadi [3], a voice dataset from the UCI Repository was utilized for the classification of PD. The dataset comprised voice recordings from 42 patients, totaling 5875 instances. They proposed a model that combined the Random Forest (RF) and Rotation Forest algorithms to classify the predictions into two categories: severe or non-severe. The accuracy results for the total Unified PD Rating Scale (UPDRS) and motor UPDRS using this model were found to be 76.09% and 79.49%, respectively. In another study conducted by Mohammed et al. [4], a multi-agent approach was employed to filter and identify the most relevant features that could enhance PD classification accuracy while reducing training time. They utilized a dataset consisting of 31 human voice recordings, 23 of which were diagnosed with PD. Initially, the dataset contained 22 features, which were reduced to 14 after the filtering process. Eleven different classification algorithms were applied to the selected features, and the results were evaluated. This approach achieved an accuracy of 96.6%.

Recently, Velmurugan and Dhinakaran [5] proposed an approach known as the Ensemble Stacking Learning Algorithm (ESLA) for PD classification. The ESLA method integrated the linear regression and Adaboost ensemble techniques with the RF and Extreme Gradient Boosting (XGBoost) algorithms to effectively identify individuals with PD. The dataset employed was collected from 188 PD patients. Initially, basic models were developed using the RF and XGBoost algorithms for prediction. Subsequently, the outputs of these prediction models were utilized as inputs in the next step to fine-tune parameters and create models with enhanced accuracy. The top models for the RF and XGBoost were then chosen. The RF model's accuracy increased from 84.21% to 84.86%, while the XGBoost model's accuracy increased from 88.15% to 88.85%. The proposed ESLA method leveraged the stacking technique to create four stacked models, combining RF, XGBoost, logistic regression, Adaboost, and multilayer perceptron (MLP), to further enhance the classification performance. This method outperformed the individual classifiers, yielding an accuracy of 90.13%.

The study by Sharma et al. [6] proposed a binary version of the Rao algorithm to overcome the problem of feature selection. The Rao algorithm was applied to four public PD datasets using the kNN classifier for PD classification. The highest accuracy of the classifications obtained from the four datasets was 99.25%.

In their study, Sabeena et al. [7] proposed a novel framework for feature selection and classification to identify individuals with PD. The dataset used consisted of speech samples from 188 PD patients and 64 healthy individuals. An optimization-based ensemble feature selection method was employed. It involved three different approaches for selecting the optimized subsets of features. The results from these approaches were combined using an ensemble technique. The selected features were then utilized in various classifiers, which yielded accuracies ranging from 83.66% to 98.77%. In another study by Ul Haq et al. [8], a dataset of 196 voice samples with 23 attributes was utilized. Among the 31 individuals in the dataset, 23 were diagnosed with PD, and eight were considered healthy. Relief-ant-colony optimization (ACO), and Relief-ACO methods were employed to select subsets of features. The selected feature subsets were then used with the SVM classifier. The results showed that when the Relief-ACO feature selection method was combined with SVM using the radial basis function (RBF) kernel, an accuracy of 98.20% was achieved, outperforming other feature selection methods. Similarly, when used with SVM using the linear kernel, the Relief-ACO feature selection method achieved a high accuracy of 99.50% compared to other feature selection methods.

In the study conducted by Sarankumar et al. [9], a dataset of voice data collected from 42 patients was analyzed. The dataset contained a total of 5875 audio files. After preprocessing the dataset, a clustering process was performed using wavelet cleft fuzzy. Next, feature selection was carried out from the clustering step using the firming bacteria foraging algorithm. The selected features were then employed to predict PD patients using the Deep Brooke inception net classification algorithm, resulting in an accuracy of 99.88%. In another study by Pahuja and Nagabhushan [10], a free voice dataset of PD patients from the UCI repository was used. This dataset had six recordings for each patient. Classification algorithms ANN, SVM, and kNN, were employed and achieved accuracies of 95.89%, 88.21%, and 72.31%, respectively.

The research conducted by Yücelbaş [11] used a dataset comprising voice recordings of 252 individuals. The dataset employed 188 patients with PD and 64 healthy individuals, with three recordings for each person, resulting in a total of 756 recordings. The study proposed an information gain algorithm-based KNN hybrid model (IGKNN) for feature selection analysis. The proposed IGKNN method, using 22 selected features, achieved an accuracy of 98%. Pramanik et al. [12] used a publicly available dataset from the UCI machine learning repository in a different study. This dataset included 752 acoustic features for 252 people, including 188 PD patients and 64 healthy people. A total of 21 baseline features (BF), 22 vocal fold features (VFF), and 11-time frequency features (TFF) were extracted from this dataset. A collaborative feature bank was built to evaluate the performance of PD detection using three feature selection techniques: Correlated Feature Selection (CFS), Fisher Score Feature Selection (FSFS), and Mutual Information-based Feature Selection (MIFS). The Naïve Bayes classifier was used in the evaluation. The best accuracy obtained from utilizing the three feature selection strategies was 78.97%.

The study conducted by Salmanpour, Shamsaei, Saberi, et al. [13] aimed to categorize PD into its distinct subtypes. To achieve this, the researchers compiled 30 datasets over a period of four years from 885 individuals diagnosed with Parkinson's Progressive Marker and 163 healthy individuals. These datasets encompassed a combination of non-imaging, imaging, and radiomic features extracted from DAT-SPECT images. The study used 16 algorithms for feature reduction, eight algorithms for clustering, and 16 classifiers. The radiomics features aided in generating a consistent cluster structure, enabling the subdivision of PD into three distinct subtypes: mild, intermediate, and severe.

The study by Nahar et al. [14] was based on 44 acoustic features extracted from a dataset of 80 people, 40 of whom were PD patients and 40 who were healthy. The feature selection was performed using three different methods: Boruta, Recursive Feature Elimination (RFE), and RF. Gradient Boosting, Extreme Gradient Boosting, Bagging, and an Extra Tree Classifier were employed. The classifier results were examined using the original 44 features, and the Extreme Gradient Boosting classifier achieved a good accuracy of 78.08%. Furthermore, the classification results were analyzed after using the three feature selection methods, and an accuracy of 82.35% was achieved using the RFE feature selection method and the Bagging classifier.

While previous studies have explored PD diagnosis using voice analysis, significant gaps remain. Concerns related to generalizability and accuracy have been highlighted due to the inadequate dataset diversity and feature selection methodologies. This study tackles these limitations by combining two independent datasets and offering a fusion of filter feature selection with ensemble learning and genetic selection.

3. Materials and Methods

3.1. Datasets

Two distinct biomedical voice datasets were employed in this study for the assessment of PD.

The first dataset encompasses a compilation of biomedical voice measurements obtained from 31 individuals, 23 of whom were diagnosed with PD. Each row in the dataset corresponds to a voice recording from these individuals, while each column represents a specific voice measure. This dataset was expertly curated through a collaborative effort between Max Little of the University of Oxford and the National Center for Voice and Speech in Denver, Colorado, entailing the meticulous recording of speech signals [15].

The dataset contains 195 sustained vowel phonations, encompassing a range of time since diagnosis spanning from 0 to 28 years. The subjects' ages vary from 46 to 85 years, with a mean age of 65.8 and a standard deviation of 9.8. For each subject, an average of six phonations were captured, varying in duration from one to 36 s. These phonations were recorded within an IAC sound-treated booth, utilizing a head-mounted microphone (AKG C420) positioned 8 cm away from the lips. The calibration of the microphone involved a Class 1 sound level meter (B&K 2238) situated 30 cm from the speaker. The voice signals were directly recorded onto a computer through CSL 4300B hardware (Kay Elemetrics), sampled at 44.1 kHz, and with a 16 bit resolution. To ensure the robustness of the algorithms, all samples underwent digital amplitude normalization prior to the computation of the metrics. The details of the subjects are given in Table 2.

Table 2. List of subjects with sex, age, Parkinson's stage, and number of years since diagnosis [1]. Entries labeled "n/a" for healthy subjects for whom Parkinson's stage and years since diagnosis are not applicable. "H&Y" refers to the Hoehn and Yahr PD stage, where higher values indicate a greater level of disability.

Subject Code	Sex	Age	Stage (H&Y)	Years Since Diagnosis
S01	M	78	3.0	0
S34	F	79	2.5	$\frac{1}{4}$
S44	M	67	1.5	1
S20	M	70	3.0	1
S24	M	73	2.5	1
S26	F	53	2.0	$1\frac{1}{2}$
S08	F	48	2.0	2
S39	M	64	2.0	2
S33	M	68	2.0	3
S32	M	50	1.0	4
S02	M	60	2.0	4
S22	M	60	1.5	$4\frac{1}{2}$
S37	M	76	1.0	5
S21	F	81	1.5	5
S04	M	70	2.5	$5\frac{1}{2}$
S19	M	73	1.0	7
S35	F	85	4.0	7
S05	F	72	3.0	8
S18	M	61	2.5	11
S16	M	62	2.5	14
S27	M	72	2.5	15
S25	M	74	3.0	23
S06	F	63	2.5	28
S10 (healthy)	F	46	n/a	n/a
S07 (healthy)	F	48	n/a	n/a
S13 (healthy)	M	61	n/a	n/a
S43 (healthy)	M	62	n/a	n/a
S17 (healthy)	F	64	n/a	n/a
S42 (healthy)	F	66	n/a	n/a
S50 (healthy)	F	66	n/a	n/a
S49 (healthy)	M	69	n/a	n/a

[1] Adapted from [15].

The second dataset utilized in this study was built by Sakar et al. [16] for their study, which comprised a comparative analysis of speech signal processing algorithms for PD classification and the use of the tunable Q-factor wavelet transform. This dataset was collected at the Department of Neurology in the Cerrahpaşa Faculty of Medicine, Istanbul University. It entailed the comprehensive data of 188 PD patients (107 men and 81 women) spanning an age range of 33 to 87 years (mean age: 65.1 ± 10.9). Additionally, a control group consisting of 64 healthy individuals (23 men and 41 women) with ages ranging from 41 to 82 years (mean age: 61.1 ± 8.9) was included. During the data collection process, voice recordings were captured using a microphone set to a frequency of 44.1 KHz. Specifically, sustained phonation of the vowel /a/ was necessary to collect from each subject with three repetitions. Subsequently, a comprehensive set of speech signal processing algorithms, including Time-Frequency features, Mel Frequency Cepstral Coefficients (MFCCs), Wavelet Transform-based features, Vocal Fold features, and TWQT features, were diligently applied to the speech recordings of PD patients.

3.2. Filter Feature Selection

The goal of feature selection in machine learning and data mining is to identify and maintain a subset of important features from the original dataset. The motivation for feature selection stems from its ability to increase model performance, reduce computational complexity, and improve model interpretability. Filter methods have received substantial attention among the various approaches to feature selection due to their simplicity, efficiency, and capacity to evaluate feature significance independently of any specific learning algorithm.

Filter feature selection methods attempt to prioritize and choose features based on their unique properties and association with the target variable without considering the learning process of the specific model. In this study, we focus on the importance of filtering quasi-constant features as a crucial step in the filter feature selection process. Quasi-constant features refer to those with minimal variance or almost constant values across the dataset, providing limited or negligible discriminatory information.

Identifying and removing quasi-constant features reduces dimensionality and improves model generalization. By eliminating these features, we may reduce noise, improve computational performance, and promote more meaningful dataset exploration. However, to effectively filter out quasi-constant features, it is essential to set an appropriate threshold that determines the acceptable level of variance below which a feature is considered quasi-constant and subsequently removed.

3.3. Genetic Algorithm

Genetic Algorithms (GAs), members of the evolutionary algorithm family, have emerged as a popular and robust solution to addressing the limitations encountered by conventional optimization techniques in terms of efficiency and effectiveness. They are inspired by concepts of natural selection and genetics, imitating the process of evolution to find optimal solutions within a specific area.

The concept of a population-based search is at the heart of GAs, in which a set of potential solutions, referred to as individuals or chromosomes, undergo iterative refinement to explore the solution space. GAs enable the propagation of desirable features and the examination of new solution regions by utilizing genetic operators such as selection, crossover, and mutation. This population-centric method enables GAs to tackle complicated optimization problems with high dimensionality, non-linearity, and multimodality effectively.

GAs work by iteratively generating new populations, with each population being evaluated based on a fitness function that assesses the quality of individual solutions. They promote convergence towards optimal or near-optimal solutions across generations by repeatedly applying selection, crossover, and mutation operators. This repeated exploration and exploitation approach enables them to navigate the solution space with ease, exceeding local optima and delivering strong solutions.

3.4. Methods

Two distinct methods were employed in the experiments to evaluate the effectiveness of the filtering approach.

The selection of the quasi-constant threshold for filtering features was performed using a trial-and-error method. After careful evaluation of different threshold values, it was found that the best results were achieved when the threshold was set to 0.0001. However, both lower and higher threshold values yielded decreased accuracy in our experiments.

A combination of filter feature selection and ensemble learning methods was employed in the first method. First, quasi-constant features with a threshold value of 0.0001 were identified and subsequently removed from the dataset, resulting in a refined dataset. This refined dataset was then subjected to five different classification algorithms: Gaussian Naïve Bayes classifier, Support Vector Machine (SVM), Decision Tree, Random Forest, and XGBoost.

The performance evaluation revealed that among the tested classification algorithms, Decision Tree, Random Forest, and XGBoost exhibited the highest classification accuracy and predictive power. Building upon this finding, further analysis was conducted by employing ensemble learning methods: stacking and voting, using the three best-performing algorithms. Additionally, bagging was also applied to the three selected algorithms to explore potential performance enhancements and model robustness. The first method is summarized in Figure 1.

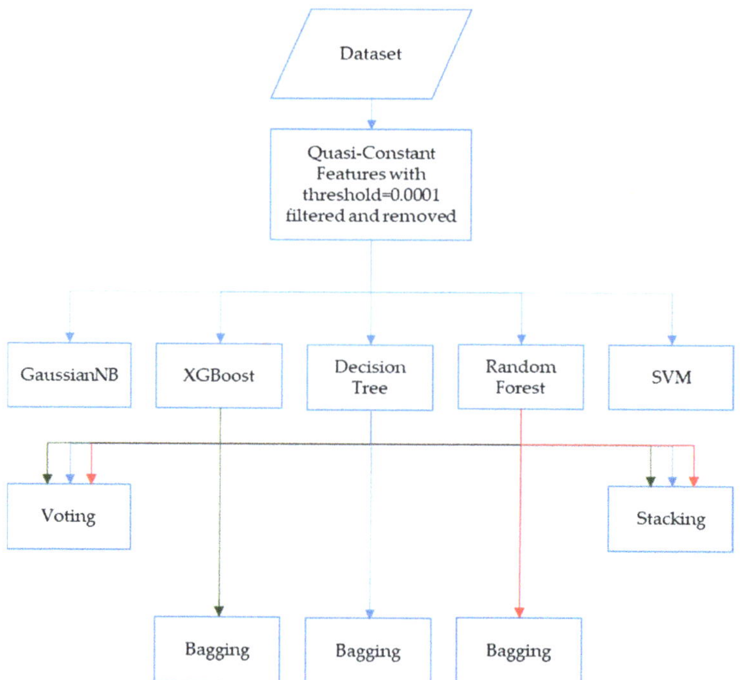

Figure 1. Method 1, where filter feature selection was applied to five different classification algorithms followed by ensemble learning methods.

In the second method, after filtering out the quasi-constant features, a genetic algorithm was utilized to further optimize the feature selection process for the same set of classification algorithms: GaussianNB, SVM, Decision Tree (with entropy and Gini index), XGBoost, Random Forest, and additionally, logistic regression. A pictorial representation of the second method is shown in Figure 2.

The genetic selection was performed after 40 generations of populations with 50 individuals. The crossover probability was 0.5, and the mutation probability was 0.2. The crossover independent probability was set to 0.5 and the mutation independent probability to 0.05. The tournament size was set to three, and the number of generations after which the optimization is terminated when the best individual has not changed in all the previous generations (n_gen_no_change) was set to 10.

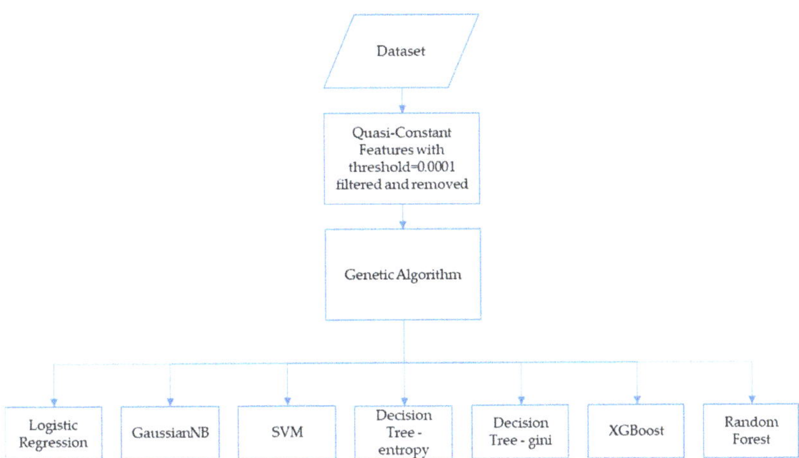

Figure 2. Method 2, where filter feature selection is followed by a genetic algorithm before applying to the classification models.

4. Results and Discussion

With the quasi-constant threshold set to 0.0001, five of the twenty-four features in Dataset 1 and 188 of the 754 features in Dataset 2 were identified as quasi-constant and subsequently eliminated. This was a significant number of features in both datasets and enabled streamlining the feature space to enhance the efficiency and effectiveness of subsequent modeling tasks.

Different classification models were then tested on the filtered datasets. The accuracies of these models are given in Table 3.

Table 3. Results on applying filter feature selection.

Classification Model	Accuracy (in %) after Filter Feature Selection	
	Dataset 1	**Dataset 2**
GuassianNB	67.34	75.13
SVM	85.71	76.72
Decision Tree-entropy	100.00	76.72
Random Forest	97.95	92.06
XGboost	100.00	86.24

Among the various models tested, Decision Tree, Random Forest, and XGBoost demonstrated notably higher accuracies compared to the others on both datasets. Therefore, ensemble methods, namely voting, stacking, and bagging, were applied to these three models. Both hard voting and soft voting were employed. The models were stacked to leverage the strengths of multiple classifiers. Bagging was applied independently to each of the three models, utilizing 5-fold cross-validation and a total of 500 trees. The resulting accuracies obtained from these ensemble approaches on Dataset 1 and Dataset 2 are presented in Tables 4 and 5, respectively.

Voting with both hard and soft voting classifiers attained perfect accuracy (100%) on Dataset 1. Stacking also displayed good results, with a 96.2% accuracy on Dataset 1 and a 90.06% accuracy on Dataset 2. Bagging had lower accuracy than voting and stacking.

Ensemble approaches take advantage of the diversity and complementary features of individual models, resulting in higher accuracy. The perfect accuracy achieved by voting on Dataset 1 suggests strong agreement among the models, contributing to accurate classification. The relatively high accuracy of the stacked models verifies the efficiency of combining the predictions of the base models to produce greater performance. Bagging decreases the variance and instability of classification models by training individual models on diverse subsets of the dataset and aggregating their predictions. The slight decrease in accuracy from bagging could have resulted from the intrinsic randomness introduced during the resampling process, which may result in a minor trade-off between accuracy and model stability.

Table 4. Results on applying ensemble learning methods to Dataset 1.

Ensemble Learning Method		Accuracy (in %)
Voting	Hard voting	100.00
	Soft voting	100.00
Stacking	Stacking	96.20
Bagging	Decision Tree-entropy	91.05
	Random Forest	89.70
	XGBoost	92.40

Table 5. Results on applying ensemble learning methods to Dataset 2.

Ensemble Learning Method		Accuracy (in %)
Voting	Hard voting	91.53
	Soft voting	88.89
Stacking	Stacking	90.06
Bagging	Decision Tree-entropy	88.34
	Random Forest	86.56
	XGBoost	87.76

In the second method, the filtered dataset was subjected to further feature refinement using genetic selection. A genetic algorithm investigates several feature combinations to determine an optimal subset that achieves the maximum classification accuracy. The genetic selection process begins with the generation of an initial population of potential feature subsets, each of which represents a unique combination of features. These subsets were then analyzed using the same classification models in addition to logistic regression. The results thus obtained are summarized in Table 6.

Table 6. Results from applying genetic selection.

Classification Model	Accuracy (in %) after Filter Feature Selection	
	Dataset 1	Dataset 2
GuassianNB	91.83	77.63
SVM	81.63	77.63
Decision Tree-entropy	81.63	76.65
Decision Tree-Gini	81.63	74.34
Random Forest	83.67	74.34
XGboost	87.75	76.97
Logistic Regression	89.79	76.97

It can be observed that the accuracy of the Guassian Naïve Bayes classifier improved to 91.83% for Dataset 1 and 77.63% for Dataset 2 after genetic selection. This indicates that genetic algorithms were effective in choosing the relevant features that boosted the performance of the classifier. However, with the SVM classifier, the accuracy declined to 81.63% for Dataset 1 and improved only slightly for Dataset 2 with 77.63% accuracy. The accuracy of the decision tree model, measured using both entropy and the Gini index, also failed to improve significantly with genetic selection. The same was true for random forest and XGBoost classifiers. Logistic regression also produced similar results to the rest, with an accuracy of 89.79% with Dataset 1 and 76.97 with Dataset 2. In summary, genetic selection had varying effects on the accuracy of the different classification models. This implies that the effectiveness of genetic algorithms may also be dependent on the properties of the classification model.

Precision may also be an important evaluation metric for the detection of PD. Precision is a performance metric that quantifies the accuracy of a classification model's positive predictions. It determines the proportion of true positive predictions (positive instances correctly identified) out of all predicted positive instances (true positives + false positives). By focusing on precision, we can ensure that the models accurately identify actual PD patients while also reducing the risks of misclassifying individuals in good health as having the disease, as that can lead to unnecessary fear, stress, and even medical interventions. A high precision score gives reliability and greater confidence to employ the models in PD diagnosis. The precision of the predictions made by the models with filter feature selection and genetic selections on both datasets is given in Tables 7 and 8, respectively.

Table 7. Precision in applying filter feature selection.

Classification Model	Precision (in %) after Filter Feature Selection			
	Dataset 1		Dataset 2	
	PD Patient	Healthy	PD Patient	Healthy
GuassianNB	100	41	83	48
SVM	84	100	77	100
Decision Tree-entropy	100	100	86	51
Random Forest	100	92	94	86
XGboost	100	100	88	79

Table 8. Precision in applying genetic algorithms.

Classification Model	Precision (in %) after Genetic Selection			
	Dataset 1		Dataset 2	
	PD Patient	Healthy	PD Patient	Healthy
GuassianNB	90	100	78	73
SVM	85	62	78	78
Decision Tree-entropy	84	67	76	62
Decision Tree-Gini	85	62	77	50
Random Forest	89	64	74	0
XGBoost	94	69	78	64
Logistic Regression	92	80	79	61

It is notable that the decision tree and XGBoost classifiers achieved perfect precision in identifying both PD patients and healthy individuals. The Gaussian Naïve Bayes and random forest classifiers attained perfect precision in identifying PD patients in Dataset 1, whereas the SVM classifier showed perfect precision in detecting healthy individuals in both datasets. It is also noteworthy that SVM was the only classifier that achieved perfect precision in identifying at least one category (PD patients or healthy individuals) in Dataset 2.

After genetic selection on Dataset 1, all the models achieved relatively high precision in identifying PD patients, ranging from 84% to 94%. The Gaussian Naïve Bayes classifier was 100% precise in identifying healthy individuals. However, all the other models showed less precision in identifying healthy individuals (ranging from 62% to 80%) than PD patients. The same can also be observed in the case of Dataset 2 after genetic selection. All models showed higher precision in identifying PD patients (ranging from 74% to 79%) than in identifying healthy individuals (ranging from 0% to 78%). This suggests that identifying PD patients may be easier than identifying healthy people from the selected datasets. One possible reason for this could be the unequal distribution of PD patients and healthy individuals in both datasets. Both datasets contained information from a higher number of PD patients than healthy people, which made the models more proficient in learning the patterns and characteristics associated with PD. This imbalance in class distribution may have led to a bias towards PD patients during the training process, potentially resulting in higher precision in identifying PD cases.

The overall results for Dataset 1 were better than those for Dataset 2. This disparity may be due to the difference in the number of features between the two datasets. Initially, Dataset 1 had only 24 features, which is substantially fewer than Dataset 2, which had 754 features. Even after applying the filter feature selection technique, a relatively large number of features (566 features) were preserved in Dataset 2 compared to Dataset 1. The presence of a larger feature space in Dataset 2 might have introduced additional complexity and made it more challenging for the models to discern the meaningful patterns associated with PD. This demonstrates that having a greater number of features may not necessarily translate to better results and may even generate noise or redundancy, resulting in poor model performance.

While previous research has established the efficacy of ensemble techniques [5,7], a comparative analysis with the current literature demonstrates a remarkable outperformance of ensemble learning methods, as exemplified by the perfect accuracy (100%) achieved by both hard and soft voting on Dataset 1. Moreover, the hard voting classifier achieved an accuracy of 91.53% on Dataset 2, surpassing the performance reported in the related literature [5]. The introduction of genetic selection is a novel approach. While certain models responded differently to genetic selection, this nuanced approach illustrates the complicated interplay between feature selection strategies and classification outcomes. Following genetic selection, the GaussianNB classifier achieved the best accuracy for both datasets, with an accuracy of 91.83% for Dataset 1 and 77.63% for Dataset 2. The emphasis on precision ensures that PD patients are accurately identified while minimizing the risk of misclassifying healthy individuals, a crucial aspect for real-world clinical applications. Filter feature selection led to perfect (100%) precision in the predictions of decision trees and XGBoost classifiers. With genetic selection, there was an average precision of 88.42% in identifying PD patients and 72% in identifying healthy individuals in Dataset 1. In Dataset 2, these values were 77.14% for PD patients and 55.43% for healthy individuals. This holistic viewpoint illustrates the depth and breadth of this research, effectively establishing its relevance and impact on improving patient care and prognosis. Overall, this research not only benchmarks favorably against the prior literature but also offers a novel strategy for enhancing the accuracy and reliability of PD detection through voice data analysis.

5. Conclusions and Future Scope

This study aimed to develop an efficient method for the detection of PD from voice clips. A combination of filter feature selection, ensemble learning, and genetic selection was employed. The results of the study demonstrated the effectiveness of filter feature selection in streamlining the feature space and enhancing the efficiency of subsequent modeling tasks. By eliminating quasi-constant features, a significant number of irrelevant features were successfully removed, leading to high model accuracy. The application of ensemble learning techniques, such as voting, stacking, and bagging, further explored the classification performance of these models. Additionally, the genetic selection approach analyzed the precision of the classification models in identifying PD patients and healthy individuals. The models exhibited relatively high precision in identifying PD patients, while the precision in identifying healthy individuals was comparatively lower. Moreover, the comparison between Dataset 1 and Dataset 2 demonstrated the effect of feature space on model performance. Dataset 1, with a smaller number of features, yielded better results compared to Dataset 2, which had a larger feature space even after filter feature selection.

While this study contributes significantly to the field of PD detection, a few limitations warrant careful consideration. The precision analysis performed in this study reveals a potential bias toward recognizing PD patients more accurately than healthy people. This bias stems from the inherent class imbalance within the datasets, where PD patients are overrepresented compared to healthy individuals. This discrepancy could lead to a skewed learning process, affecting the models' generalizability when applied to larger, more balanced populations. Furthermore, the variation in performance between Dataset 1 and Dataset 2 underscores the sensitivity of model outputs to the dimensionality of the feature space. The larger feature set of Dataset 2, even after filter feature selection, suggests the possibility of increased noise or redundancy, thereby affecting model robustness and performance.

Future studies could explore the use of sampling techniques, such as oversampling or undersampling, to balance the datasets. This would help in achieving better performance and addressing the bias towards the majority class. The current study utilized specific datasets for model development and evaluation. Future research could involve testing the developed models on external datasets or real-world data to assess their generalizability and robustness. This would provide insights into the practical applicability of the proposed methods and their performance across different populations. By addressing these future research areas, we can further advance the field of PD detection from voice data and contribute to the development of accurate, reliable, and clinically applicable diagnostic tools.

Author Contributions: Conceptualization, A.M.A. and F.S. (Faisal Saeed); methodology, F.S. (Farsana Salim) and F.S. (Faisal Saeed); software, F.S. (Farsana Salim); validation, A.M.A. and F.S. (Faisal Saeed); formal analysis, F.S. (Farsana Salim) and F.S. (Faisal Saeed); investigation, F.S. (Farsana Salim) and F.S. (Faisal Saeed); resources, A.M.A.; data curation, F.S. (Farsana Salim); writing—original draft preparation, A.M.A. and F.S. (Farsana Salim); writing—review and editing, A.M.A. and F.S. (Faisal Saeed); visualization, F.S. (Farsana Salim); supervision, F.S. (Faisal Saeed); project administration, A.M.A.; funding acquisition, A.M.A. All authors have read and agreed to the published version of the manuscript.

Funding: This project was funded by the Deanship of Scientific Research (DSR) at King Abdulaziz University, Jeddah, under grant no. (G: 479-611-1442). The authors, therefore, acknowledge with thanks DSR for technical and financial support.

Institutional Review Board Statement: Not applicable.

Informed Consent Statement: Not applicable.

Data Availability Statement: The datasets are available online in [15,16].

Acknowledgments: This Project was funded by the Deanship of Scientific Research (DSR) at King Abdulaziz University, Jeddah, under grant no. (G: 479-611-1442). The authors, therefore, acknowledge with thanks DSR for technical and financial support.

Conflicts of Interest: The authors declare no conflict of interest.

References

1. Tjaden, K.; Lam, J.; Wilding, G. Vowel Acoustics in Parkinson's Disease and Multiple Sclerosis: Comparison of Clear, Loud, and Slow Speaking Conditions. *J. Speech Lang. Hear. Res.* **2013**, *56*, 1485–1502. [CrossRef] [PubMed]
2. Tsanas, A.; Little, M.; McSharry, P.; Ramig, L. Accurate telemonitoring of Parkinson's disease progression by non-invasive speech tests. *Nat. Preced.* **2009**, *57*, 884–893. [CrossRef]
3. Sheikhi, S.; Kheirabadi, M.T. An Efficient Rotation Forest-Based Ensemble Approach for Predicting Severity of Parkinson's Disease. *J. Healthc. Eng.* **2022**, *2022*, e5524852. [CrossRef] [PubMed]
4. Mohammed, M.A.; Mohamed, E.; Abdulkareem, K.H.; Mostafa, S.A.; Maashi, M.S. A Multi-agent Feature Selection and Hybrid Classification Model for Parkinson's Disease Diagnosis. *Assoc. Comput. Mach.* **2021**, *17*, 1–22. [CrossRef]
5. Velmurugan, T.; Dhinakaran, J. A Novel Ensemble Stacking Learning Algorithm for Parkinson's Disease Prediction. *Math. Probl. Eng.* **2022**, *2022*, 9209656. [CrossRef]
6. Sharma, S.R.; Singh, B.; Kaur, M. Classification of Parkinson disease using binary Rao optimization algorithms. *Expert Syst.* **2021**, *38*, e12674. [CrossRef]

7. Sabeena, B.; Sivakumari, S.; Teressa, D.M. Optimization-Based Ensemble Feature Selection Algorithm and Deep Learning Classifier for Parkinson's Disease. *J. Healthc. Eng.* **2022**, *2022*, e1487212. [CrossRef]
8. Ul Haq, A.; Li, J.; Memon, M.H.; Ali, Z.; Abbas, S.Z.; Nazir, S. Recognition of the parkinson's disease using a hybrid feature selection approach. *J. Intell. Fuzzy Syst.* **2020**, *39*, 1319–1339. [CrossRef]
9. Sarankumar, R.; Vinod, D.; Anitha, K.; Manohar, G.; Vijayanand, K.S.; Pant, B.; Sundramurthy, V.P. Severity Prediction over Parkinson's Disease Prediction by Using the Deep Brooke Inception Net Classifier. *Comput. Intell. Neurosci.* **2022**, *2022*, 7223197. [CrossRef] [PubMed]
10. Pahuja, G.; Nagabhushan, T.N. A Comparative Study of Existing Machine Learning Approaches for Parkinson's Disease Detection. *IETE J. Res.* **2018**, *67*, 4–14. [CrossRef]
11. Yücelbaş, C. A new approach: Information gain algorithm-based k-nearest neighbors hybrid diagnostic system for Parkinson's disease. *Phys. Eng. Sci. Med.* **2021**, *44*, 511–524. [CrossRef] [PubMed]
12. Pramanik, M.; Pradhan, R.; Nandy, P.; Qaisar, S.M.; Bhoi, A.K. Assessment of Acoustic Features and Machine Learning for Parkinson's Detection. *J. Healthc. Eng.* **2021**, *2021*, 9957132. [CrossRef] [PubMed]
13. Salmanpour, M.R.; Shamsaei, M.; Saberi, A.; Hajianfar, G.; Soltanian-Zadeh, H.; Rahmim, A. Robust identification of Parkinson's disease subtypes using radiomics and hybrid machine learning. *Comput. Biol. Med.* **2021**, *129*, 104142. [CrossRef] [PubMed]
14. Nahar, N.; Ara, F.; Neloy, M.A.I.; Biswas, A.; Hossain, M.S.; Andersson, K. Feature Selection Based Machine Learning to Improve Prediction of Parkinson Disease. In *Brain Informatic*; Springer: Berlin/Heidelberg, Germany, 2021; pp. 496–508. [CrossRef]
15. Little, M.A.; McSharry, P.E.; Hunter, E.J.; Spielman, J.; Ramig, L.O. Suitability of Dysphonia Measurements for Telemonitoring of Parkinson's Disease. *IEEE Trans. Biomed. Eng.* **2009**, *56*, 1015–1022. [CrossRef]
16. Sakar, C.O.; Serbes, G.; Gunduz, A.; Tunc, H.C.; Nizam, H.; Sakar, B.E.; Tutuncu, M.; Aydin, T.; Isenkul, M.E.; Apaydin, H. A comparative analysis of speech signal processing algorithms for Parkinson's disease classification and the use of the tunable Q-factor wavelet transform. *Appl. Soft Comput.* **2019**, *74*, 255–263. [CrossRef]

Article

A New Deep-Learning-Based Model for Breast Cancer Diagnosis from Medical Images

Salman Zakareya [1], **Habib Izadkhah** [1,2,*] and **Jaber Karimpour** [1]

1 Department of Computer Science, University of Tabriz, Tabriz 5166616471, Iran; selman33111@hotmail.com (S.Z.); karimpour@tabrizu.ac.ir (J.K.)
2 Research Department of Computational Algorithms and Mathematical Models, University of Tabriz, Tabriz 5166616471, Iran
* Correspondence: izadkhah@tabrizu.ac.ir

Abstract: Breast cancer is one of the most prevalent cancers among women worldwide, and early detection of the disease can be lifesaving. Detecting breast cancer early allows for treatment to begin faster, increasing the chances of a successful outcome. Machine learning helps in the early detection of breast cancer even in places where there is no access to a specialist doctor. The rapid advancement of machine learning, and particularly deep learning, leads to an increase in the medical imaging community's interest in applying these techniques to improve the accuracy of cancer screening. Most of the data related to diseases is scarce. On the other hand, deep-learning models need much data to learn well. For this reason, the existing deep-learning models on medical images cannot work as well as other images. To overcome this limitation and improve breast cancer classification detection, inspired by two state-of-the-art deep networks, GoogLeNet and residual block, and developing several new features, this paper proposes a new deep model to classify breast cancer. Utilizing adopted granular computing, shortcut connection, two learnable activation functions instead of traditional activation functions, and an attention mechanism is expected to improve the accuracy of diagnosis and consequently decrease the load on doctors. Granular computing can improve diagnosis accuracy by capturing more detailed and fine-grained information about cancer images. The proposed model's superiority is demonstrated by comparing it to several state-of-the-art deep models and existing works using two case studies. The proposed model achieved an accuracy of 93% and 95% on ultrasound images and breast histopathology images, respectively.

Keywords: medical image; breast cancer diagnoses; machine learning; deep learning; classification

check for updates

Citation: Zakareya, S.; Izadkhah, H.; Karimpour, J. A New Deep-Learning-Based Model for Breast Cancer Diagnosis from Medical Images. *Diagnostics* **2023**, *13*, 1944. https://doi.org/10.3390/diagnostics13111944

Academic Editors: Mugahed A. Al-antari and Gary J. Whitman

Received: 9 April 2023
Revised: 15 May 2023
Accepted: 28 May 2023
Published: 1 June 2023

1. Introduction

Breast cancer is the most commonly diagnosed form of cancer worldwide and the second leading cause of cancer-related deaths. In 2020, breast cancer was diagnosed in 2.3 million women globally, resulting in 685,000 fatalities. Additionally, as of the end of 2020, 7.8 million women had received a breast cancer diagnosis within the last five years [1]. Clinical studies have demonstrated that early detection is crucial for effective treatment and can significantly improve the survival rate of breast cancer patients [2].

Computer-aided detection and diagnosis (CAD) software systems have been developed and clinically used since the 1990s to support radiologists in screening, improve predictive accuracy, and prevent misdiagnosis due to fatigue, eye strain, or lack of experience [3].

The rapid progress of machine learning in both application and efficiency, especially deep learning, has increased the interest of the medical community in using these techniques to improve the accuracy of cancer screening from images. Machine learning can play an essential role in helping medical professionals in the early detection of cancerous lesions. Despite the benefits of using these techniques, cancer screening is associated with a high risk of false positives and false negatives. However, early detection of cancer can contribute to up to a 40% decrease in the mortality rate [2].

Deep-learning networks employ a deeply layered architecture that enables hierarchical learning and progressive extraction of features from data, starting from simple to progressively more complex abstractions. By autonomously learning the maximum possible set of features, the deep-learning algorithm can deliver the most precise results, rendering these networks highly effective for medical image classification and the identification of features such as lesions [4,5].

The performance of convolutional neural networks (CNNs) is hindered by several challenges, and the most popular CNNs attempt to improve their performance by simply stacking convolution layers deeper and deeper. However, the significant areas in an image can vary considerably in size, making it difficult to select the appropriate kernel size for the convolution operation due to the extreme variability in the information's location. For information that is more locally distributed, a smaller kernel is recommended, whereas a larger kernel is chosen for information distributed more widely. Very deep networks are more prone to overfitting, and gradient updates are challenging to share throughout the entire network, in addition to being computationally expensive to naively stack large convolution processes [6,7].

Granular computing is a type of computing that focuses on the use of granules, which are small, discrete pieces of knowledge that can be combined, manipulated, or analyzed to solve complex problems. It is a form of computing that is based on the idea of breaking down complex problems into smaller, more manageable pieces. Granular computing has been used in various areas, such as data mining, decision support systems, and knowledge discovery. Granular computing can be used in image classification by dividing the image into smaller regions or sub-regions, also known as granules, and extracting features from them [8].

This study proposes a novel deep-learning model designed to enhance the accuracy of breast cancer detection while simultaneously reducing the network's parameters to improve training time. Inspired by GoogLeNet and the residual block, the proposed model considers both the depth and width of the network. Multiple filters of varying sizes operate at the same level, and their outputs are concatenated and transmitted to the next module. Additionally, the use of shortcut connections and two learnable activation functions, as opposed to traditional activation functions, is expected to reduce time consumption and improve diagnostic accuracy, thereby potentially alleviating the workload of medical professionals. We also propose a granular computing-based algorithm for capturing more detailed and fine-grained information about cancer images. We will apply granular computing to the dataset before starting the training process.

The paper's contributions can be outlined as follows:

1. The proposed model has the highest diagnostic accuracy compared to existing breast cancer methods;
2. This paper is the first study to use the granular computing concept in disease diagnosis;
3. Utilizing wide and depth networks, shortcut connections, and intermediate classifiers, we design a new deep network to improve the detection of breast cancer;
4. An attention mechanism is proposed to highlight the important features in the input image, thereby resulting in improved accuracy of the classifier;
5. Two learnable activation functions are developed and utilized instead of traditional activation functions. Learnable activation functions provide a flexible framework that can be fine-tuned during training for optimal performance on specific tasks.

The following is the structure of this work: Section 2 addresses the related works, Section 3 presents the model and implementation, Section 4 presents the results and discussion, and ultimately, the conclusion is presented.

2. Related Work

Breast cancer ranks among the most prevalent cancers affecting women worldwide. Early detection and accurate diagnosis are crucial factors for effective treatment and improved patient outcomes [9]. Ultrasound imaging is a widely used method for breast cancer screening and diagnosis, but it requires skilled radiologists to interpret the images accurately [10]. According to the National Breast Cancer Foundation's 2020 report, AI has been successfully used to diagnose more than 276,000 breast cancer cases. By analyzing breast cancer images using AI, breast lumps (masses), mass segmentation, breast density, and breast cancer risk can be identified. In the majority of patients, lumps in the breast are the most common sign of breast cancer [9]; therefore, their detection is an essential step used in CAD.

A review of deep-learning applications in breast tumor diagnosis utilizing ultrasound and mammography images is provided in [11]. Moreover, the research summarizes the latest progressions in computer-aided diagnosis/detection (CAD) systems that rely on deep-learning methodologies to automatically recognize breast images, ultimately enhancing radiologists' diagnostic precision. Remarkably, the classification process underpinning the novel deep-learning approaches has demonstrated significant usefulness and effectiveness as a screening tool for breast cancer.

Recent studies have explored the use of deep-learning techniques, particularly convolutional neural networks (CNNs), for automated breast ultrasound image classification [12–15]. These studies have shown encouraging results. Several convolutional neural networks (CNNs) models are used for breast cancer image classifications including AlexNet, VGGNet, GoogLeNet, ResNet, and Inception. In their study, the authors of [5] categorized breast lesions as either benign or malignant. They developed a CNN model to remove speckle noise from the ultrasound images and then proposed another CNN model for classifying the ultrasound images. The study [16] discriminates benign cysts from malignant masses in US images.

In the study presented in [17], various deep-learning models were employed to classify breast cancer ultrasound images based on their benign, malignant, or normal status. A dataset comprising a total of 780 images was utilized, and data augmentation and preprocessing techniques were applied. Three models were evaluated for classification. Specifically, ResNet50 achieved an accuracy of 85.4%, ResNeXt50 achieved 85.83%, and VGG16 achieved 81.11%.

The study [18] introduced a novel ensemble deep-learning-enabled clinical decision support system for the diagnosis and classification of breast cancer based on ultrasound images. The study presented an optimal multilevel thresholding-based image segmentation technique for identifying tumor-affected regions. Additionally, an ensemble of three deep-learning models was developed to extract features, and an optimal machine-learning classifier was utilized to detect breast cancer.

In the study [14], the authors proposed a system to classify breast masses into normal, benign, and malignant. Ten well-known, pre-trained CNNs classification models were compared, and the best model was Inception ResNetV2. In [19], a vector-attention network (BVA Net) was proposed to classify benign and malignant mass tumors in the breast.

In [20], the authors proposed a CNN-based CAD system for breast ultrasound image classification (benign and malignant lesions). The study [21] developed a deep-learning model based on ResNet18 CNN architecture for breast ultrasound image classification. In addition, the study [22] compared the performance of different deep-learning models, including CNNs, recurrent neural networks (RNNs), and hybrid models, for breast cancer diagnosis on ultrasound images. In addition to binary classification, some studies have also explored multiclass classification of breast ultrasound images. For example, the study [23] proposed a CNN-based CAD system that can classify breast lesions into four categories: benign, malignant, cystic, or complex cystic-solid. The system achieved an overall accuracy of 87% on a dataset of 1000 images.

Gao et al. have devised a computer-aided diagnosis (CAD) system geared toward screening mammography readings, which demonstrated an accuracy rate of approximately 92% [24]. Similarly, in several studies [25,26], multiple convolutional neural networks (CNNs) were employed for mass detection in mammographic and ultrasound images.

The study conducted by [3] provides a comprehensive review of the techniques used for the diagnosis of breast cancer in histopathological images. The state-of-the-art machine-learning approaches employed at each stage of the diagnosis process, including traditional methods and deep-learning methods, are presented, and a comparative analysis between the different techniques is provided. The technical details of each approach and their respective advantages and disadvantages are discussed in detail.

Lee et al. [27] conducted a study utilizing a deep-learning-based computer-aided prediction system for ultrasound (US) images. The research involved a total of 153 women with breast cancer, comprising 59 patients with lymph node metastases (LN+) and 94 patients without (LN−). Multiple machine-learning algorithms, including logistic regression, support vector machines (SVMs), XGBoost, and DenseNet, were trained and evaluated on the US image data. The study found that the DenseNet model exhibited the best performance, achieving an area under the curve (AUC) of 0.8054. This study highlights the potential of deep-learning techniques in the development of accurate and efficient prediction systems for breast cancer diagnosis using US imaging.

Sun et al. [28] conducted a study utilizing a convolutional neural network (CNN) trained and tested on ultrasound images of 169 patients. The training dataset consisted of 248 US images from 124 patients, while the testing dataset comprised 90 US images from 45 patients. The results of the study revealed a somewhat inferior performance, with an AUC of 0.72 (SD 0.08) and an accuracy of 72.6% (SD 8.4). Notably, the validation process did not include cross-validation or bootstrapping methods. These findings suggest that further research is necessary to improve the performance of CNNs in breast cancer diagnosis using ultrasound imaging.

In a study by [29], a comparison was made between convolutional neural networks (CNNs) and traditional machine-learning (ML) methods, specifically random forests, in the context of breast cancer diagnosis. The study utilized a dataset of 479 breast cancer patients, comprising 2395 breast ultrasound images. The research also focused on different regions of the ultrasound images, including intratumoral, peritumoral, and combined regions, to train and evaluate the models. The study found that CNNs outperformed random forests in all modalities ($p < 0.05$), and the combination of intratumoral and peritumoral regions provided the best result, with an AUC of 0.912 [0.834–99.0]. While confidence intervals were provided, the method used to determine them was not mentioned. These results highlight the potential of CNNs in breast cancer diagnosis using ultrasound imaging and the importance of considering different regions of the image in the analysis.

The study proposed by [30] implemented the multilevel transfer-learning (MSTL) algorithm using three pre-trained models, namely EfficientNetB2, InceptionV3, and ResNet50, along with three optimizers, which included Adam, Adagrad, and stochastic gradient descent (SGD). The study utilized 20,400 cancer cell images, 200 ultrasound images from Mendeley, and 400 from the MT-Small dataset. This approach has the potential to reduce the need for large ultrasound datasets to realize powerful deep-learning models. The results of this study demonstrate the effectiveness of the MSTL algorithm in breast cancer diagnosis using ultrasound imaging.

The study [31] presents a review of studies investigating the ability of deep-learning (DL) approaches to classify histopathological breast cancer images. The article evaluates current DL applications and approaches to classify histopathological breast cancer images based on papers published by November 2022. The study findings indicate that convolutional neural networks, as well as their hybrids, represent the most advanced DL approaches currently in use for this task. The authors of the study defined two categories of classification approaches, namely binary and multiclass solutions, in the context of DL-based classification of histopathological breast cancer images. Overall, this review provides insights into the current state of the art in DL-based classification of histopathological breast cancer images and highlights the potential of advanced DL approaches to improve the accuracy and efficacy of breast cancer diagnosis.

The study [32] proposed a breast cancer classification technique that leverages a transfer-learning approach based on the VGG16 model. To preprocess the images, a median filter was employed to eliminate speckle noise. The convolution layers and max pooling layers of the pre-trained VGG16 model were utilized as feature extractors, while a two-layer deep neural network was devised as a classifier.

The vision transformer (ViT) architecture has been proven to be advantageous in extracting long-range features and has thus been employed in various computer vision tasks. However, despite its remarkable performance in traditional vision tasks, the ViT model's supervised training typically necessitates large datasets, thereby posing difficulties in domains where it is challenging to amass ample data, such as medical image analysis. In [33], the authors introduced an enhanced ViT architecture, denoted as ViT-Patch, and investigated its efficacy in addressing a medical image classification problem, namely, identifying malignant breast ultrasound images.

In summary, these studies showcase the capability of deep-learning techniques in automating breast image classification and underscore the significance of devising precise CAD systems to support radiologists in detecting breast cancer. The majority of current approaches employ pre-existing deep-learning architectures for detecting breast cancer. In the following, we introduce a novel architecture that surpasses all previous methods.

3. Methodology

Inspired by GoogLeNet [34] and residual block [35] and adding several other features, in this paper, we developed a new deep architecture for breast cancer detection from images. GoogLeNet and residual block are based on convolutional neural network (CNN) architecture. GoogLeNet is a deep convolutional neural network architecture developed by Google's research team in 2014. It was the winner of the ImageNet Large Scale Visual Recognition Challenge (ILSVRC) in 2014 and achieved state-of-the-art performance on a variety of computer vision tasks.

The GoogLeNet architecture consists of a 22-layer deep neural network with a unique "Inception" module that enables the network to efficiently capture spatial features at different scales using parallel convolutional layers. The Inception module combines multiple convolutional filters of different sizes and concatenates their outputs, allowing the network to capture both fine-grained and high-level features. On the other hand, a residual block is a building block used in deep neural networks that helps to address the problem of vanishing gradients during training. A residual block consists of two or more stacked convolutional layers followed by a shortcut connection that bypasses these layers. The shortcut connection allows the gradient to be directly propagated to earlier layers, allowing for better optimization and deeper architectures.

This study introduces a novel deep-learning-based architecture for breast cancer detection that stands out from existing architectures in four significant aspects, resulting in superior performance.

1. Proposing a granular computing-based algorithm aiming to extract more detailed and fine-grained information from breast cancer images, leading to improved accuracy and performance;
2. Utilizing wide and deep modules, shortcut connections, and intermediate classifiers simultaneously in the architecture;
3. Designing an attention mechanism; the attention mechanism in CNNs provides a powerful tool for selectively focusing on relevant features in the input data, enabling the network to achieve better accuracy and efficiency;
4. Designing two learnable activation functions and using them instead of traditional activation functions.

Figure 1 depicts the overall process proposed in this paper. The input of the proposed method is a breast cancer image. If the size of the images is different, they are resized to a pre-determined size. After that, the pixels of the image are normalized between [0,1]. Resizing and normalization are preprocessing steps. After the preprocessing step, we use granular computing to highlight important features of an image. The output of the previous step is used to train the proposed deep-learning model. These steps are used for all images in the dataset. In the following, we will describe each of these cases.

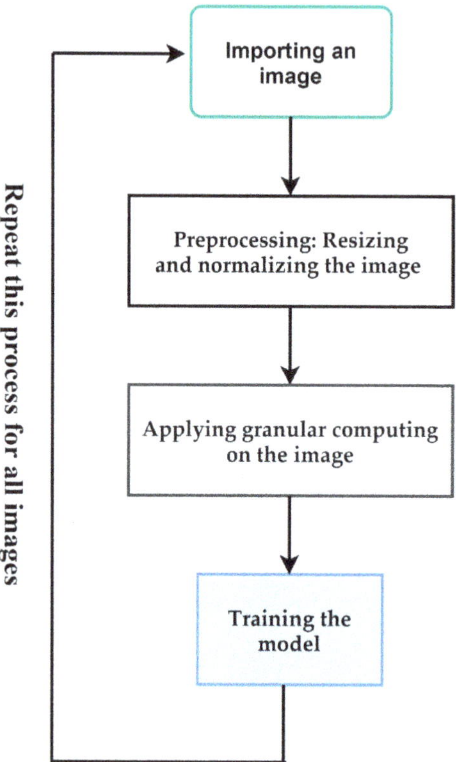

Figure 1. The overall process of the proposed method. All steps in this figure are repeated for all images.

3.1. Granular Computing

Granular computing can be used in image classification by dividing the image into smaller regions or sub-regions, also known as granules, and extracting features from them. This approach can improve the accuracy of the classification task by capturing more detailed and fine-grained information about the image [8].

Here, we propose general steps by which granular computing can be applied to image classification in deep learning:

Input: An image to be classified.

Output: A label representing the class of the image.

Preprocessing: Resize the image to a fixed size and normalize the pixel values to a range between 0 and 1.

Granulation: Divide the image into smaller regions or sub-regions, known as granules. This can be performed by using techniques such as windowing, tiling, or segmentation.

Feature Extraction: For each granule, extracts features using techniques such as local binary patterns, histograms of oriented gradients, or CNN. This will result in a set of feature vectors, one for each granule.

Feature Aggregation: Combine the feature vectors obtained from the granules and use them to classify the image. This can be performed by using techniques such as mean pooling or max pooling.

Figure 2 shows the proposed steps for granular computing used in this paper. Considering the above steps, we propose Algorithm 1 for applying granular computing in this paper. In this algorithm, we have used the pre-trained VGG16 architecture to extract features for each granularity. The size of each granular is considered to be 32×32. We apply granular computing to the dataset before starting the training process.

Algorithm 1. Extracting more detailed features by granular computing

---

Repeat the following steps for all images
img = Load the image

Preprocessing step: resizing and normalization
img = resize(img, (224, 224))
img = normalize(img/255.0)

Granulation step: split the image in windows of size 24 * 24
granules = []
for i in range(0, 224, 32):
for j in range(0, 224, 32):
granule = img[i:i + 32, j:j + 32,:]
granules.append(granule)

Feature Extraction step:
model = VGG16(weights= 'imagenet', include_top = False, input_shape = (32, 32, 3))
features = []
for granule in granules:
feature = model.predict(np.expand_dims(granule, axis = 0)).squeeze()
features.append(feature)

Feature Aggregation step:
features = np.array(features)
aggregated_features = np.mean(features, axis = 0)

3.2. Learnable Activation Function

Sigmoid, ReLU, and tanh are widely used activation functions in artificial neural networks, and they perform satisfactorily in many cases. However, these functions have some limitations that make their use suboptimal, thus necessitating the development of learnable activation functions.

Learnable activation functions in artificial neural networks are like functions that can adapt and modify themselves based on the input received by the neural network. These functions are characterized by a group of parameters that can be fine-tuned by the neural network during the training process. The primary advantage of learnable activation functions is their flexibility and adaptability in adjusting to different types of input data being processed. There are two primary types of learnable activation functions in artificial neural networks:

- Parametric activation functions: These functions have a fixed form (such as sigmoid or ReLU), but they add extra learnable parameters to them. The neural network changes these parameters to adjust the activation function appropriately on training data;
- Adaptive activation functions: These functions do not have any fixed form or formula. Instead, they rely on a neural network structure such as RNN or LSTM to learn appropriate activation functions.

We, here, develop two learnable activation functions named LAF_sigmoid and LAF_relu. To develop a new parametric learnable activation function, we start by defining a general form of the function that has learnable parameters. Let us call this function "Learnable Activation Function", or LAF for short:

$$LAF(x; W) = a * F(x; W) + b \qquad (1)$$

Here, a and b are adjustable parameters that are learned during training, and $F()$ is a non-linear function that defines the shape of the activation function. The weight matrix W contains learnable values that determine the shape of $F()$, and it is optimized through backpropagation.

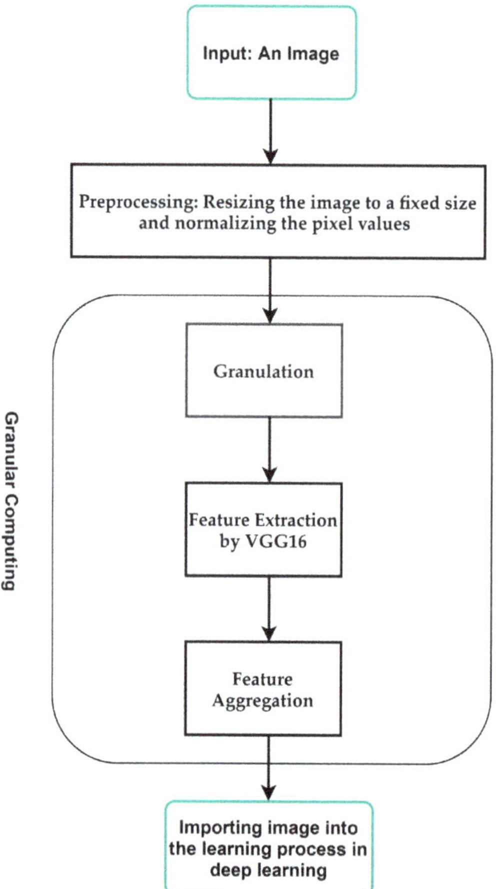

Figure 2. The granular computing process proposed in this paper.

To further develop the LAF, we can choose a suitable non-linear function $F()$. One possible choice is the sigmoid function:

$$F(x; W) = 1/(1 + \exp(-Wx)) \qquad (2)$$

The sigmoid function is a common choice for activation functions due to its smoothness and boundedness, which is important for the stable training of neural networks. The LAF with the Sigmoid function becomes:

$$LAF\_sigmoid(x; W, a, b) = a * (1/(1 + \exp(-Wx))) + b \qquad (3)$$

This activation function will be used in the dense layers of the network.

Another possible choice for $F()$ is the ReLU function:

$$F(x; W) = max(0, Wx) \tag{4}$$

The ReLU function is preferred for some tasks because of its simplicity and computational efficiency. The learnable parameters a and b can be added to shift and scale the ReLU function, resulting in the LAF with the ReLU function:

$$LAF\_relu(x; W, a, b) = a * max(0, Wx) + b \tag{5}$$

This activation function will be used in the convolutional layers of the network.

The values of a, b, and W can be trained through backpropagation using gradient descent or other optimization algorithms. The choice of the initial values and number of hidden units are important factors that can affect the success of training the neural network using LAFs. There are several advantages of using learnable activation functions in artificial neural networks:

1. Improved performance: By incorporating learnable activation functions, the neural network performance can be improved significantly. This is because the activation function adapts to the input data, allowing for a more accurate representation of complex relationships between features;
2. Non-linear mapping: Learnable activation functions allow for non-linear mappings between input and output, which can capture more complex patterns in the data;
3. Flexibility: With traditional activation functions, the network architecture is fixed. However, using learnable activation functions allows for more flexibility in the network architecture, as the activation function can be modified according to the specific task;
4. Reduced overfitting: Learnable activation functions can also help reduce overfitting, as they can adapt to the input data and generalize better to new data that has not been seen before;
5. Efficient training: The use of learnable activation functions can also make the training process more efficient by allowing gradients to be propagated through the network more smoothly. This can lead to faster convergence and improved performance.

3.3. The Attention Mechanism

The attention mechanism in convolutional neural networks (CNNs) is a tool that enables networks to selectively focus on specific parts of input data that are crucial for making decisions. Mimicking the process of human attention, this mechanism allows neural networks to attend selectively to relevant features in input data. The attention mechanism can be incorporated into various parts of a network, including the input layer, the convolutional layer, or the dense layer. Typically, the attention mechanism is added on top of the convolutional layer as an auxiliary module.

The attention mechanism takes as input the output features of the preceding layer and computes a set of attention weights for each feature. These weights reflect the importance of each feature for the current task. To compute these attention weights, a sub-network called the attention module, which comprises one or more layers, is employed. The attention weights are then multiplied element-wise with the output features of the preceding layer to obtain a weighted sum of the characteristics. This weighted sum is then passed to the next layer of the network. By doing this, the attention mechanism selectively focuses on key parts of the input data and enhances their impact on the output.

The attention mechanism can be trained end-to-end using backpropagation, and the attention weights can be learned jointly with the neural network parameters. During training, the attention mechanism learns to weigh the importance of various features based on their relevance to the task. This enables the network to selectively attend to the most informative parts of the input data and ignore irrelevant or noisy features.

In this section, we propose an attention mechanism to apply to the output layer (top layer). The attention mechanism in CNNs can highlight the salient regions in images that are significant for the classification task.

In the case of breast cancer detection, this can be useful since certain regions of the breast image may contain more relevant features for cancer detection compared to others. Here, we develop an attention mechanism for CNN:

1. Start with a standard convolutional layer with filters of size (k, k) and stride s;
2. Add a second convolutional layer with filters of size 1×1 and stride 1. This layer will compute a scalar attention weight for each pixel in the input image;
3. Apply a Softmax activation function to the output of the attention layer to ensure that the weights sum up to 1 for each pixel;
4. Multiply the attention weight maps element-wise with the input image to obtain the attended input image;
5. Feed the attended input image into the next layer of the CNN.

The idea behind this attention mechanism is that the second convolutional layer learns to compute a scalar attention weight for each pixel in the input image, based on its relevance to the task at hand. The softmax activation function ensures that the attention weights sum up to one for each pixel, making them interpretable as a probability distribution over the pixels. The element-wise multiplication of the attention weight maps and input image highlights or downplays certain pixels, improving the accuracy of the CNN on the given task.

3.4. Wide and Depth Networks, Short Connections, and 1×1 Convolutional Layers

Our developed network takes advantage of the features of wide and deep networks, short connections, and 1×1 convolutional layers. In the following, we will examine each one.

Wide and deep neural networks such as GoogLeNet offer improved accuracy, higher capacity, faster convergence, and better regularization. They have demonstrated impressive performance in various tasks, including image recognition, speech recognition, and natural language processing. Their advantages make them an attractive choice when designing neural networks.

In neural networks, short connections, which are used in ResNet and DenseNet networks, are a type of connection between the neurons that bypass one or more layers in the network. These connections allow information to flow between two layers that are not directly connected in the network architecture. Short connections, also known as skip connections, in neural networks can be represented mathematically as an element-wise summation or concatenation operation between the input to a layer and the output of that layer. In other words, the output of a layer is added to or concatenated with the input to that layer or a previous layer. For example, in a convolutional neural network (CNN), a short connection can be introduced between two convolutional layers by adding the output of the first convolutional layer to the input of the second convolutional layer. This can be added as follows:

$$x1 = \text{Convolutional\_layer\_1(input)}$$

$$x2 = \text{Convolutional\_layer\_2(x1 + input)}$$

where "+" denotes element-wise summation.

Short connections or residual connections in neural networks have several advantages:

1. Improved gradient flow: By adding short connections, the gradient can flow through the network more effectively, which eliminates the vanishing gradient problem. The gradient can be propagated directly to earlier layers, allowing the network to train deeper architectures;

2. Improved training speed: The use of short connections reduces the number of layers in the critical learning path, which can speed up the training process. The reduced depth also means that less computation is required, resulting in a more efficient model;

3. Improved accuracy: Short connections enable the learning of more complex functions by allowing the network to make use of the information from earlier layers. This can result in higher accuracy in tasks such as image recognition and speech processing;

4. Reduced overfitting: Short connections can help reduce overfitting by providing a regularization mechanism. They allow the network to learn simpler representations for the input data, which leads to better generalization.

In CNNs, 1×1 convolutional layers are utilized as a type of layer that executes a convolution operation by convolving the input tensor with a kernel of size 1×1. Despite their small size, 1×1 convolutional layers have various advantages in CNNs:

1. Dimensionality reduction: 1×1 convolutional layers can be used to reduce the dimensionality of feature maps, which can be useful in reducing the computational complexity of CNNs while maintaining their accuracy. By using 1×1 convolutional layers, the number of parameters can be reduced while still retaining the important features;

2. Non-linear transformations: Even though it has a kernel of size 1×1, this layer applies non-linear transformations to the input feature maps. The non-linear activation function applied after the convolution operation contributes to this non-linearity;

3. Improved model efficiency: By reducing the number of parameters, 1×1 convolutional layers reduce the computational cost of the model. This can, in turn, improve the efficiency of the implementation of the model, allowing it to be run on smaller devices or with fewer computational resources;

4. Feature interaction: A 1×1 convolutional layer can act as a feature interaction layer and induce correlations between features, which can further enhance the representation power of the network.

These advantages make 1×1 convolutional layers an important building block in CNNs, especially in deeper networks where computational cost and memory usage are of key concern.

3.5. The Designed Architecture

To reduce the high volume of processing in the GoogLeNet network, we changed the architecture from fully connected to sparsely connected network architectures within the convoluted layers. The Inception layer, which was inspired by the Hebbian principle of human learning, is critical to this sparsely connected architecture. For example, a deep-learning model for recognizing a particular pattern (e.g., face) in an image might have a layer that focuses on individual parts of an image. The next layer then focuses on the overall pattern in the image and identifies the various objects in it. To this end, the layer requires appropriate filter sizes to detect these objects. The Inception layer is crucial in this scenario, allowing internal layers to determine which filter size is relevant for learning the required information. Therefore, even if the size of the pattern in the image is different, the layer can recognize it accordingly.

The proposed system is shown in Figure 3, which shows the block diagram used to diagnose medical images. The proposed system has the same structure as the simple GoogLeNet network. We replaced the Inception module with a new module named X-module. The structure of the new deep neural network consists of the following:

1. Input layer;
2. A convolutional-based attention layer;
3. Convolution layer;
4. Two X modules with different filter sizes followed by a down-sample module;
5. An auxiliary classifier with a learnable Softmax classifier;
6. Three X modules with different filter sizes followed by a down-sample module;
7. An auxiliary classifier with a learnable Softmax classifier;

8. An X-module followed the Average pool, dropout layer;
9. A dense layer-based attention layer;
10. Learnable Softmax classifier as output layer.

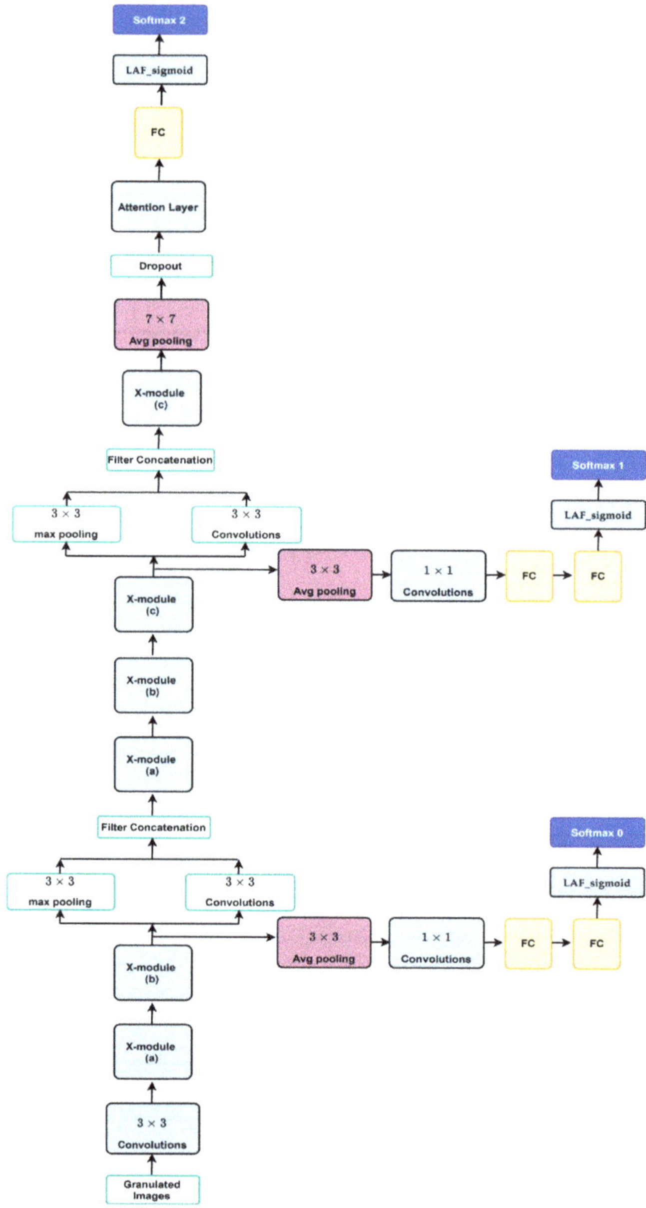

Figure 3. The new CNN system proposed to diagnose medical images.

The details are explained as follows:

Input Layer: In this step, the medical image is entered into the system.

Convolutional-based attention layer: This layer allows for the selective focus on specific parts of the input data that hold significance in determining an outcome.

Convolution Layer: This layer uses convolution operations to produce new feature maps.

X-Module: This module considers both the depth and width of the network, with multiple filters of varying sizes operating at the same level. The outputs of these filters are concatenated before being transmitted to the subsequent module. The main unit in X-module is a sub-block called R-block (Figure 4), which is inspired by the residual block. The main difference is that the designed block uses learnable activation functions. The block uses a shortcut connection; the input is added to the output of the block to pass gradient updates through the entire network easily and reduce overfitting. The R-block consists of two convolutional layers stacked on top of each other, with the first layer being succeeded by a batch normalization layer and a learnable activation function that is dependent on parameters. Using a parameter learnable activation function helps to reduce time consumption and better learning. Three types of R-blocks are implemented upon the filter size. The filter size of the two convolutional layers in the first R-block is 3×3. In the second R-block, the filter size of the two convolutional layers is 5×5. The first convolutional layer in the third R-block has a 3×3 filter size and the second convolutional layer has a 5×5 filter size. We have reduced the number of parameters and computational costs in the X-module by incorporating an additional 1×1 convolution in the initial layer, preceding the 3×3 and 5×5 convolutions. An extra 1×1 convolution is also utilized after the max pooling layer.

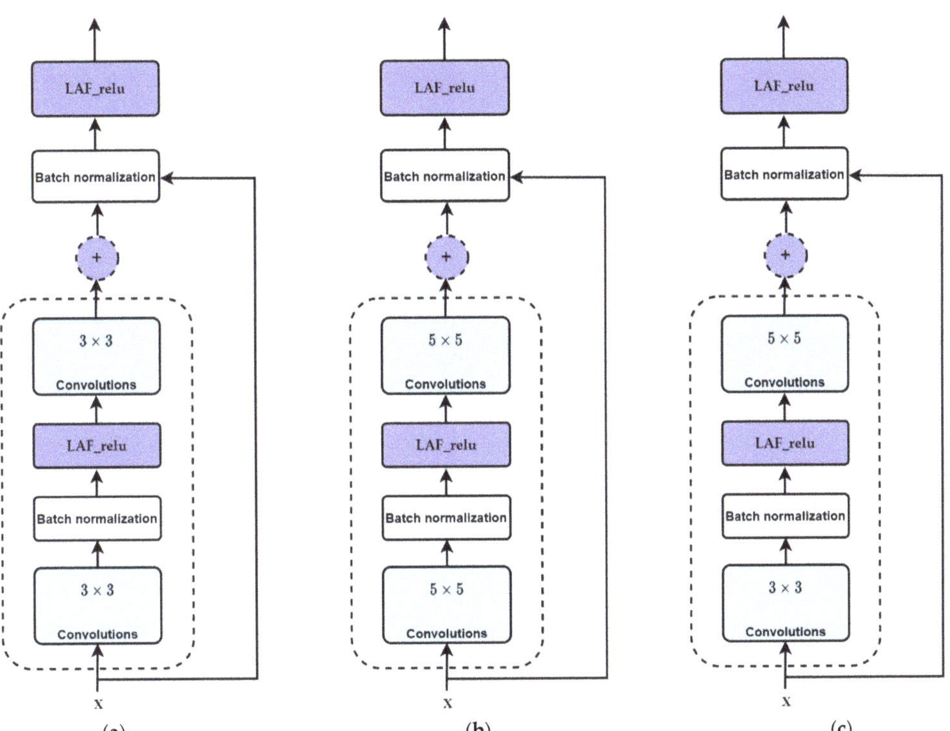

Figure 4. Structure of R-block (the main units of X-module): (**a**) R-block 3×3. (**b**) R-block 5×5. (**c**) R-block 3×5.

Different configurations of the X-module are implemented to build different deep neural network (DNN) models. In the first X-module, the first R-block is 3×3, the second is 5×5, and the third R-block has a 3×5 convolution filter size. In the second model, three R-blocks of filter sizes 3×3 are utilized. In the third model, three R-blocks of filter sizes 5×5 are utilized. The structure of the R-block and X-module are shown in Figures 4 and 5, respectively.

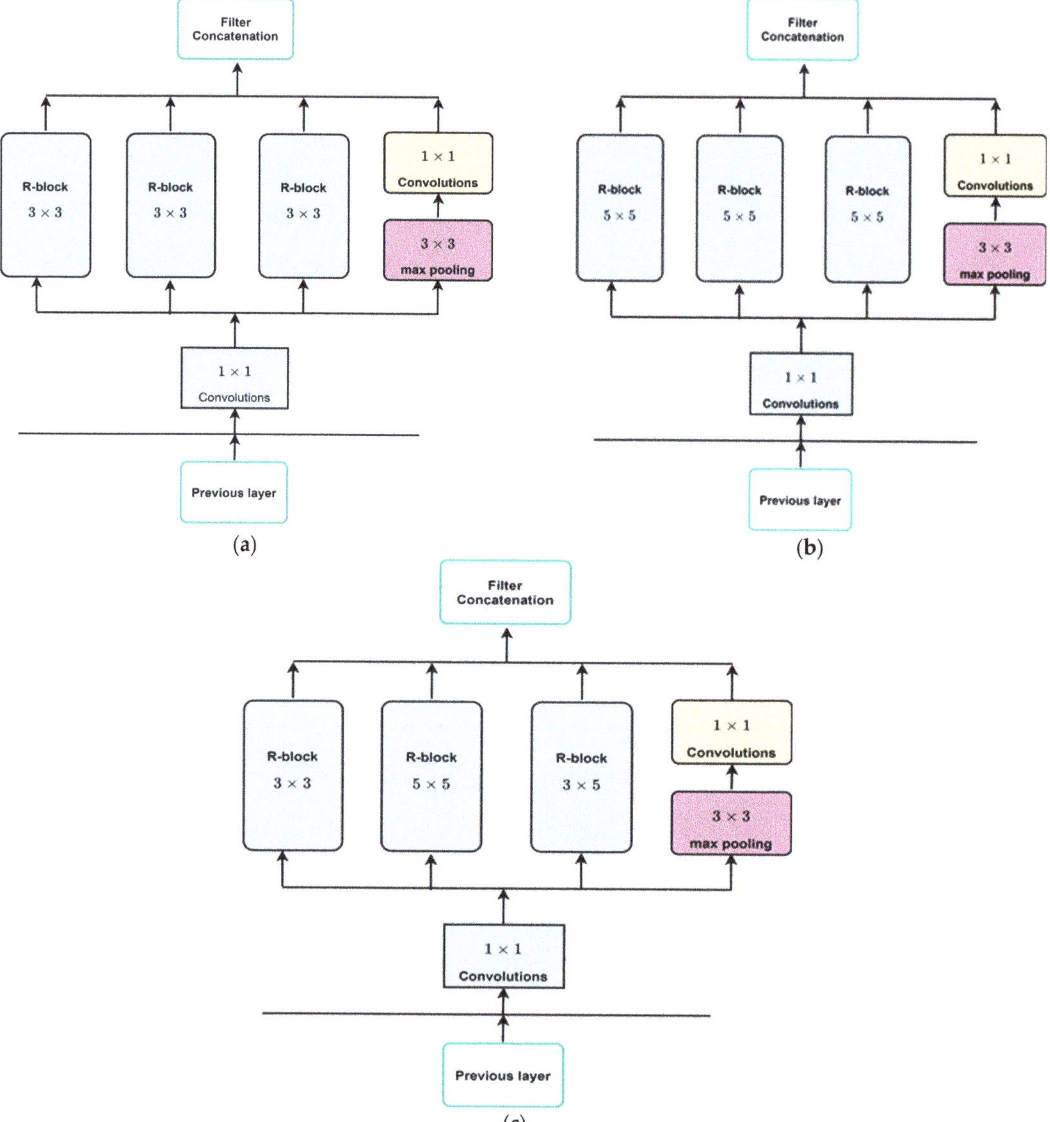

Figure 5. The different used structures into X-module: (**a**) X-module includes three R-blocks of 3×3, called DNN R3_R3_R3. (**b**) X-module includes three R-blocks of 5×5, called DNN R5_R5_R5. (**c**) X-module includes three R-blocks of 3×3, 5×5, and 3×5, called DNN R3_R5_R35.

Downsampling module: Following the X-module, this module is implemented to decrease both the size of the feature map and the number of network parameters. The pooling max function is concatenated with a 3×3 convolutional layer. As stated before, deep neural networks are computationally expensive. To make it cheaper, the number of input channels is limited by adding a 1×1 convolution before the 3×3 convolution.

Dense layer-based attention layer: This attention takes as input a 3D tensor representing the output features of the previous layer and outputs a 2D tensor of attention scores, where each score represents the relevance of a specific feature.

Output layer: The output in this step will be normal or abnormal.

4. Result and Discussion

The experiments were conducted using the Cairo University ultrasound images dataset and the breast histopathology images dataset for training and testing. The code was written in Python, and the experiments were performed on Kaggle, leveraging the power of their hardware, including GPUs. To train the model, the following settings were employed: Adam optimizer with a learning rate set to 0.0001, 100 epochs, a batch size of 32, and a dropout rate of 50%. The number of epochs in a CNN is one of the hyperparameters that can be tuned to improve the performance of the model. The number of epochs refers to the number of times the entire training dataset is passed through the CNN during the training phase. To tune the number of epochs, we used the early stopping technique. Early stopping is a method used to prevent overfitting by monitoring the validation loss during training and stopping the training process once the validation loss starts to increase. After applying this technique, we set the number of epochs to 100.

In this study, the proposed approach is assessed in terms of several performance metrics, including accuracy, loss, precision, recall, and F1 score. These metrics are defined as follows:

Accuracy: This metric measures the overall performance of the model by calculating the percentage of correctly predicted labels to the total number of samples in the test dataset. Mathematically, it can be expressed as:

$$\text{Accuracy} = (\text{Number of Correct Predictions})/(\text{Total Number of Predictions}) \quad (6)$$

Loss: This metric represents the error between the predicted output and the actual output. The loss function is typically defined during the training phase of the model, and it is used to optimize the model parameters by minimizing the difference between its predictions and the true values. The most commonly used loss function in deep learning is the mean squared error (MSE), which measures the average of the squared differences between the predicted and true values.

Precision: This metric measures the proportion of true positives (samples that were correctly classified as positive) to the total number of positive predictions made by the model. It can be calculated as:

$$\text{Precision} = \text{True Positives}/(\text{True Positives} + \text{False Positives}) \quad (7)$$

Recall: This metric measures the proportion of true positives to the total number of true positives and false negatives in the dataset.

F1: This metric calculates the harmonic mean of precision and recall.

Different DNN models are implemented using different configurations of the X-module, as depicted in Figure 5. The first model has three R-blocks of 3×3 convolution filters. The second model utilized three R-blocks of 5×5. In the third model, the first R-block is 3×3, the second is 5×5, and the third R-block has a 3×3 filter size in the first convolutional layer, and the second convolutional layer has a 5×5 filter size. We have utilized various filters and kernels (kernel size). By specifying multiple values for the kernel parameter within a filter, our model can effectively identify patterns that occur at different scales within an image. The incorporation of multiple kernels also assists in reducing overfitting and improving the generalization of the model. This is because including filters with varying kernel sizes compels the network to learn more diverse and robust feature representations, leading to an improved ability to generalize the model to new images.

The Cairo University ultrasound images dataset was collected in 2018 that consists of 780 images with an average image size of 500×500 pixels [36]. The images are categorized into three classes, which are normal (133 images), benign (487 images), and malignant (210 images). The data collected at baseline include breast ultrasound images among women in ages between 25 and 75 years old. The number of patients is 600 female patients. Because the number of images in different classes is unbalanced, this may cause the model to learn some classes better than others, and this can cause the model to perform inappropriately during use or testing. To prevent this from happening, we randomly selected an equal number of images from each class. Before starting the training process of the model, due to the lack of data, we start the data augmentation process. We resize the dataset using cubic interpolation to fit the input requirements of the model. For augmentation, we applied width and height shifts of 0.1 and a horizontal flip, which tripled the size of the dataset. With this technique, we tripled the number of data for each class. After this process, we split the dataset into training and test sets, allocating approximately 80% for training and 20% for testing. Of course, we have not used these new images for testing. We maintain the sequence of each image so that every image appears only once in each of the aforementioned sets.

Figure 6 depicts the accuracy and loss diagrams for the three proposed models on the Cairo University ultrasound dataset.

On the Cairo University ultrasound images dataset, the performance of the three proposed CNN models in the testing phase is summarized in Table 1. The third model, DNN R3_R5_R35, which uses a different size of convolutional filters, achieves the best accuracy and low loss. It achieved 93% accuracy performance.

Table 2 is the resultant confusion matrix of the DNN R3_R5_R35 model. This table shows promising results so that the proposed model has correctly diagnosed the presence or absence of cancer in most cases, and it has not been able to correctly diagnose only five cases out of 68 cases.

Proposed DNN R3_R5_R35

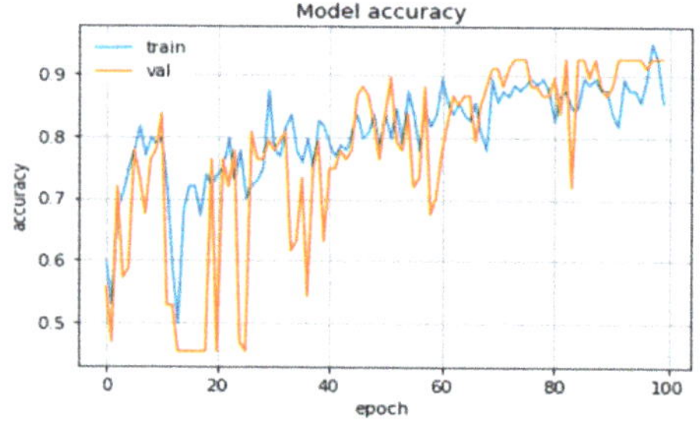
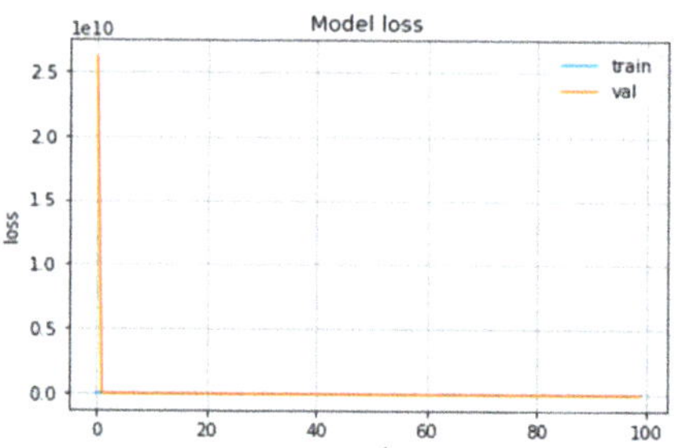

Proposed DNN R3_R3_R3

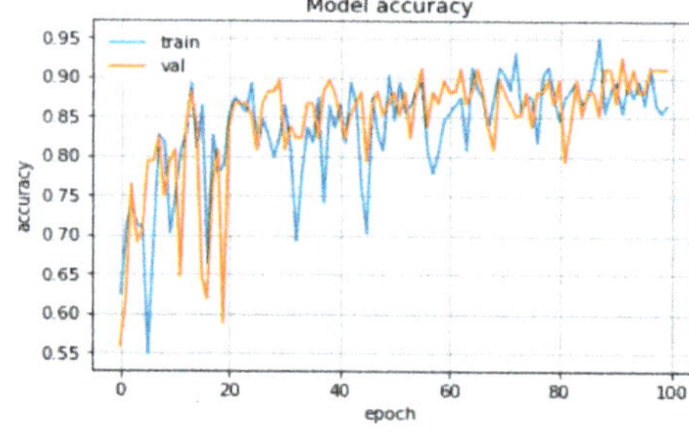
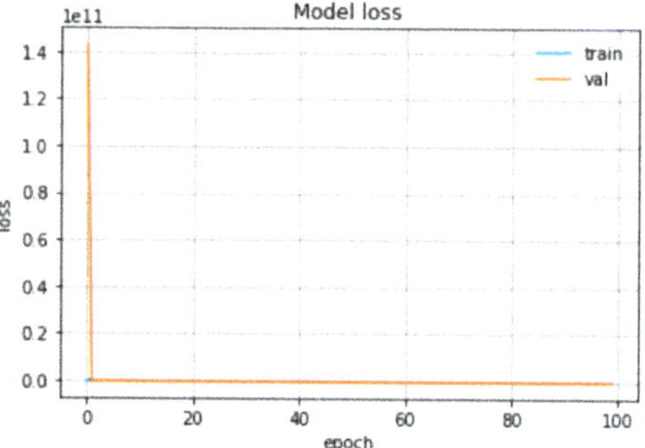

Figure 6. *Cont.*

Proposed DNN R5_R5_R5

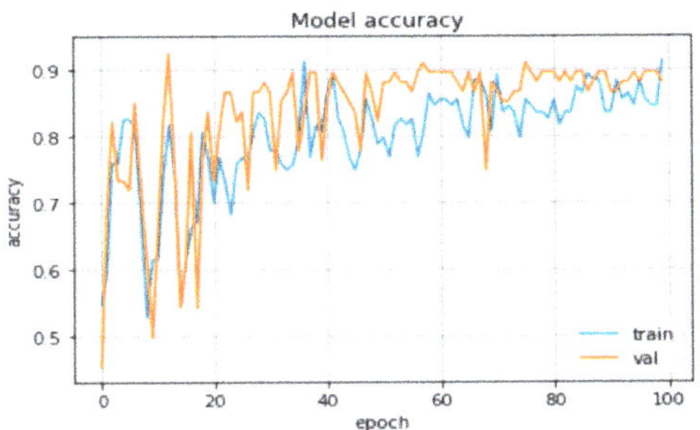

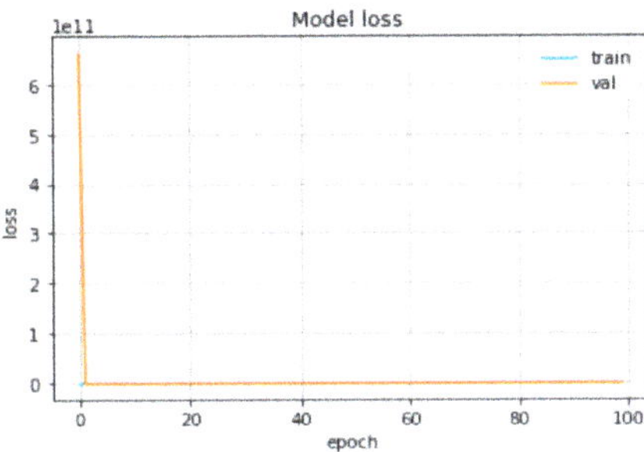

Figure 6. Accuracy and loss diagrams for the three proposed models.

Table 1. The newly developed model results on Cairo University ultrasound images dataset (test part).

Deep-Learning Model	Precision	Recall	F_1-Score	Loss	Accuracy
Proposed DNN R5_R5_R5	0.88	0.88	0.88	0.2919	0.88
Proposed DNN R3_R3_R3	0.91	0.91	0.91	0.2445	0.91
Proposed DNN R3_R5_R35	0.93	0.93	0.93	0.2103	0.93

Table 2. Confusion matrix for the test dataset. (**0** indicating no breast cancer and **1** indicating existing breast cancer, one of the benign and malignant in the image.).

		Predicted	
		0	1
Actual	0	33	4
	1	1	30

As we used some of the features of the state-of-the-art GoogLeNet and ResNet architectures in the design of the new architecture, we compared the proposed architecture, i.e., DNN R3_R5_R35, with these architectures in Table 3. To perform this comparison, we have utilized GoogLeNet with 22 layers and ResNet with 50 layers for comparison. The results indicate that ResNet outperformed GoogLeNet in two critical evaluation metrics: accuracy and F1 score. Inspection of the table containing the results reveals that the proposed method has surpassed both of these models and yielded a higher detection accuracy than these two state-of-the-art architectures.

Table 3. Comparison of the proposed model against GoogLeNet and ResNet on the test dataset.

Deep-Learning Model	Precision	Recall	F_1-Score	Loss	Accuracy
GoogLeNet	0.86	0.87	0.85	0.59	0.87
ResNet50 (Residual Network)	0.87	0.85	0.86	0.51	0.88
DNN R3_R5_R35	0.93	0.93	0.93	0.2103	0.93

The performance of GoogLeNet during the training phase is shown in Figure 7. A comparison between Figures 6 and 7 demonstrates that there is a suitable performance for the proposed models and no overfitting compared to GoogLeNet. The DNN R3_R5_R35 model that uses different sizes of convolutional filters achieves the best performance. Here, we compare the proposed architecture with the state-of-the-art image processing architectures in terms of prediction accuracy and loss on test data. Table 4 shows these comparisons. The proposed model is superior to all existing image processing architectures in terms of prediction accuracy on breast cancer images.

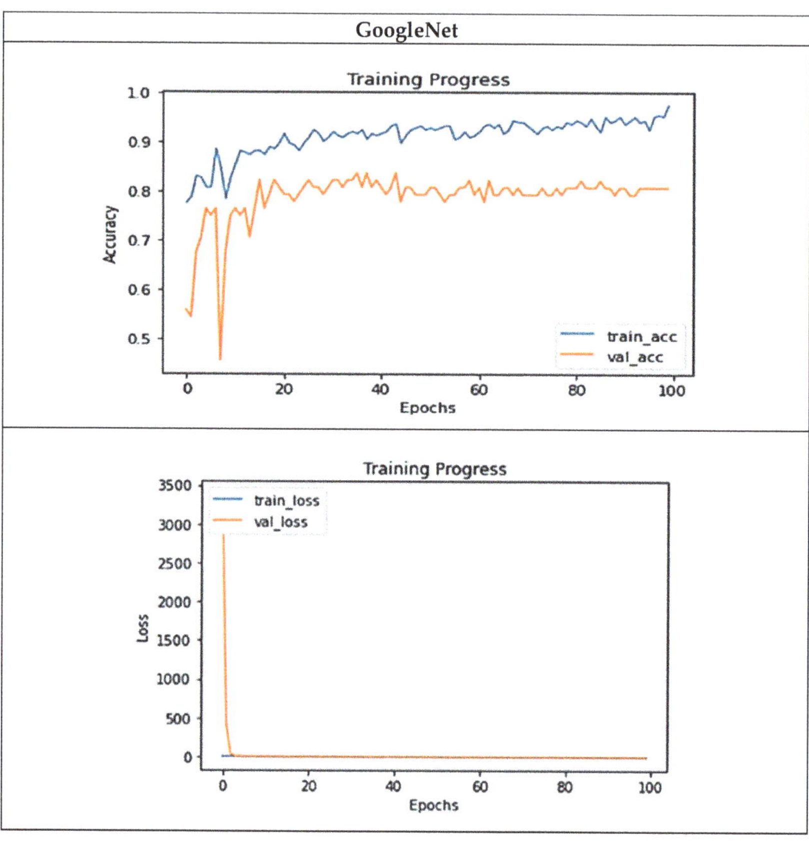

Figure 7. Accuracy and loss diagrams for the GoogLeNet.

Table 4. Comparison of the proposed model against nine state-of-the-art image processing models on the Cairo University ultrasound images dataset.

Deep-Learning Model	Loss (%)	Accuracy (%)
AlexNet	66	69
ZFNet	64	69
VGG-16	6	73
Inception v4	46	85
MobileNet	55	85
WideResNet	39	88
GoogLeNet	59	87
ResNet34	61	83
ResNet50	51	88
Proposed DNN R3_R5_R35	**21**	**93**

We also applied the proposed model to the breast histopathology images dataset to further evaluate it. The original dataset consisted of 162 whole-mount slide images of breast cancer specimens scanned at $40\times$. From that, 277,524 patches of size 50×50 were extracted (198,738 IDC-negative and 78,786 IDC-positive). Invasive Ductal Carcinoma (IDC) is the most common subtype of all breast cancers. We choose 40,000 images in total from the dataset: 20,000 random images from both classes. We split the dataset into training and test sets, allocating 80% for training and 20% for testing. The comparison results of the proposed DNN R3_R5_R35 model against existing approaches are listed in Table 5. The best results are highlighted in bold. It is evident from the table that the proposed model has outperformed the other models in the two assessed criteria.

Table 5. Comparative analysis on breast histopathology images dataset.

Ref.	Year	Method/Model	Accuracy (%)	F_1-Score (%)
[32]	2023	Transfer learning with VGG16	91	89
[33]	2023	ViT-Patch 32	89	-
[37]	2021	CNN architecture	87	87
[38]	2021	DenseNet121 Model	86	87
[38]	2021	DenseNet169	85	86
[38]	2021	MobileNet	86	85
[38]	2021	ResNet50	84	83
[38]	2021	VGG19	84	80
[38]	2021	VGG16	85	85
[38]	2021	EfficientNetB0	84	84
[38]	2021	EfficientNetB4	82	82
[38]	2021	EfficientNetB5	84	84
[39]	2020	Residual learning-based CNN	84	83
[40]	2020	CNN	85	82
[41]	2021	Patch-Based Deep-Learning Modeling	85	85
Proposed Study		Proposed DNN R3_R5_R35 Model	**95**	**93**

The objective of this study was to enhance the accuracy of breast cancer detection through the application of deep-learning techniques in the development of computer-aided detection systems. The proposed model, utilizing various filter sizes, demonstrated 93% and 95% accuracy on two distinct datasets: ultrasound images and breast histopathology images, respectively. The second goal was to decrease the parameters of the network, aiming to improve the training time. The time issue during the training process for any deep-learning model is still challenging and depends on the facilities that be used. Training the model on ultrasound and histopathology images takes less than two hours and less than six hours, respectively, which is suitable compared to other DNN models.

A short review above, we can conclude the main findings of this paper as follows:

1. The granular computing technique used in this paper, by breaking down images into smaller, more granular components, can effectively extract features from images, allowing for more accurate and efficient image analysis. This leads to increased efficiency by reducing the computational complexity of image analysis tasks. Moreover, breaking down images into smaller, granular computing can improve the accuracy of image analysis tasks, leading to more reliable results;

2. Activation functions with learnable parameters offer greater flexibility and adaptability compared to traditional activation functions with fixed parameters. This allows the network to better adapt to different types of data and tasks. These functions can also improve the flow of gradients through the network during training, making it easier to optimize the network and reduce the risk of vanishing gradients. Better regularization is another advantage of these functions. Learnable activation functions can be used as a form of regularization, helping to prevent overfitting by constraining the network's capacity and reducing the risk of memorization;

3. In this study, a range of filters and kernels of varying sizes were employed to effectively identify patterns at multiple scales within an image. By incorporating multiple kernels within a filter, the network was able to learn diverse and robust feature representations, which helped to reduce overfitting and improve the generalization of the model. This approach enabled the model to consider a wider range of input features,

leading to higher accuracy in complex tasks compared with a model that employs a single filter and kernel. The use of multiple kernels within a filter, therefore, represents an effective strategy for improving the ability of a neural network to generalize to new images by facilitating the learning of more sophisticated features across a range of spatial scales;

4. Utilizing a wide and depth network, shortcut connections, attention layers, auxiliary classifiers, and using a learnable activation function improves the accuracy of diagnosis and consequence and decreases the load on doctors. In addition, using 1×1 convolutions reduces time consumption in the model. Compared to existing breast cancer methods, the proposed model achieves the highest diagnostic accuracy.

There are two potential limitations to the presented model:

1. The extraction of some patterns from the image may be dependent on the granularity size. In the proposed granulation, the granularity size was set to 32×32 pixels, regardless of the image size. Consequently, some patterns may not be extracted, weakening the effectiveness of granulation. However, the model's overall performance demonstrates that the proposed granulation method outperformed state-of-the-art models for the datasets under consideration;

2. Incorporating granularity in a model requires additional time before the training process can commence. It is worth noting, however, that once these granules have been established, they can be reused multiple times.

5. Conclusions and Future Work

By automating the diagnosis process, healthcare professionals can focus on providing personalized treatment plans for patients. This not only improves patient outcomes but also reduces healthcare costs by minimizing unnecessary procedures and tests. The use of machine-learning and deep-learning techniques in breast cancer detection has the potential to revolutionize the way breast cancer is diagnosed and managed. Incorporating these tools into healthcare systems has the potential to lower cancer-related mortality rates and enhance the overall management of breast cancer. This paper proposed a novel granular computing-based deep-learning model for breast cancer detection, which is evaluated under ultrasound images and breast histopathology images datasets. The proposed model has used some effective features of GoogLeNet and ResNet architectures (such as wide and depth modules, 1×1 convolutional filters, auxiliary classifiers, and skip connection) and has added some new features such as granular computing, activation functions with learnable parameters, and attention layer to the new architecture. Granular computing can extract the important features of the image and create a new image with the important features highlighted before sending it to the training process. This feature makes the model require fewer images than in the case where granularity is not used. The proposed model achieved an accuracy improvement compared to state-of-the-art models. In particular, the deep-learning model based on granular computing exhibited an accuracy of 93% and 95% on two real-world datasets, ultrasound images and breast histopathology images, respectively. The model delivered promising results on the datasets. The findings may encourage radiologists and physicians to leverage the model in the early detection of breast cancer, leading to improved diagnosis accuracy, reduced time consumption, and eased workload of doctors. It has been confirmed that granular computing has a positive effect on the performance of problems where the number of available images is small, such as breast cancer.

For future work, the following items can be performed: (1) The framework can be further optimized for real-time performance. Taking inspiration from the MobileNet architecture, e.g., separable convolutions feature, the number of parameters of the proposed architecture can be reduced so that the accuracy does not decrease much. (2) Instead of granular computing, one can use fuzzy clustering. This method uses fuzzy logic to group pixels in an image into clusters based on their similarity. (3) There are usually very few medical photos. The number of these photos can be increased with techniques such as

Sketch2Photo [42] before starting the learning process. This will increase the accuracy of the model. (4) Over the last 30 years, hyperspectral imagery (HSI) has gained prominence for its ability to discern anomalies from natural ground objects based on their spectral characteristics. The importance of HSI has been recognized in a variety of remote sensing applications, including but not limited to object classification, hyperspectral unmixing, anomaly detection, and change detection [43]. We can use this technique to identify breast cancer. (5) The primary challenge in content-based image retrieval (CBIR) systems is the presence of a semantic gap that must be narrowed for effective retrieval. To address this issue, various techniques, such as those outlined in [44], can be employed to incorporate semantic considerations. (6) To reduce the processing time, it is suggested to use distributed and parallel similarity retrieval techniques, such as [45], on large CT image sequences. (7) The proposed framework can be extended to include other types of cancer detection, such as lung or prostate cancer. This would enable the development of a comprehensive cancer detection system that can be integrated into existing healthcare systems.

Author Contributions: Conceptualization, S.Z. and H.I.; methodology, S.Z. and H.I.; software, S.Z.; validation, S.Z., H.I. and J.K.; formal analysis, S.Z., H.I. and J.K.; investigation, S.Z., H.I. and J.K; resources, S.Z., H.I. and J.K; data curation, S.Z., H.I. and J.K; writing—original draft preparation, S.Z.; writing—review and editing, S.Z., H.I. and J.K.; visualization, S.Z.; supervision, H.I. and J.K; project administration, H.I. All authors have read and agreed to the published version of the manuscript.

Funding: This research received no external funding.

Institutional Review Board Statement: Not applicable.

Informed Consent Statement: Informed consent was obtained from all subjects involved in the study.

Data Availability Statement: Publicly available datasets were analyzed in this study. Ultrasound images dataset: https://www.data-in-brief.com/article/S2352-3409(19)31218-1/fulltext (accessed on 1 September 2022). Breast Histopathology Images: https://www.kaggle.com/datasets/paultimoth ymooney/breast-histopathology-images (accessed on 1 September 2022.).

Conflicts of Interest: There is no conflict of interest.

References

1. World Health Organization: Breast Cancer Web Site. Available online: https://www.who.int/newsroom/fact-sheets/detail/brea st-cancer (accessed on 30 September 2022).
2. Mahmood, T.; Li, J.; Pei, Y.; Akhtar, F.; Imran, A.; Rehman, K.U. A brief survey on breast cancer diagnostic with deep learning schemes using multi-image modalities. *IEEE Access* **2020**, *8*, 165779–165809. [CrossRef]
3. Liew, X.Y.; Hameed, N.; Clos, J. A review of computer-aided expert systems for breast cancer diagnosis. *Cancers* **2021**, *13*, 2764. [CrossRef]
4. Almajalid, R.; Shan, J.; Du, Y.; Zhang, M. Development of a deep-learning-based method for breast ultrasound image segmentation. In Proceedings of the 2018 17th IEEE International Conference on Machine Learning and Applications (ICMLA), Orlando, FL, USA, 17–20 December 2018; pp. 1103–1108.
5. Latif, G.; Butt, M.O.; Al Anezi, F.Y.; Alghazo, J. Ultrasound image despeckling and detection of breast cancer using deep CNN. In Proceedings of the 2020 RIVF International Conference on Computing and Communication Technologies (RIVF), Ho Chi Minh, Vietnam, 14–15 October 2020; pp. 1–5.
6. Zhu, W.; Xiang, X.; Tran, T.D.; Hager, G.D.; Xie, X. Adversarial deep structured nets for mass segmentation from mammograms. In Proceedings of the 2018 IEEE 15th International Symposium on Biomedical Imaging (ISBI 2018), Washington, DC, USA, 4–7 April 2018; pp. 847–850.
7. Al-Antari, M.A.; Al-Masni, M.A.; Choi, M.T.; Han, S.M.; Kim, T.S. A fully integrated computer-aided diagnosis system for digital X-ray mammograms via deep learning detection, segmentation, and classification. *Int. J. Med. Inform.* **2018**, *117*, 44–54. [CrossRef] [PubMed]
8. Yao, J.T.; Vasilakos, A.V.; Pedrycz, W. Granular computing: Perspectives and challenges. *IEEE Trans. Cybern.* **2013**, *43*, 1977–1989. [CrossRef] [PubMed]
9. Lee, J.; Kang, B.J.; Kim, S.H.; Park, G.E. Evaluation of computer-aided detection (CAD) in screening automated breast ultrasound based on characteristics of CAD marks and false-positive marks. *Diagnostics* **2022**, *12*, 583.
10. Cheng, H.D.; Shan, J.; Ju, W.; Guo, Y.; Zhang, L. Automated breast cancer detection and classification using ultrasound images: A survey. *Pattern Recognit.* **2010**, *43*, 299–317. [CrossRef]

11. Jiménez-Gaona, Y.; Rodríguez-Álvarez, M.J.; Lakshminarayanan, V. Deep-learning-based computer-aided systems for breast cancer imaging: A critical review. *Appl. Sci.* **2020**, *10*, 8298. [CrossRef]
12. Wang, S.; Huang, J. Breast Lesion Segmentation in Ultrasound Images by CDeep3M. In Proceedings of the 2020 International Conference on Computer Engineering and Application (ICCEA), Guangzhou, China, 18–20 March 2020; pp. 907–911.
13. Wei, K.; Wang, B.; Saniie, J. Faster Region Convolutional Neural Networks Applied to Ultrasonic Images for Breast Lesion Detection and Classification. In Proceedings of the 2020 IEEE International Conference on Electro Information Technology (EIT), Romeoville, IL, USA, 31 July–1 August 2020; pp. 171–174.
14. Badawy, S.M.; Mohamed, A.E.N.A.; Hefnawy, A.A.; Zidan, H.E.; GadAllah, M.T.; El-Banby, G.M. Classification of Breast Ultrasound Images Based on Convolutional Neural Networks—A Comparative Study. In Proceedings of the 2021 International Telecommunications Conference (ITC-Egypt), Alexandria, Egypt, 13–15 July 2021; pp. 1–7.
15. Tang, P.; Yang, X.; Nan, Y.; Xiang, S.; Liang, Q. Feature Pyramid Nonlocal Network With Transform Modal Ensemble Learning for Breast Tumor Segmentation in Ultrasound Images. *IEEE Trans. Ultrason. Ferroelectr. Freq. Control* **2021**, *68*, 3549–3559.
16. Xiao, T.; Liu, L.; Li, K.; Qin, W.; Yu, S.; Li, Z. Comparison of transferred deep neural networks in ultrasonic breast masses discrimination. *BioMed Res. Int.* **2018**, *2018*, 4605191. [CrossRef]
17. Uysal, F.; Köse, M.M. Classification of Breast Cancer Ultrasound Images with Deep Learning-Based Models. *Eng. Proc.* **2022**, *31*, 8.
18. Ragab, M.; Albukhari, A.; Alyami, J.; Mansour, R.F. Ensemble deep-learning-enabled clinical decision support system for breast cancer diagnosis and classification on ultrasound images. *Biology* **2022**, *11*, 439. [CrossRef]
19. Xing, J.; Chen, C.; Lu, Q.; Cai, X.; Yu, A.; Xu, Y.; Huang, L. Using BI-RADS stratifications as auxiliary information for breast masses classification in ultrasound images. *IEEE J. Biomed. Health Inform.* **2020**, *25*, 2058–2070. [CrossRef]
20. Ragab, D.A.; Sharkas, M.; Marshall, S.; Ren, J. Breast cancer detection using deep convolutional neural networks and support vector machines. *PeerJ* **2019**, *7*, e6201. [CrossRef]
21. Yu, X.; Wang, S.H. Abnormality diagnosis in mammograms by transfer learning based on ResNet18. *Fundam. Inform.* **2019**, *168*, 219–230. [CrossRef]
22. Islam, M.M.; Haque, M.R.; Iqbal, H.; Hasan, M.M.; Hasan, M.; Kabir, M.N. Breast cancer prediction: A comparative study using machine learning techniques. *SN Comput. Sci.* **2020**, *1*, 1–14. [CrossRef]
23. Alzubaidi, L.; Al-Shamma, O.; Fadhel, M.A.; Farhan, L.; Zhang, J.; Duan, Y. Optimizing the performance of breast cancer classification by employing the same domain transfer learning from hybrid deep convolutional neural network model. *Electronics* **2020**, *9*, 445. [CrossRef]
24. Gao, Y.; Geras, K.J.; Lewin, A.A.; Moy, L. New frontiers: An update on computer-aided diagnosis for breast imaging in the age of artificial intelligence. *AJR Am. J. Roentgenol.* **2019**, *212*, 300. [CrossRef]
25. Yap, M.H.; Pons, G.; Marti, J.; Ganau, S.; Sentis, M.; Zwiggelaar, R.; Marti, R. Automated breast ultrasound lesions detection using convolutional neural networks. *IEEE J. Biomed. Health Inform.* **2017**, *22*, 1218–1226. [CrossRef]
26. Moon, W.K.; Lee, Y.W.; Ke, H.H.; Lee, S.H.; Huang, C.S.; Chang, R.F. Computer-aided diagnosis of breast ultrasound images using ensemble learning from convolutional neural networks. *Comput. Methods Programs Biomed.* **2020**, *190*, 105361. [CrossRef]
27. Lee, Y.W.; Huang, C.S.; Shih, C.C.; Chang, R.F. Axillary lymph node metastasis status prediction of early-stage breast cancer using convolutional neural networks. *Comput. Biol. Med.* **2021**, *130*, 104206. [CrossRef]
28. Sun, Q.; Lin, X.; Zhao, Y.; Li, L.; Yan, K.; Liang, D.; Li, Z.C. Deep learning vs. radiomics for predicting axillary lymph node metastasis of breast cancer using ultrasound images: Don't forget the peritumoral region. *Front. Oncol.* **2020**, *10*, 53. [CrossRef] [PubMed]
29. Sun, S.; Mutasa, S.; Liu, M.Z.; Nemer, J.; Sun, M.; Siddique, M.; Ha, R.S. Deep learning prediction of axillary lymph node status using ultrasound images. *Comput. Biol. Med.* **2022**, *143*, 105250. [CrossRef] [PubMed]
30. Ayana, G.; Park, J.; Jeong, J.W.; Choe, S.W. A novel multistage transfer learning for ultrasound breast cancer image classification. *Diagnostics* **2022**, *12*, 135. [CrossRef]
31. Yusoff, M.; Haryanto, T.; Suhartanto, H.; Mustafa, W.A.; Zain, J.M.; Kusmardi, K. Accuracy Analysis of Deep Learning Methods in Breast Cancer Classification: A Structured Review. *Diagnostics* **2023**, *13*, 683. [CrossRef] [PubMed]
32. Hossain, A.A.; Nisha, J.K.; Johora, F. Breast Cancer Classification from Ultrasound Images using VGG16 Model based Transfer Learning. *Int. J. Image Graph. Signal Process.* **2023**, *13*, 12. [CrossRef]
33. Feng, H.; Yang, B.; Wang, J.; Liu, M.; Yin, L.; Zheng, W.; Liu, C. Identifying Malignant Breast Ultrasound Images Using ViT-Patch. *Appl. Sci.* **2023**, *13*, 3489. [CrossRef]
34. Szegedy, C.; Liu, W.; Jia, Y.; Sermanet, P.; Reed, S.; Anguelov, D.; Rabinovich, A. Going deeper with convolutions. In Proceedings of the IEEE Conference on Computer Vision and Pattern Recognition, Boston, MA, USA, 7–12 June 2015; pp. 1–9.
35. He, K.; Zhang, X.; Ren, S.; Sun, J. Deep residual learning for image recognition. In Proceedings of the IEEE Conference on Computer Vision and Pattern Recognition, Las Vegas, NV, USA, 27–30 June 2016; pp. 770–778.
36. Al-Dhabyani, W.; Gomaa, M.; Khaled, H.; Fahmy, A. Dataset of breast ultrasound images. *Data Brief* **2020**, *28*, 104863. [CrossRef]
37. Alanazi, S.A.; Kamruzzaman, M.M.; Islam Sarker, M.N.; Alruwaili, M.; Alhwaiti, Y.; Alshammari, N.; Siddiqi, M.H. Boosting breast cancer detection using convolutional neural network. *J. Healthc. Eng.* **2021**, *2021*, 1–11. [CrossRef]
38. Seemendra, A.; Singh, R.; Singh, S. Breast cancer classification using transfer learning. In *Evolving Technologies for Computing, Communication and Smart World: Proceedings of ETCCS*; Springer: Singapore, 2020; pp. 425–436.

39. Gour, M.; Jain, S.; Sunil Kumar, T. Residual learning based CNN for breast cancer histopathological image classification. *Int. J. Imaging Syst. Technol.* **2020**, *30*, 621–635. [CrossRef]

40. Shahidi, F.; Daud, S.M.; Abas, H.; Ahmad, N.A.; Maarop, N. Breast cancer classification using deep learning approaches and histopathology image: A comparison study. *IEEE Access* **2020**, *8*, 187531–187552. [CrossRef]

41. Hirra, I.; Ahmad, M.; Hussain, A.; Ashraf, M.U.; Saeed, I.A.; Qadri, S.F.; Alfakeeh, A.S. Breast cancer classification from histopathological images using patch-based deep learning modeling. *IEEE Access* **2021**, *9*, 24273–24287. [CrossRef]

42. Liu, H.; Xu, Y.; Chen, F. Sketch2Photo: Synthesizing photo-realistic images from sketches via global contexts. *Eng. Appl. Artif. Intell.* **2023**, *117*, 105608. [CrossRef]

43. Wang, S.; Hu, X.; Sun, J.; Liu, J. Hyperspectral anomaly detection using ensemble and robust collaborative representation. *Inf. Sci.* **2023**, *624*, 748–760. [CrossRef]

44. Zhuang, Y.; Chen, S.; Jiang, N.; Hu, H. An Effective WSSENet-Based Similarity Retrieval Method of Large Lung CT Image Databases. *KSII Trans. Internet Inf. Syst.* **2022**, *16*, 7.

45. Zhuang, Y.; Jiang, N.; Xu, Y.; Xiangjie, K.; Kong, X. Progressive Distributed and Parallel Similarity Retrieval of Large CT Image Sequences in Mobile Telemedicine Networks. *Wirel. Commun. Mob. Comput.* **2022**, *2022*, 6458350. [CrossRef]

 *diagnostics*

Article

An Adaptive Early Stopping Technique for DenseNet169-Based Knee Osteoarthritis Detection Model

Bander Ali Saleh Al-rimy [1,*], Faisal Saeed [2], Mohammed Al-Sarem [3], Abdullah M. Albarrak [4] and Sultan Noman Qasem [4]

1 Department of Computer Science, Faculty of Computing, Universiti Teknologi Malaysia, Johor Bahru 81310, Malaysia
2 DAAI Research Group, Department of Computing and Data Science, School of Computing and Digital Technology, Birmingham City University, Birmingham B4 7XG, UK; faisal.saeed@bcu.ac.uk
3 College of Computer Science and Engineering, Taibah University, Medina 41477, Saudi Arabia; msarem@taibahu.edu.sa
4 Computer Science Department, College of Computer and Information Sciences, Imam Mohammad Ibn Saud Islamic University (IMSIU), Riyadh 11432, Saudi Arabia; amsbarrak@imamu.edu.sa (A.M.A.); snmohammed@imamu.edu.sa (S.N.Q.)
* Correspondence: bander@utm.my

Abstract: Knee osteoarthritis (OA) detection is an important area of research in health informatics that aims to improve the accuracy of diagnosing this debilitating condition. In this paper, we investigate the ability of DenseNet169, a deep convolutional neural network architecture, for knee osteoarthritis detection using X-ray images. We focus on the use of the DenseNet169 architecture and propose an adaptive early stopping technique that utilizes gradual cross-entropy loss estimation. The proposed approach allows for the efficient selection of the optimal number of training epochs, thus preventing overfitting. To achieve the goal of this study, the adaptive early stopping mechanism that observes the validation accuracy as a threshold was designed. Then, the gradual cross-entropy (GCE) loss estimation technique was developed and integrated to the epoch training mechanism. Both adaptive early stopping and GCE were incorporated into the DenseNet169 for the OA detection model. The performance of the model was measured using several metrics including accuracy, precision, and recall. The obtained results were compared with those obtained from the existing works. The comparison shows that the proposed model outperformed the existing solutions in terms of accuracy, precision, recall, and loss performance, which indicates that the adaptive early stopping coupled with GCE improved the ability of DenseNet169 to accurately detect knee OA.

Keywords: knee OA detection; DenseNet169; early stopping; self-adaptive; GCE

 check for updates

Citation: Al-rimy, B.A.S.; Saeed, F.; Al-Sarem, M.; Albarrak, A.M.; Qasem, S.N. An Adaptive Early Stopping Technique for DenseNet169-Based Knee Osteoarthritis Detection Model. *Diagnostics* **2023**, *13*, 1903. https://doi.org/10.3390/diagnostics13111903

Academic Editor: Mugahed A. Al-antari

Received: 11 March 2023
Revised: 23 May 2023
Accepted: 25 May 2023
Published: 29 May 2023

1. Introduction

Osteoarthritis is a degenerative joint disorder that affects millions of people worldwide, leading to pain and loss of mobility in the affected joints [1,2]. Early detection and diagnosis of osteoarthritis are crucial for effective treatment, but traditional diagnostic methods can be time-consuming and invasive. In recent years, deep learning-based techniques have shown great potential in the early detection of knee osteoarthritis using medical imaging [2,3]. This type of approach can automate the analysis of radiographic images, reducing the dependence on subjective interpretations, and increasing the accuracy and consistency of diagnosis [4,5].

Deep learning is a powerful tool used for image analysis, pattern recognition, and decision making. It is based on the use of artificial neural networks, which are modelled after the human brain, and can learn from data [6]. In the context of knee osteoarthritis detection, deep learning algorithms can be trained to recognize patterns and features that are indicative of the disease in radiographic images [7,8]. The ability to automatically extract and analyze these features can provide a more objective and accurate diagnosis than

146

traditional methods [8,9]. Recent studies have shown that deep learning-based techniques can achieve high accuracy in the detection of knee osteoarthritis using X-ray and MRI images [9,10]. These results demonstrate the potential of deep learning-based techniques for the early detection of knee osteoarthritis and pave the way for the development of more accurate and efficient diagnostic tools.

However, it is worth noting that knee osteoarthritis detection using deep learning is still in the development phase, and more research is required to validate and improve these solutions before they become widely adopted in clinical practice [11]. One of the main challenges in deep learning-based solutions for osteoarthritis detection is overfitting, which occurs when a model is trained on a limited dataset and performs well on the training data but performs poorly on new, unseen data [12,13]. According to [14], the limited size and diversity of available datasets can lead to poor generalization and overfitting of the models. This can result in a high accuracy on the training set but a low accuracy on the test set and real-world data [15].

Increasing the number of epochs in a deep learning model can also lead to overfitting and affect the accuracy of the model in various ways [16]. Generally speaking, as the number of training epochs increases, the model will continue to learn from the training data, and the accuracy of the training set will typically increase [17]. However, as the model continues to learn, it may start to overfit the training data, resulting in a decrease in the accuracy on the test set or real-world data [18].

In the case of a DenseNet169 model, which is a type of convolutional neural network, increasing the number of epochs can lead to an improvement in the accuracy on the training set. The model will be able to learn more from the training data, and the weights and biases of the network will be adjusted to better fit the data. However, after a certain number of epochs, the accuracy on the validation set may start to decrease, indicating that the model has started to overfit [19].

It is worth noting that the optimal number of epochs will depend on the specific data, the architecture of the model, and the task at hand [20]. One way to determine the optimal number of epochs is to use techniques such as early stopping, which involves monitoring the performance of the model on a validation set and stopping the training when the performance starts to degrade [21]. By stopping the training before the model starts to overfit, early stopping can help to prevent overfitting and improve the accuracy of the model on new, unseen data.

Recent research has proposed various early stopping methods to improve the accuracy of DenseNet169-based models. For example, Ref. [21] proposed an early stopping strategy that monitors the performance of the model on the validation set and stops the training when the performance starts to degrade. In addition, Refs. [22,23] employed the early stopping regularization that monitors the performance of the model on the validation set and stops the training when the performance starts to degrade. In their study, Ref. [13] investigated knee OA early detection, and OA grading identification using deep learning. The researchers developed a new approach to classify data in deep learning models using the Laplace distribution-based strategy (LD-S) and created an aggregated multiscale dilated convolution network (AMD-CNN) to extract features from multivariate data of knee osteoarthritis (KOA) patients. They combined the AMD-CNN and LD-S to create a new KOA-CAD method that achieves three objectives in computer-aided diagnosis. Similarly, Ref. [24] introduced a new method for identifying knee osteoarthritis (KOA) in its early stages, which involves using deep learning to extract features from data and classify it. The algorithm being suggested utilizes X-ray images to both train and test the results. It uses hybrid feature descriptors, which extract features through combinations of CNN with HOG and CNN with LBP. The system employs three multi-classifiers to categorize diseases based on the KL grading system using KNN, RF, and SVM. However, there are several current research issues related to early stopping methods in deep learning models including the DenseNet169. One issue is determining the optimal stopping point [23]. The optimal stopping point will depend on the specific data, the architecture of the model, and the task at hand, and it is not clear yet how to determine it in an automated and general way that works well across different datasets and tasks [25,26]. Another issue is related to the trade-off between the accuracy and the generalization of the model. While early stopping can prevent overfitting, it can also result in underfitting if the model is stopped too early; this can lead to decreased accuracy on the test set or real-world data [27]. Therefore, there is a need to strike a balance between preventing overfitting and ensuring that the model has enough capacity to generalize well to new data. Additionally, the definition of performance degradation can vary depending on the dataset and the task, which makes it difficult to generalize early stopping methods across different datasets and tasks [28].

To this end, this study is devoted to investigating the applicability and efficacy of a novel adaptive early stopping technique in DenseNet169 in the context of knee OA detection. Our early stopping mechanism sets a patience threshold for early stopping by calculating the running average of the validation loss. In such a way, our technique can avoid arbitrary termination of the training. The proposed technique also embeds a novel gradual cross-entropy coefficient for accurate loss estimation during the early stopping of model training. The contribution of this paper is three-fold, as follows:

1. An adaptive early stopping technique was proposed for DenseNet169 that dynamically adjusts the number of epochs and the batch size during the training, based on the contribution of each batch to the accuracy of the model.
2. A gradual loss estimation method based on cross-entropy was proposed for measuring the dissimilarity between the predicted class probabilities and the true class labels.
3. An improved DenseNet169-based knee OA detection model which incorporates the techniques in (1) and (2) was developed and experimentally evaluated using the Knee Osteoarthritis Severity Grading dataset.

The rest of this paper is structured as follows. Section 2 provides the details on the methodology design and techniques proposed in this study. Section 3 describes the dataset and experimental environment used to carry out the model evaluation. It also explains the results and analytically discusses the findings from the experimental evaluation. Section 4 concludes the paper with suggestions for further research suggestions.

Related Works

The evolution in detecting and assessing the severity level of knee OA has seen a transition from traditional methods to the utilization of advanced machine learning and deep learning techniques. These include the use of complex network theory [11], circular Fourier filters [2], and deep learning algorithms to analyze radiographic knee X-ray images and aid in the early detection and diagnosis of the disease. A pivotal development in this research is the usage of a deep learning-based algorithm to automatically assess and grade the OA severity, often achieving comparable accuracy with expert radiologists [7,8]. In some instances, utilizing deep learning techniques on properly preprocessed images, such as through image sharpening, has resulted in improved accuracy rates [24]. Similarly, a semi-automatic model based on deep Siamese convolutional neural networks has been used to detect OA lesions according to the KL scale [29]. Furthermore, transfer learning has been deployed to aid the classification performance of models trained on imbalanced datasets.

With the advancement in deep learning architectures, new methodologies for OA severity assessment are introduced. A variety of deep-learning models have been proposed in the literature for diagnosing the severity of knee OA. For example, Ref. [30] leveraged a fully convolutional network (FCN) to locate knee joints and a deep convolutional neural network (CNN) to differentiate various stages of knee OA severity. Likewise, Ref. [31] introduced a technique using deep Siamese CNNs for automatic grading of knee OA severity following the KL grading scale, treating knee OA as a multi-class problem based on KL grades. Moreover, Ref. [32] presented a Discriminative Regularized Auto-Encoder (DRAE) for the early detection of knee OA, specifically differentiating between non-OA and minimal OA. The DRAE combines a discriminative loss function with the standard auto-encoder training criterion to improve the identification of knee OA.

Pre-trained deep learning models such as DenseNet and ResNet were also used in several studies for the knee OA severity level assessment. In particular, DenseNet201 was employed in [33] to develop knee OA grading. The model trains the DenseNet201 architecture on knee radiographic images from the OAI dataset. Using the Kellgren and Lawrence (KL) grading system, the model classifies the severity from grade 0 through grade 4. Similarly, the knee OA model proposed in [34] utilized DenseNet169. The model involves training the DenseNet169 using a balanced combination of two loss functions, categorical cross-entropy and mean squared error. This model inherently enables the prediction of knee OA severity on both an ordinal scale (0, 1, 2, 3, 4) and a continuous scale (0–4).

The study [35] proposed two novel learning structures, Deep Hybrid Learning-I (DHL-I) and Deep Hybrid Learning-II (DHL-II), both devised for efficient knee osteoarthritis (OA) severity classification based on Kellgren-Lawrence (KL) grades. DHL-I, based on a convolutional neural network (CNN), introduces a five-class prediction structure. This model is trained on knee X-ray images, then extracts features, applies principal component analysis (PCA) for dimensionality reduction, and then uses support vector machines (SVMs) for classification. DHL-II follows the same process, but the pre-trained CNN developed for DHL-I is fine-tuned using the concept of transfer learning to classify knee OA into four, three, and two classes.

When training a deep neural network model for assessing knee OA severity level, setting the appropriate number of training epochs and batch size per epoch often poses a challenge [36]. Overfitting might occur if too many epochs are used, while underfitting may result from too few epochs [37]. Training a neural network involves finding the right balance to avoid overfitting the training data. While adjusting the number of training epochs can help, it is computationally intensive and is not guaranteed to find an optimal value. Early stopping offers a more efficient solution [36]. This strategy involves training the model for many epochs, then halting the training when the model's performance on a validation dataset starts to decline, ensuring optimal generalization performance [37]. This can be achieved by setting a potentially large number of training epochs initially, and then halting the training process when there is no further improvement in the model's performance on the validation dataset.

Several studies have adopted early stopping for knee OA severity level assessment. A convolutional neural network with ResU-Net architecture (ResU-Net-18) was used in [38] to develop a Multiple-JSW for knee OA severity and progression. The model segments the knee X-ray images, and the minimum and multiple joint space widths (JSW) were estimated from this segmentation and verified against radiologist measurements. During ResU-Net-18 training, the early stopping mechanism was implemented. This technique ends the training if there is no reduction in the loss for 10 consecutive epochs, serving as a preventive measure against overfitting.

The study conducted by [39] developed a fully automated deep-learning model for assessing the severity of knee osteoarthritis (OA) using the Kellgren–Lawrence (KL) grading system. The algorithm was developed to use posterior–anterior (PA) and lateral (LAT) views of knee radiographs for this assessment. Early stopping was employed to halt the training before the model overfitted. The early stopping parameter was set to 20, which stops the training after 20 epochs. Nonetheless, identifying the ideal number of epochs poses a significant challenge. Similar to methods not employing early stopping, this approach could lead to overfitting if the early stopping criteria are set too high, and to underfitting if the criteria are set too low.

A pre-trained CNN model was also used in [29], which developed a semi-automatic computer-aided diagnosis (CAD) for detecting knee OA based on ResNet-34. The model used deep Siamese convolutional neural networks and a fine-tuned ResNet-34 to detect OA lesions in both knees based on the Kellgren and Lawrence (KL) scale. In order to balance the prevention of overfitting with maintaining model accuracy, an early stopping criterion was implemented. This stopped training when there was not any improvement in the validation accuracy observed after 50 epochs.

Although some of the existing knee OA grading models employ early stopping, they rely on statically set patience parameter values based on the number of epochs and the batch size. Such a static approach makes the model rigid and unable to adapt to the varying characteristics of the OA in X-ray images. If the patience value is set statically, it may not be optimal, as a value that is too high may lead to overfitting. This is because the model could continue training beyond the point of optimal generalization, learning the noise in the training data. On the other hand, a value that is too low may stop the training prematurely, leading to an underfit model that does not capture the underlying patterns in the data. Hence, it is crucial to find a balance and possibly employ dynamic strategies in setting the patience parameter for early stopping.

2. Materials and Methods

The methodology section of this paper describes the methods and procedures used to develop and evaluate the deep learning-based knee osteoarthritis (OA) detection method using X-ray images. We first present the dataset used in this study and the pre-processing steps applied to the images. Next, we describe the DenseNet169 architecture, and the fine-tuning process used to adapt the model to the knee OA detection task. We also describe the implementation details of the early stopping methods used to prevent overfitting and improve the accuracy of the model. Finally, we present the evaluation metrics and statistical analysis used to assess the performance of the proposed method. This section provides a detailed description of the steps taken to achieve the results and conclusions of this study, allowing for replication and further research on the topic.

2.1. X-ray Images Pre-Processing

The model development was carried out in three phases, i.e., pre-processing, training, and fine-tuning. At pre-processing phase, X-ray images underwent several procedures, namely, embedding, data augmentation, transposition, and flipping. Furthermore, vision transformer (ViT) was used to divide the input images into fixed-size patches and then positionally embed them into the transformer's encoder (TE). This step reduces the overhead on the model as it replaces the convolutions while maintaining a high level of accuracy. Concretely, the ViT takes an image, $x \in R(H \times W \times C)$, as input and turns it into a sequence of patches , $x_p \in R(N \times P \times P \times C)$, where (H, W) denotes the hight and width of the original image, C denotes the number of channels, (P, P) denotes the patch resolution, and N is the number of patches, which is calculated as follows.

$$N = \frac{WH}{P^2} \tag{1}$$

Then, the generated patches are embedded linearly into the TE. The TE uses a multi-head self-attention layer to control the embedding and generates a richer representation of image data. In particular, the self-attention layer consolidates the ability of TE to relate the sequence of inputs with each other.

2.2. An Adaptive Early Stopping for DenseNet169-Based Knee OA Detection Model

The model adopts the DenseNet169 architecture in which each layer is connected to every other layer [40,41]. The rationale behind this choice is that DenseNet169 architecture has far fewer trainable parameters compared to other architectures. Therefore, DenseNet169 helps to increase the depth of deep CNNs while avoiding information vanishing, which happens when the path between input and output layers becomes too big. By reducing the number of parameters, DenseNet169 gets rid of redundant feature maps, which, in turn, reduces the number of filters as well [42]. Figure 1 shows the architecture of DenseNet169.

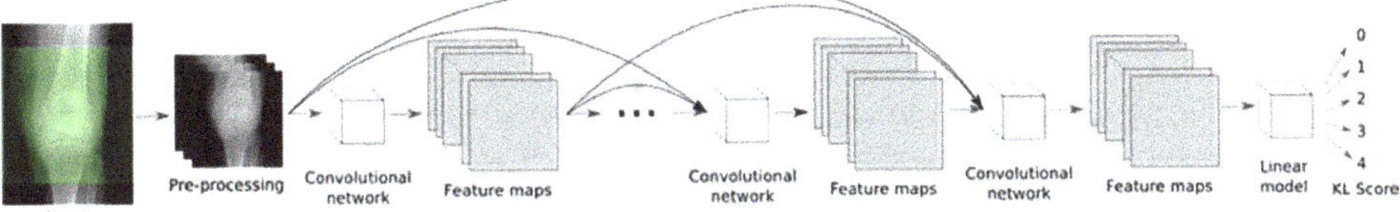

Figure 1. The architecture of DenseNet169.

The DenseNet169 architecture is composed of several types of layers including convolutional, maxpool, dense, and transition layers [41]. Moreover, the architecture uses two activation functions, namely, Relu and SoftMax. The former is used throughout the architecture, except for the final layer, in which SoftMax is used instead. The purpose of convolutional layers is to apply multiple filters to the X-ray image and generate a feature map that describes the intensity of the extracted features. Concretely, if we take an image input with L × N size followed by a convolutional layer and apply an m × m filter, the output of the convolution will be an (l − m + 1) × (l − n + 1).

The maxpool layer in DenseNet169 is then used to decrease the feature map size. To achieve that, a pooling filter is applied over the feature map, which aggregates the features in the area covered by the filter region. Concretely, a feature map with (n_h, n_w, n_c) dimensions can be reduced by applying the MaxPool technique as described in Equation 1 as follows:

$$\text{MaxPool} = \frac{n_c \times (n_h - f + 1) \times (n_w - f + 1)}{s^2} \tag{2}$$

where h denotes the height, w denotes the width, c denotes the channel of the feature map, and f denotes the size of the filter.

The dense layer in DenseNet169 architecture consists of nodes (neurons) that receive inputs from all nodes in the preceding layer. Those inputs undergo matrix–vector multiplication. Concretely, it is assumed that M is an $x \times y$ matrix, p is a $(1 \times y)$ vector, and the matrix λ of parameters of the preceding layer was learned using the backpropagation. Therefore, the weights (φ^{l_y}) and biases (η^{l_y}) associated with layer l_y can be calculated as follows:

$$\varphi^{l_y} = \varphi^{l_y} - \alpha \times d\varphi^{l_y} \tag{3}$$

$$\eta^{l_y} = \eta^{l_y} - \alpha \times d\eta^{l_y} \tag{4}$$

The $d\varphi^{l_y}$ and $d\eta^{l_y}$ are the partial derivatives of the loss function of φ and η. Finally, the transition layer decreases the model complexity by reducing the number of channels using 1 × 1 convolution. Table 1 shows the layered architecture of DenseNet169. It details the information for each layer, including the kernel size, tensor size, and used parameters. From the table, we can observe that Relu and SoftMax are used as activation functions. Moreover, the stride value (which determines the number of pixels that shift over the input matrix) was set to 2 in all convolutions, pooling, and transition layers. In addition, the dropout that helps in preventing overfitting and reduces the variance is set to 0.2 for all dense layers. We can also observe that the tensor size decreases by half when moving toward the output layer.

Table 1. DenseNet169 layered architecture.

Layer	Kernel Size	Parameters	Tensor Size
Convolution	$Conv = 7 \times 7$	Stride = 2, Relu	112 × 112
Pooling	$MaxPool = 3 \times 3$	Stride = 2	56 × 56
Dense1	$Conv = 1 \times 1 \times 6$ $Conv = 3 \times 3 \times 6$	Dropout = 0.2	56 × 56

Table 1. *Cont.*

Layer	Kernel Size	Parameters	Tensor Size
Transition 1	$Conv = 1 \times 1$ $AvgPool = 2 \times 2$	Stride = 2	56×56 28×28
Dense 2	$Conv = 1 \times 1 \times 12$ $Conv = 3 \times 3 \times 12$	Dropout = 0.2	28×28
Transition 2	$Conv = 1 \times 1$ $AvgPool = 2 \times 2$	Stride = 2	28×28 14×14
Dense 3	$Conv = 1 \times 1 \times 32$ $Conv = 3 \times 3 \times 32$	Dropout = 0.2	14×14
Transition 2	$Conv = 1 \times 1$ $AvgPool = 2 \times 2$	Stride = 2	14×14 7×7
Dense 4	$Conv = 1 \times 1 \times 32$ $Conv = 3 \times 3 \times 32$	Dropout = 0.2	7×7
Classification	$AvgPool = 1 \times 1$ 1000D (fully connected SoftMax)		1×1

2.3. An Adaptive Early Stopping Technique

Unlike classical sequential models, our model dynamically adjusts the number of epochs as well as the number of steps per epoch (batch size) during the training, based on the contribution of each batch to the accuracy of the model. An early stopping mechanism was incorporated into the feedforward and backpropagation during the model training. This mechanism solves the issue of identifying an appropriate number of training epochs as well as batch size per epoch. The early stopping allows the model to start with arbitrary values for both parameters and stops the training when no further improvement happens at both levels. During the model's training, the early stopping mechanism monitors one or more performance measures based on which the training can be aborted. In our study, we monitor the loss on the validation set. The model stops the training when no further decrement is achieved in the validation loss. To avoid immature early stopping, we set a patience threshold as a baseline value calculated using the running average of the loss difference (ε). Equation (5) was used to calculate the value of the patience parameter. The equations show that the value is updated at every step within the epoch based on the average of previous values, which avoids arbitrary stopping.

$$Patience = \frac{avg\left(\varepsilon_{t_{i-1}}\right) + \varepsilon_{t_i}}{i+1} \tag{5}$$

$$\varepsilon = t_i - t_{i-1} \tag{6}$$

where t_i denotes the ith value of the observed measure.

The model waits until the threshold's value is satisfied, then triggers the early stopping. Such a controlling mechanism relies on two parameters, a global parameter (macro controller), and a local parameter (micro controller). On the one hand, the macro controlling parameter aborts the training when a set of preceding and current epochs make no improvement to the accuracy. On the other hand, the micro-controlling parameter aborts the running epoch at the time when it detects that no further improvement to the accuracy is made during that epoch. Therefore, model training takes less time and uses fewer resources.

However, dropping part of the data on the macro and micro level could deprive the model of valuable data located that would have been used in later epochs. To mitigate such drawback of early stopping, an improved loss function technique with the ability to compensate for potentially lost data was developed.

2.4. A Gradual Cross-Entropy Loss Estimation Technique

As pointed out above, the existing loss function techniques rely on the entire data allocated for the epoch to calculate the loss. However, the early stopping aborts the epoch execution and drops a portion of training data. Consequently, the loss estimation is negatively affected. To address the effect of early stopping on the accuracy of the loss estimation, our study proposes a gradual cross-sntropy (GCE) technique which improves the loss estimation at the micro (epoch) level. Unlike the existing loss calculation techniques that consider all epoch data, the GCE calculates the loss based on only the portion of the data that was consumed in the epoch until the moment of abortion. Intuitively, the early stopping at the micro level discards the remaining data in the batch allocated for the current epoch. Therefore, it is necessary to exclude the discarded data from the loss calculation. Concretely, Equation (7) shows that the entropy value is divided by the total number of examples (N) in the epoch.

$$J(w) = -\frac{1}{N}\sum_{i=1}^{N}\sum_{c=1}^{C} 1_{y_{i\epsilon}C_c} \log_{p_{model}}[y_i\epsilon C_c] \tag{7}$$

where C denotes the category (class) and $1_{y_{i\epsilon}C_c}$ denotes the ith observation that belongs to the cth category. Such a calculation overlooks the effect of early stopping. To rectify such a drawback, our study introduces a gradual weighting coefficient δ into the loss function as shown in Equation (8). The calculation of δ value is shown in Equation (9).

$$J(w) = -\frac{\delta}{N}\sum_{i=1}^{N}\sum_{c=1}^{C} 1_{y_{i\epsilon}C_c} \log_{p_{model}}[y_i\epsilon C_c] \tag{8}$$

$$\delta = \frac{N}{l} \tag{9}$$

where l denotes the number of examples that were consumed so far during the current epoch. The coefficient δ reduces the weight of N according to the actual number of training examples that were used during the epoch. If an epoch stops early, the loss calculation is carried out based on the consumed data only. Therefore, the accuracy of loss estimation is improved, which consequently improves the accuracy of the model. The GCE is integrated into the detection model and used during the training phase to support the feed-forward and backpropagation.

2.5. Dataset Description

In this study, the Knee Osteoarthritis Severity Grading dataset is used to train and evaluate the performance of the proposed model. It contains knee X-ray images for OA detection and KL grading. Five gradings constitute the dataset labels as follows: healthy knee image (grade 0), doubtful joint space narrowing (JSN) with possible OA (grade 1 or healthy), confirmed OA and possible joint space narrowing (grade 2 or minimal), multiple moderate OA with confirmed JSN and mild sclerosis (grade 3 or moderate), and large OA with significant JSN and severe sclerosis (grade 4 or severe). The data are distributed based on the grades, such that there are 604 images for grade 0, 275 images for grade 1, 403 files for grade 2, 200 images for grade 3, and 44 images for grade 4. Figure 2 shows samples of images with various labels.

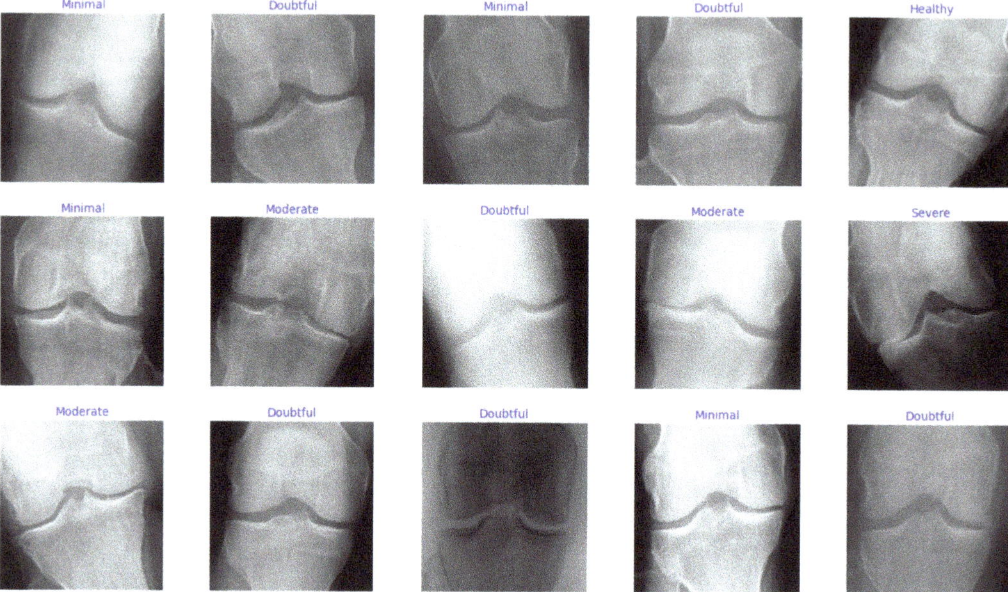

Figure 2. Samples of images in the dataset.

From the data distribution illustrated above, it can be seen that the dataset has a class imbalance, which might lead to classification bias toward the majority label. To mitigate this drawback, data augmentation was used to balance the training set so that each class contains 500 samples. Several techniques were employed to conduct the augmentation, including flipping, rotation, shifting, and zooming. Table 2 shows the augmentation parameter customization used in this study, which was determined experimentally.

Table 2. Augmentation optimization parameters.

Parameter	Value
horizontal_flip	True
rotation_range	25
width_shift_range	0.22
height_shift_range	0.23
zoom_range	0.25

The augmented samples were then added to the dataset and used for training the detection model. The dataset was divided into three subsets, training, validation, and testing, using the cross-validation method by which the data were sampled randomly, and the five labels were represented in all subsets. Table 3 shows the data distribution among the three subsets after conducting the K-fold cross-validation split.

Table 3. Data distribution among the training, validation, and testing subsets.

Subset Name	Number of Samples
Training set	2500
Validation set	826
Testing set	1656

2.6. Development and Evaluation Environment

This experiment was carried out in an MS Windows 10 machine with 16 GB RAM, 12th Gen Intel Core i 7, 4.7 GHz, and NVIDIA 1050 Ti GPU. Python with several libraries such as Tensor flow, Keras, Pandas, Numpy, Matplotlib, and Sci-kit learn was used to develop the DenseNet169 DL model.

2.7. Evaluation Metrics

This study makes use of the confusion matrix to evaluate the performance of the proposed model. On one side, the matrix shows the actual values, and on another side, it shows the predicted values. Then, the ratio of true positive, true negative, false positive, and false negative can be deduced. Several metrics were used to measure the performance of the model including accuracy, *F1* score, loss rate, precision, and recall. The following equations are used to calculate these metrics for multi-class classification:

$$Accuracy = \frac{\sum TP + \sum TN}{\sum TP + \sum TN + \sum FP + \sum FN} \tag{10}$$

$$Recall = \frac{\sum TP}{\sum TP + \sum FN} \tag{11}$$

$$Precision = \frac{\sum TP}{\sum TP + \sum FP} \tag{12}$$

$$F1\ Score = \frac{2 \times Pr \times Re}{Pr + Re} \tag{13}$$

where *TP*, *TN*, *FP*, and *FN* represent the true positive, true negative, false positive, and false negative, respectively.

3. Results and Discussion

In this section, the experimental results are detailed. The experiments were conducted on the training dataset in three rounds. During the first round, we built a multi-class classifier using the five labels in the dataset. In the second round, another multi-class classifier was built using only three class labels. To achieve this, the class labels in the dataset were categorized into three classes, including healthy, moderate, and severe. Lastly, we built a binary classifier where the labels were put under two categories, healthy and unhealthy. The purpose of such multi-round training is to investigate the effect of multi-classes on the accuracy of the model. This helps to determine whether increasing the classes affects the ability of DenseNet169 to detect the stage of OA.

Table 4 summarizes the performance of the proposed model with respect to accuracy, F1 score, precision, and recall. The model applies the adaptive early stopping when the training process does not make any further improvements, which helps to prevent overfitting. The results show that the five-class classification achieved 0.62 accuracies, 0.65 F score, 0.58 precision, and 0.61 recall. For the three-class classification, the results show an increase in the accuracy to 0.93, F1 score to 0.90, precision to 0.91, and recall to 0.91. When we used the binary class classification, the results were increased to 0.94 for accuracy, F1 score, precision, and recall. The confusion matrix for the three classification tasks was used to calculate the performance metrics (accuracy, recall, precision and F1 score). The horizontal side of those matrices represents the actual labels, while the vertical side represents the predicted labels. The intersection between the actual and predicted labels determines the performance of the model as to whether it generates true positives (TP), true negatives (TN), false positives (FP), or false negatives (FN). Based on such a prediction, the accuracy of the model was calculated.

Table 4. The performance of the proposed model with respect to the accuracy, F1 score, precision, and recall.

	Precision	Recall	F1 Score	Accuracy
2-class	0.9456	0.9469	0.9449	0.9408
3-class	0.9315	0.9058	0.9132	0.9179
5-class	0.5995	0.6220	0.6059	0.6274

Figure 3 shows the training and validation performance of the model for the three-class tasks (5-class (a), 3-class (b), and 2-class (c)) over the training epochs. It also shows the best fit where the training and validation curves intersect. It can be observed that the loss decreases in both the training and validation sets when the number of epochs increases. In the three classification tasks (i.e., five-class, three-class, and two-class classification), it can be noticed that the training loss was higher than the validation loss at the early epochs. While the training loss continues to decrease at the late epochs during the five-class training (3:a), the validation loss curve tends to flatten, which indicates that the loss does not improve by increasing the epochs. However, during the three-class and two-class classifications, both the training and validation losses overlap most of the time. The loss curves also show the effect of the adaptive early stopping technique as the training stops at 12 epochs (five-class), 13 epochs (three-class), and 14 epochs (two-class) when the model detects no more improvement on the validation set.

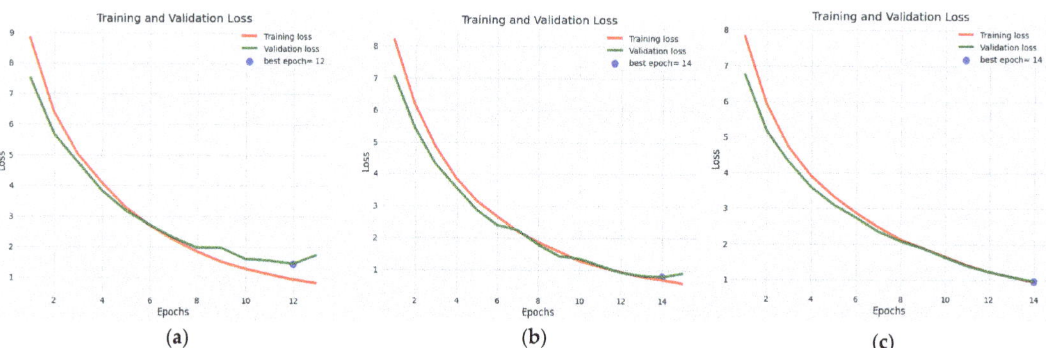

Figure 3. Training and validation performance of the model for the 5-class (**a**), 3-class (**b**), and 2-class (**c**) tasks over the training epochs.

Figure 4 shows the comparison between five-class classification, three-class classification, and the binary classification that we conducted using our proposed model. The results were taken after each epoch using the validation set. It can be observed that the binary classification achieved the highest accuracy, while the accuracy of the five-class classification was the lowest. Moreover, it can be observed that the training accuracy increases when the number of epochs increases, until the number of epochs reaches 25, where we can see that the increase becomes less gradual. Furthermore, the comparison shows that the validation accuracy was not stable and oscillated around 0.6. This indicates that the data with five-class labels negatively affect the accuracy when new data are introduced to the model. The reason behind this drop in the model's accuracy could be the overlapping between the class boundaries, which makes it difficult for the model to distinguish between the characteristics of those classes. The high training accuracy confirms such claim as the model overfits the training examples. In contrast, the validation accuracy for three-class classification and binary classification increased to around 0.9. This means that the model performed well when the target label was less granular but dropped when the labeling became more specific. This could be due to the inability of the model to explore discriminative features that represent the fine granular labels (the five-class case). One potential solution is to embed an attention layer as a feature selection mechanism into the model structure so that it can focus on a set of features relevant to the target classes.

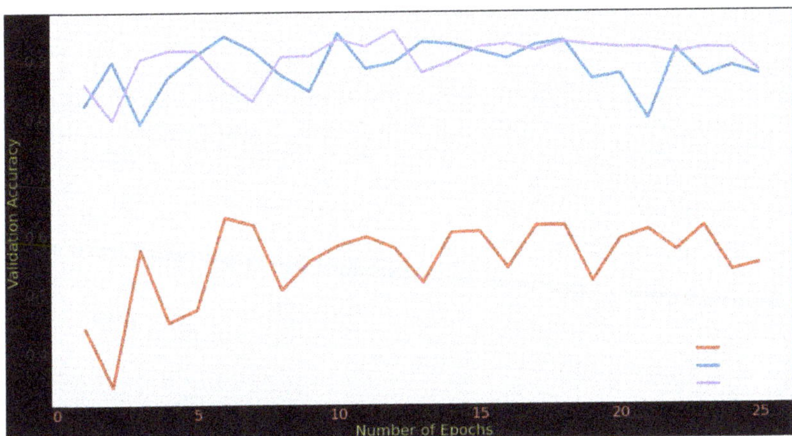

Figure 4. Adaptive DenseNet169 accuracy trending based on the number of epochs for 2-class, 3-class, and 5-class classification.

Tables 5–7 show a comparison between the proposed adaptive DenseNet169 model with the results of existing works. The comparison is also visualized in Figures 5–7. The comparison was conducted between the proposed adaptive DenseNet169 and the standard DenseNet169. We also compared the performance of the proposed model with the existing studies related to pre-trained models for knee OA detection with early stopping capabilities, namely, AMD-CNN [13], deep CNN [24], DHL-II [35], and ResNet-34 [29]. In this comparison, we used several metrics, namely, precision, recall, F1 score, and accuracy. By comparing the results obtained from our model with those obtained by related works, it can be observed that the proposed model outperformed the previous models in terms of accuracy, recall, and precision. In the comparison, the proposed model as well as the models developed by existing works were trained using the same number of epochs. In our model, the adaptive early stopping was applied at a batch level, in which the model aborts the training of the respective epoch if new instances have little or no contributions to improving the validation accuracy.

Table 5. Comparison between the performance of the proposed model with the related models for the 2-class classification.

	Standard DenseNet169	[13]	[24]	DHL II [35]	ResNet [29]	Adaptive DenseNet169
Precision	0.936	0.9298	0.9043	0.9	0.896	**0.9456**
Recall	0.9354	0.9155	0.8753	0.904	0.902	**0.9469**
F1 score	0.9371	0.9197	0.8889	0.92	0.907	**0.9449**
Accuracy	0.9354	0.9155	0.9358	0.917	0.9	**0.9408**

Table 6. Comparison between the performance of the proposed model with the related models for 3-class classification.

	Standard DenseNet169	[13]	[24]	DHL II [35]	ResNet [29]	Adaptive DenseNet169
Precision	0.9241	0.9107	0.9132	0.867	0.86	**0.9315**
Recall	0.8954	0.884	0.892	0.88	0.875	**0.9058**
F1 score	0.9075	0.9021	0.9027	0.898	0.884	**0.9132**
Accuracy	0.9054	0.8979	0.8979	0.893	0.881	**0.9179**

Table 7. Comparison between the performance of the proposed model with the related models for the 5-class classification.

	Standard DenseNet169	[13]	[24]	DHL II [35]	ResNet [29]	Adaptive DenseNet169
Precision	0.5822	0.5716	0.5873	0.581	0.583	**0.5995**
Recall	0.6059	0.5921	0.6174	0.604	0.61	**0.622**
F1 score	0.587	0.5735	0.5951	0.596	0.602	**0.6059**
Accuracy	0.6059	0.5921	0.6074	0.617	0.607	**0.6274**

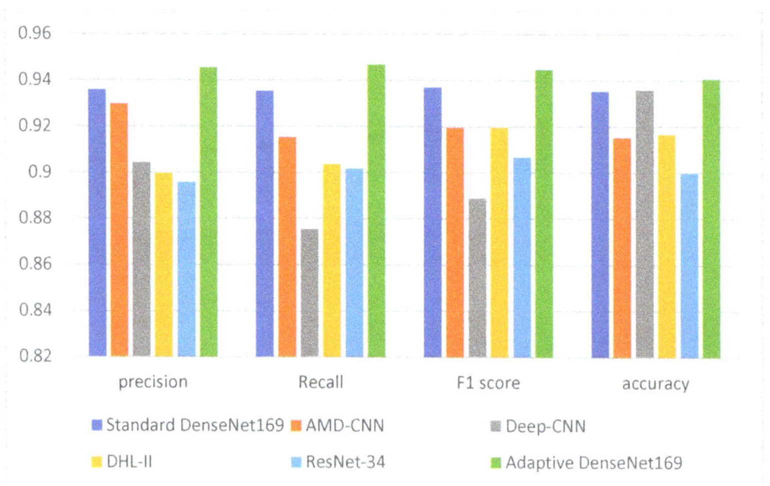

Figure 5. Comparison between the performance of the proposed model with the related models (Standard Densenet169, AMD-CNN [13], Deep-CNN [24], DHL-II [35], ResNet [29]) for the 2-class classification.

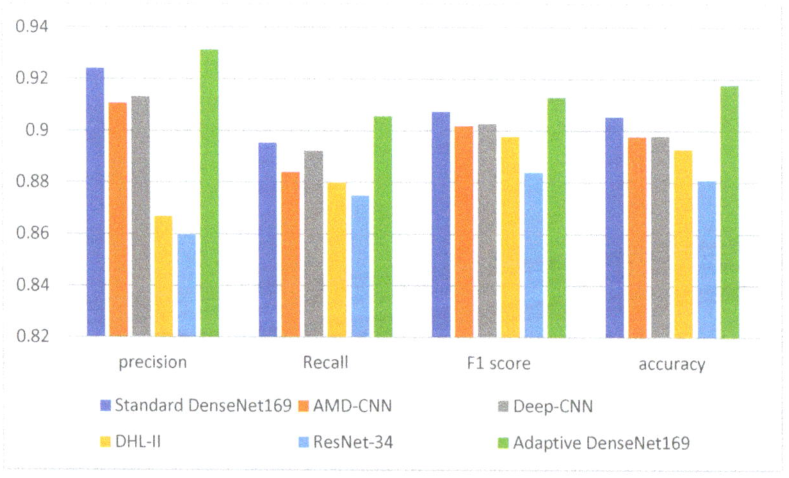

Figure 6. Comparison between the performance of the proposed model with the related models (Standard Densenet169, AMD-CNN [13], Deep-CNN [24], DHL-II [35], ResNet [29]) for the 3-class classification.

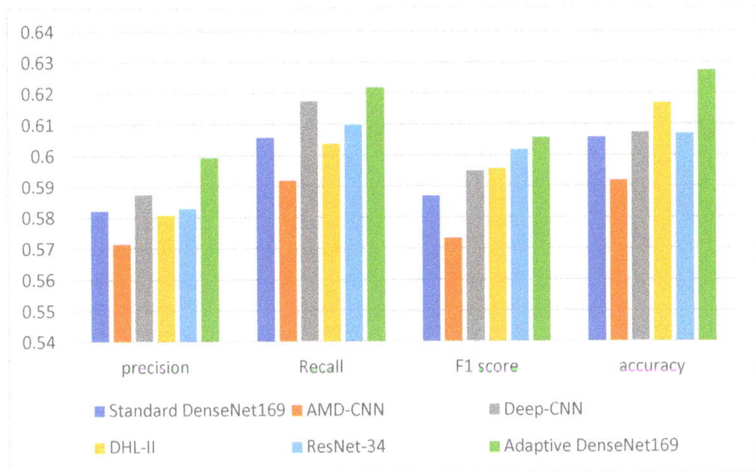

Figure 7. Comparison between the performance of the proposed model with the related models (Standard Densenet169, AMD-CNN [13], Deep-CNN [24], DHL-II [35], ResNet [29]) for the 5-class classification.

The improved performance (in terms of precision, recall, F1 score, and accuracy) that our proposed adaptive DenseNet169 model shows over existing models in the two-class, three-class, and five-class classifications can be primarily credited to the incorporation of adaptive early stopping and the use of the gradual cross-entropy (GCE) loss estimation technique. Early stopping allows our model to run an adequate number of epochs with an appropriate batch size for each classification, preventing overfitting and avoiding premature termination of training when the epochs and batch sizes are underestimated. This adaptability allows for the dynamic readjustment of the patience parameter, which ensures optimal data utilization, consequently leading to maximized accuracy.

This improved performance is attributed to the efficacy of the GCE technique, which adaptively tunes the patience parameter based on the validation loss at the epoch level. Unlike conventional methods that solely depend on the number of epochs for setting the patience parameter, our model incorporates GCE to base the loss estimate on the data processed at the epoch level before early stopping. This technique negates the influence of discarded data on the loss calculation, ensuring more precise loss estimation. Hence, these combined strategies allow our model to adapt and learn more effectively and accurately, resulting in its improved performance over the existing models.

It can also be observed that the five-class classification achieved the lowest performance by all models across all metrics (precision, recall, F1 score, and accuracy). This can stem from the class imbalance, as the class "Healthy" has the highest number of images (2286), and the class "Severe" has the least number of images (173). This creates a significant discrepancy between the classes, and, as a result, the model might become biased toward the "Healthy" class, simply because it encounters more examples of this class during training, making it less capable of accurately detecting and differentiating between the less represented "Doubtful", "Minimal", "Moderate", and "Severe" classes. To address this issue, multimodal deep learning can be an effective solution, as it leverages multiple types of data input, such as combining image data with structured clinical data. For example, the model could be trained on both X-ray images and corresponding clinical data such as patient age, weight, gender, pain levels, and other relevant health metrics. By integrating these additional data sources, the model could learn more complex representations and dependencies, leading to more accurate OA severity predictions. However, collecting and integrating diverse types of knee OA-related data (such as images, text, structured clinical data, etc.) can be challenging due to data privacy and protection regulations such as the Health Insurance

Portability and Accountability Act (HIPPA). These regulations, either mandated by local or federal jurisdictions, strictly control the access to and usage of personal health information, thereby adding a layer of complexity to the data collection process for multimodal training. In addition, ensuring accurate alignment across different types of data is essential and non-trivial. For successful and accurate multimodal training, it is critical to ensure that all data types—the images, the textual information, and the structured clinical data—correctly correspond to the same entity, such as a patient. This alignment guarantees that the integrated data maintain their contextual relevance, thereby enabling the model to develop a coherent understanding of the information. Moreover, it can be difficult to understand which modality is contributing to the predictions and how they are interacting with each other. These challenges can be investigated further in future studies. Researchers could delve deeper into these issues, developing innovative solutions to streamline the alignment process across different data types, and enhance the interpretability of multimodal models.

4. Conclusions

In this study, we present a novel approach to improve the performance of DenseNet169-based knee osteoarthritis detection using X-ray images. Our approach utilizes an adaptive early stopping technique coupled with gradual cross-entropy loss estimation. We have shown that our approach improved the accuracy of knee osteoarthritis detection when compared to traditional early stopping techniques. Our results demonstrate that the proposed approach can lead to more accurate and efficient diagnostic tools for knee osteoarthritis. This study also investigates the effect of several types of classification on detection accuracy and shows that fewer classes generate accurate predictions. It is important to note that our approach is not without limitations. Further research is needed to investigate the generalizability of our method to other types of imaging modalities and to different types of osteoarthritis. Additionally, more efforts are needed to improve the model for multi-class classification when the number of classes increases. This is crucial for diagnosing the development of OA and identifying what stage the disease is at. The incorporation of other types of information, such as clinical data, may further improve the performance of the proposed method. Despite these limitations, our results are a promising step toward the development of more effective deep learning-based diagnostic tools for knee osteoarthritis.

Author Contributions: Conceptualization, B.A.S.A.-r., F.S., A.M.A. and S.N.Q.; Methodology, B.A.S.A.-r., A.M.A. and S.N.Q.; Software, B.A.S.A.-r. and M.A.-S.; Validation, F.S. and M.A.-S.; Formal analysis, B.A.S.A.-r. and S.N.Q.; Investigation, B.A.S.A.-r., A.M.A. and S.N.Q.; Resources, F.S., A.M.A. and S.N.Q.; Data curation, M.A.-S.; Writing—original draft, B.A.S.A.-r., A.M.A. and S.N.Q.; writing—review and editing, B.A.S.A.-r., F.S. and M.A.-S.; Visualization, F.S. and M.A.-S.; Project administration, A.M.A. and S.N.Q.; Funding acquisition, A.M.A. and S.N.Q. All authors have read and agreed to the published version of the manuscript.

Funding: This paper is funded by the Deanship of Scientific Research at Imam Mohammad Ibn Saud Islamic University, Research Group No. RG-21-07-05.

Institutional Review Board Statement: Not applicable.

Informed Consent Statement: Not Applicable.

Data Availability Statement: Not applicable.

Acknowledgments: The authors extend their appreciation to the Deanship of Scientific Research at Imam Mohammad Ibn Saud Islamic University for funding this work through Research Group No. RG-21-07-05.

Conflicts of Interest: The authors declare no conflict of interest.

References

1. Lim, J.; Kim, J.; Cheon, S. A deep neural network-based method for early detection of osteoarthritis using statistical data. *Int. J. Environ. Res. Public Health* **2019**, *16*, 1281. [CrossRef] [PubMed]
2. Brahim, A.; Jennane, R.; Riad, R.; Janvier, T.; Khedher, L.; Toumi, H.; Lespessailles, E. A decision support tool for early detection of knee OsteoArthritis using X-ray imaging and machine learning: Data from the OsteoArthritis Initiative. *Comput. Med. Imaging Graph.* **2019**, *73*, 11–18. [CrossRef] [PubMed]
3. Chen, P.; Gao, L.; Shi, X.; Allen, K.; Yang, L. Fully automatic knee osteoarthritis severity grading using deep neural networks with a novel ordinal loss. *Comput. Med. Imaging Graph.* **2019**, *75*, 84–92. [CrossRef] [PubMed]
4. Kokkotis, C.; Moustakidis, S.; Papageorgiou, E.; Giakas, G.; Tsaopoulos, D.E. Machine learning in knee osteoarthritis: A review. *Osteoarthr. Cartil. Open* **2020**, *2*, 100069. [CrossRef]
5. Gan, H.S.; Ramlee, M.H.; Al-Rimy, B.A.S.; Lee, Y.S.; Akkaraekthalin, P. Hierarchical Knee Image Synthesis Framework for Generative Adversarial Network: Data From the Osteoarthritis Initiative. *IEEE Access* **2022**, *10*, 55051–55061. [CrossRef]
6. Tiulpin, A.; Saarakkala, S. Automatic grading of individual knee osteoarthritis features in plain radiographs using deep convolutional neural networks. *Diagnostics* **2020**, *10*, 932. [CrossRef]
7. Teoh, Y.X.; Lai, K.W.; Usman, J.; Goh, S.L.; Mohafez, H.; Hasikin, K.; Qian, P.; Jiang, Y.; Zhang, Y.; Dhanalakshmi, S. Discovering knee osteoarthritis imaging features for diagnosis and prognosis: Review of manual imaging grading and machine learning approaches. *J. Healthc. Eng.* **2022**, *2022*, 4138666. [CrossRef]
8. Zebari, D.A.; Sadiq, S.S.; Sulaiman, D.M. Knee Osteoarthritis Detection Using Deep Feature Based on Convolutional Neural Network. In Proceedings of the 2022 International Conference on Computer Science and Software Engineering (CSASE), Duhok, Iraq, 15–17 March 2022.
9. Ahmed, S.M.; Mstafa, R.J. A comprehensive survey on bone segmentation techniques in knee osteoarthritis research: From conventional methods to deep learning. *Diagnostics* **2022**, *12*, 611. [CrossRef]
10. Kale, R.D.; Khandelwal, S. A Review on: Deep Learning and Computer Intelligent Techniques Using X-Ray Imaging for the Early Detection of Knee Osteoarthritis. In *Machine Learning, Image Processing, Network Security and Data Sciences: Proceedings of the 4th International Conference, MIND 2022, Virtual Event, 19–20 January 2023*; Proceedings, Part I; Springer: Berlin/Heidelberg, Germany, 2023.
11. Ribas, L.C.; Riad, R.; Jennane, R.; Bruno, O.M. A complex network based approach for knee Osteoarthritis detection: Data from the Osteoarthritis initiative. *Biomed. Signal Process. Control* **2022**, *71*, 103133. [CrossRef]
12. Üreten, K.; Maraş, H.H. Automated classification of rheumatoid arthritis, osteoarthritis, and normal hand radiographs with deep learning methods. *J. Digit. Imaging* **2022**, *35*, 193–199. [CrossRef]
13. Song, J.; Zhang, R. A novel computer-assisted diagnosis method of knee osteoarthritis based on multivariate information and deep learning model. *Digit. Signal Process.* **2023**, *133*, 103863. [CrossRef]
14. Figueiredo, J.; Santos, C.P.; Moreno, J.C. Automatic recognition of gait patterns in human motor disorders using machine learning: A review. *Med. Eng. Phys.* **2018**, *53*, 1–12. [CrossRef]
15. Wambugu, N.; Chen, Y.; Xiao, Z.; Tan, K.; Wei, M.; Liu, X.; Li, J. Hyperspectral image classification on insufficient-sample and feature learning using deep neural networks: A review. *Int. J. Appl. Earth Obs. Geoinf.* **2021**, *105*, 102603. [CrossRef]
16. Perin, G.; Buhan, I.; Picek, S. Learning when to stop: A mutual information approach to fight overfitting in profiled side-channel analysis. *Cryptology ePrint Archive* **2020**. Available online: https://eprint.iacr.org/2020/058 (accessed on 10 March 2023).
17. Yaqub, M.; Feng, J.; Zia, M.S.; Arshid, K.; Jia, K.; Rehman, Z.U.; Mehmood, A. State-of-the-art CNN optimizer for brain tumor segmentation in magnetic resonance images. *Brain Sci.* **2020**, *10*, 427. [CrossRef]
18. Chang, Y.L.; Tan, T.H.; Lee, W.H.; Chang, L.; Chen, Y.N.; Fan, K.C.; Alkhaleefah, M. Consolidated convolutional neural network for hyperspectral image classification. *Remote Sens.* **2022**, *14*, 1571. [CrossRef]
19. Chauhan, T.; Palivela, H.; Tiwari, S. Optimization and fine-tuning of DenseNet model for classification of COVID-19 cases in medical imaging. *Int. J. Inf. Manag. Data Insights* **2021**, *1*, 100020. [CrossRef]
20. Gesmundo, A.; Dean, J. munet: Evolving pretrained deep neural networks into scalable auto-tuning multitask systems. *arXiv Prepr.* **2022**, arXiv:2205.10937.
21. Heckel, R.; Yilmaz, F.F. Early stopping in deep networks: Double descent and how to eliminate it. *arXiv Prepr.* **2020**, arXiv:2007.10099.
22. Kumar, S.; Janet, B. DTMIC: Deep transfer learning for malware image classification. *J. Inf. Secur. Appl.* **2022**, *64*, 103063. [CrossRef]
23. Kim, B.; Han, M.; Baek, J. A Convolutional Neural Network-Based Anthropomorphic Model Observer for Signal Detection in Breast CT Images Without Human-Labeled Data. *IEEE Access* **2020**, *8*, 162122–162131. [CrossRef]
24. Mahum, R.; Rehman, S.U.; Meraj, T.; Rauf, H.T.; Irtaza, A.; El-Sherbeeny, A.M.; El-Meligy, M.A. A Novel Hybrid Approach Based on Deep CNN Features to Detect Knee Osteoarthritis. *Sensors* **2021**, *21*, 6189. [CrossRef] [PubMed]
25. Zhang, X.; Wang, H.; Wu, B.; Zhou, Q.; Hu, Y. A novel data-driven method based on sample reliability assessment and improved CNN for machinery fault diagnosis with non-ideal data. *J. Intell. Manuf.* **2023**, *34*, 2449–2462. [CrossRef]
26. Hung, J.; Carpenter, A. Applying faster R-CNN for object detection on malaria images. In Proceedings of the IEEE Conference on Computer Vision and Pattern Recognition Workshops, Honolulu, HI, USA, 19–20 June 2017.
27. Rajpura, P.S.; Bojinov, H.; Hegde, R.S. Object detection using deep cnns trained on synthetic images. *arXiv Prepr.* **2017**, arXiv:1706.06782.

28. Liang, S.; Zhang, R.; Liang, D.; Song, T.; Ai, T.; Xia, C.; Xia, L.; Wang, Y. Multimodal 3D DenseNet for IDH genotype prediction in gliomas. *Genes* **2018**, *9*, 382. [CrossRef] [PubMed]

29. Cueva, J.H.; Castillo, D.; Espinós-Morató, H.; Durán, D.; Díaz, P.; Lakshminarayanan, V. Detection and Classification of Knee Osteoarthritis. *Diagnostics* **2022**, *12*, 2362. [CrossRef]

30. Antony, J.; McGuinness, K.; Moran, K.; O'Connor, N.E. Automatic detection of knee joints and quantification of knee osteoarthritis severity using convolutional neural networks. In *Machine Learning and Data Mining in Pattern Recognition: Proceedings of the 13th International Conference, MLDM 2017, New York, NY, USA, 15–20 July 2017, Proceedings 13*; Springer International Publishing: Berlin/Heidelberg, Germany, 2017; pp. 376–390.

31. Tiulpin, A.; Thevenot, J.; Rahtu, E.; Lehenkari, P.; Saarakkala, S. Automatic knee osteoarthritis diagnosis from plain radiographs: A deep learning-based approach. *Sci. Rep.* **2018**, *8*, 1–10. [CrossRef]

32. Nasser, Y.; Jennane, R.; Chetouani, A.; Lespessailles, E.; El Hassouni, M. Discriminative Regularized Auto-Encoder for early detection of knee osteoarthritis: Data from the osteoarthritis initiative. *IEEE Trans. Med. Imaging* **2020**, *39*, 2976–2984. [CrossRef]

33. Chaugule, S.V.; Malemath, V.S. Knee Osteoarthritis Grading Using DenseNet and Radiographic Images. *SN Comput. Sci.* **2022**, *4*, 63. [CrossRef]

34. Abd El-Ghany, S.; Elmogy, M.; Abd El-Aziz, A.A. A fully automatic fine tuned deep learning model for knee osteoarthritis detection and progression analysis. *Egypt. Inform. J.* **2023**, *24*, 229–240. [CrossRef]

35. Ahmed, S.M.; Mstafa, R.J. Identifying Severity Grading of Knee Osteoarthritis from X-ray Images Using an Efficient Mixture of Deep Learning and Machine Learning Models. *Diagnostics* **2022**, *12*, 2939. [CrossRef]

36. Song, H.; Kim, M.; Park, D.; Lee, J.G. How does early stopping help generalization against label noise? *arXiv Prepr.* **2019**, arXiv:1911.08059.

37. Bentoumi, M.; Daoud, M.; Benaouali, M.; Taleb Ahmed, A. Improvement of emotion recognition from facial images using deep learning and early stopping cross validation. *Multimed. Tools Appl.* **2022**, *81*, 29887–29917. [CrossRef]

38. Cheung, J.C.W.; Tam, A.Y.C.; Chan, L.C.; Chan, K.; Wen, C. Superiority of multiple-joint space width over minimum-joint space width approach in the machine learning for radiographic severity and knee osteoarthritis progression. *Biology* **2021**, *10*, 1107. [CrossRef]

39. Swiecicki, A.; Li, N.; O'Donnell, J.; Said, N.; Yang, J.; Mather, R.C.; Mazurowski, M.A. Deep learning-based algorithm for assessment of knee osteoarthritis severity in radiographs matches performance of radiologists. *Comput. Biol. Med.* **2021**, *133*, 104334. [CrossRef]

40. Zhang, W.; He, X.; Li, W.; Zhang, Z.; Luo, Y.; Su, L.; Wang, P. An integrated ship segmentation method based on discriminator and extractor. *Image Vis. Comput.* **2020**, *93*, 103824. [CrossRef]

41. Huang, G.; Liu, Z.; Van Der Maaten, L.; Weinberger, K.Q. Densely connected convolutional networks. In Proceedings of the IEEE Conference on Computer Vision and Pattern Recognition, Honolulu, HI, USA, 19–20 June 2017.

42. Rubin, J.; Parvaneh, S.; Rahman, A.; Conroy, B.; Babaeizadeh, S. Densely connected convolutional networks for detection of atrial fibrillation from short single-lead ECG recordings. *J. Electrocardiol.* **2018**, *51*, S18–S21. [CrossRef]

Article

Detection of Monkeypox Disease from Human Skin Images with a Hybrid Deep Learning Model

Fatih Uysal

Department of Electrical and Electronics Engineering, Faculty of Engineering and Architecture, Kafkas University, Kars TR 36100, Turkey; fatih.uysal@kafkas.edu.tr; Tel.: +90-534-022-6128

Abstract: Monkeypox, a virus transmitted from animals to humans, is a DNA virus with two distinct genetic lineages in central and eastern Africa. In addition to zootonic transmission through direct contact with the body fluids and blood of infected animals, monkeypox can also be transmitted from person to person through skin lesions and respiratory secretions of an infected person. Various lesions occur on the skin of infected individuals. This study has developed a hybrid artificial intelligence system to detect monkeypox in skin images. An open source image dataset was used for skin images. This dataset has a multi-class structure consisting of chickenpox, measles, monkeypox and normal classes. The data distribution of the classes in the original dataset is unbalanced. Various data augmentation and data preprocessing operations were applied to overcome this imbalance. After these operations, CSPDarkNet, InceptionV4, MnasNet, MobileNetV3, RepVGG, SE-ResNet and Xception, which are state-of-the-art deep learning models, were used for monkeypox detection. In order to improve the classification results obtained in these models, a unique hybrid deep learning model specific to this study was created by using the two highest-performing deep learning models and the long short-term memory (LSTM) model together. In this hybrid artificial intelligence system developed and proposed for monkeypox detection, test accuracy was 87% and Cohen's kappa score was 0.8222.

Keywords: artificial intelligence; deep learning; image classification; monkeypox disease

Citation: Uysal, F. Detection of Monkeypox Disease from Human Skin Images with a Hybrid Deep Learning Model. *Diagnostics* **2023**, *13*, 1772. https://doi.org/10.3390/diagnostics13101772

Academic Editor: Fabiano Bini

Received: 20 April 2023
Revised: 5 May 2023
Accepted: 16 May 2023
Published: 17 May 2023

1. Introduction

Monkeypox is a type of zootonic virus that first emerged through transmission from animals to humans. There appear to be two different lineages of this virus, a west African lineage and a central African lineage. There are animal species that are susceptible to this double-stranded DNA virus. These include tree and rore squirrels, dormice and Gambian pouched rats. Monkeypox is a serious global health problem, affecting the rest of the world in addition to West and East Africa, where its genetic lineage is found. Although it originated in animals, it can also be transmitted from person to person through respiratory secretions and skin lesions during travel. So far, monkeypox has been reported in many countries including Nigeria, Israel, Singapore, Singapore, the United States and the United Kingdom, in addition to Africa, where it first emerged. With monkeypox, it generally takes between 6 and 13 days after infection for symptoms to appear. The infection is divided into two parts: invasion period and skin eruption. In the invasion period, back pain, intense headache, fever, etc., are observed and this lasts between 0 and 5 days. In skin eruption, the appearance of fever varies between 1 and 3 days. Depending on factors such as the health status of the patient and the duration of exposure to the virus, the duration of symptoms in monkeypox, where severe cases are mostly seen in children, is between 2 and 4 weeks. Case fatality rates are observed to be between 3% and 6% [1].

In addition to monkeypox, chickenpox and measles are among the diseases caused by the virus on the skin. This study uses a 4-class open source dataset of skin images and performs monkeypox detection by multi-class classification with a hybrid artificial intelligence system. The main contributions of this study are listed below.

- Since the open source dataset used in this study, which consists of normal, monkeypox, measles, and chickenpox classes, initially had an unbalanced structure, a balanced dataset was created by equalizing the amount of data in each class with data preprocessing and data augmentation operations.
- In the new augmented dataset, the dataset was randomly divided into 80% train, 10% validation and 10% test for training the deep learning models to be used for monkeypox detection.
- In order to analyze the classification results more accurately, augmentation was performed on the train dataset, while no augmentation was performed on the validation and test datasets.
- First, the classification process was performed using state-of-the-art deep learning models, CSPDarkNet, InceptionV4, MnasNet, MobileNetV3, RepVGG, SE-ResNet and Xception.
- Then, in order to improve the classification results and to develop a unique model, a hybrid deep learning model was created by combining the two models with the highest results from these deep learning models and the long short-term memory (LSTM) model.
- In order to further improve monkeypox detection, a unique hybrid artificial intelligence system was developed with a convolutional neural network (CNN)-based model and a LSTM encoder network.

2. Related Works

There are various artificial intelligence studies on the detection of monkeypox disease in the literature. Abdelhamid et al. developed a hybrid algorithm to optimize deep neural networks on a monkeypox-related dataset shared openly on the Kaggle platform. By using transfer learning with deep learning models such as AlexNet, VGG, ResNet, and GoogLeNet, they achieved the highest classification accuracy of 98.8% [2]. Almutairi optimized the hyperparameters of the VGG, Xception and MobileNet deep learning models with the metaheuristic Harris Hawks optimizer algorithm using open source, multi-class and two different datasets including monkoypex, and then performed classification with various machine learning classifiers and obtained the highest accuracy values of 98.09% and 97.75% for the two datasets [3]. Dwivedi et al. used the ResNet and EfficientNet-based deep learning models for monkeypox skin lesion detection and found the highest accuracy value was 87% with the EfficientNetB3 model [4]. Gairola and Kumar obtained an accuracy of 95.55%, one of the highest accuracy values in monkeypox detection using the AlexNet, GoogleNet and VGG deep learning models and various machine learning classifiers on an open source monkeypox dataset [5]. Irmak et al. obtained 91.38% as the highest accuracy value in classification processes using pretrained MobileNetV2 and two VGG deep learning models with different number of layers on open source monkeypox skin image dataset [6]. Using an open source dataset for monkeypox image classification, Khafaga et al. obtained 98.83% accuracy in monkeypox detection using deep convolutional neural network optimized with the AL-Biruni Earth radius stochastic fractal search algorithm in addition to the VGG19, ResNet50, GoogleNet, and AlexNet deep learning models [7]. On a two-class dataset consisting of normal and monkeypox classes, Singh and Songare used the deep learning models InceptionV3, GoogLeNet, ResNet50 and VGG16 and found the highest accuracy value of 88.27% in the GoogLeNet model [8]. Sitaula and Shahi first performed classification with 13 different deep learning models on the monkeypox dataset, and then obtained the best accuracy value of 87.13% for multi-class classification with ensemble learning using Xception and DenseNet169, which are the two best-performing models among these models [9]. Sahin et al. obtained the highest classification accuracy of 91.11% in Mo-bileNetV2 model for monkeypox detection for different epoch values using the ResNet18, MobileNetV2, EfficientNet, NasNetMobile, GoogLeNet, and ShuffleNet pretrained deep learning models. They also developed an Android mobile application with Android Studio using Android SDK 12 and the Java programming language [10].

Ahsan et al. first performed data augmentation on a very small amount of normal and monkeypox images, and then obtained a wide range of accuracy values in many different classification processes with the ResNet, VGG, Xception, NasNet, and EfficientNet deep learning models using three different optimizers [11]. Altun et al. obtained the best results with the hybrid MobileNetV3, which was optimized with an f1 score of 0.98 and an accuracy of 96%, in classification processes performed with the ResNet50, DenseNet121, Efficient-NetV2, MobileNetV3, Xcception, and VGG19 deep learning models on a two-class dataset containing monkeypox images [12]. Özşahin et al. used the proposed convolutional neural network model, AlexNet, VGG16 and VGG19 in their detection process on two datasets associated with monkeypox and chickenpox and found the best classification accuracy of 99.6% in the proposed deep learning model [13]. Saleh and Rabie used the binary chimp optimization algorithm on the data collected over the internet and obtained a 98.48% classification accuracy in monkeypox operations with an ensemble model consisting of weighted naive bayes, weighted k-nearest neighbors and long short-term memory deep learning model [14]. Almufareh et al. obtained the highest accuracy of 93% by using the model they proposed and the InceptionV3, ResNet, MobileNetV2, EfficientNet deep learning models on two different open source monkeypox skin image datasets [15]. Using the open source monkeypox dataset by Al-rusaini, the highest accuracy value was obtained in the VGG16 model with 96% in the classification processes performed with the support vector machine, ResNet50, SqueezeNet, VGG16 and InceptionV3 models [16]. In the classification process performed by Ariansyah et al. using a dataset containing monkeypox, measles and normal classification, the highest accuracy in the VGG models with the proposed convolutional neural network was achieved in the VGG16 model [17]. VGG16, ResNet50, MobileNetV1, InceptionV3, Xception models were used both alone and as feature extractors in various machine learning classifiers for classification operations on a dataset consisting of normal, monkeypox, measles and chickenpox classes by Bala et al. and also a model called MonkeyNet has been proposed within the scope of this study [18]. Çelik and Özkan performed many classification operations with pretrained VGG, EfficientNet, MobileNet and GoogleNet models on a multi-class dataset, including monkeypox images, and achieved the highest accuracy in the EfficientNet model with the original dataset and in the MobileNet model in the augmented dataset [19]. The highest accuracy value was obtained as 98.8% with Xception, VGG16, VGG19 and modified fine-tuned ResNet50 models for monkeypox detection by Gupta et al. and a secured blockchain-enabled framework was proposed [20]. For monkeypox detection, 93.39% accuracy was achieved by Pramanik et al., by proposing beta normalization-based ensemble learning framework using the InceptionV3, Xception and DenseNet169 deep learning models [21]. Thieme et al. developed a web-based app for the classification of skin lesions caused by monkeypox virus infection using a large number of monkeypox datasets, and 0.91 sensitivity and 0.898 specificity values were obtained in the test dataset with the pretrained ResNet34 deep learning model [22]. On an open-source monkeypox dataset, Velu et al. performed classification with the EfficientNet model and then compared with the reinforcement learning approach Policy Gradient, Actor–Critic, Deep Q-learning network and Double Deep Q-learning network, the highest accuracy was achieved as 0.985 [23]. For the detection of monkeypox disease by Yasmin et al., using DenseNet201, EfficientNetB7, Inception-ResNetV2, InceptionV3, VGG16, and ResNet50 models, the highest accuracy was obtained in the InceptionV3 model, and a fine-tuned version of this model was recommended, and 100% accuracy in the new model called PoxNet22 was achieved [24].

It is observed that studies in the literature often use deep learning models such as AlexNet, VGG, and ResNet for monkeypox detection on multi-class, mostly two-class, datasets and also use machine learning models for classification. This study develops a novel hybrid artificial intelligence system for monkeypox detection on an open source, four-class dataset using state-of-the-art deep learning models and the LSTM model, which has not been used so far in the literature.

In Section 3, the details of the monkeypox dataset used in this study, the data augmentation and data preprocessing applied to this dataset, and the deep learning models used for classification are described. Section 4 describes the proposed hybrid model, evaluation metrics and detailed classification results. In Section 5, the results obtained for monkeypox detection are analyzed and interpreted, the main contributions of this study and its differences from the literature are emphasized, and what improvements could be made in the future following the current study are stated.

3. Materials and Methods

The dataset used in this study for monkeypox detection is an open source shared dataset through the Kaggle platform [25]. The dataset consists of normal, monkeypox, measles and chickenpox classes. It is understood that the distribution in the dataset is unbalanced. However, in artificial intelligence models used in classification problems, the class distribution should be as balanced as possible in order to fully realize network training. For this reason, various data augmentation operations were first performed on the dataset. These augmentations are equalize, horizontal flip, random brightness contrast, hue saturation value, shift scale rotate and RGB shift. The parameters and values of the data augmentations are given in Table 1. Additionally, Figures 1 and 2 show the first version of the dataset and the new version after augmentation, respectively.

Table 1. Data augmentation types and parameters (p = probability).

Types	Parameters	Types	Parameters
Equalize	$p = 0.5$		shift_limit = 0.1
Horizontal Flip	$p = 0.7$	Shift Scale Rotate	scale_limit = 0.05
Random Brightness Contrast	brightness_limit = 0.1		rotate_limit = 60
	contrast_limit = 0.5		$p = 0.7$
	$p = 0.5$		r_shift_limit = 5
Hue Saturation Value	hue_shift_limit = 20	RGB Shift	g_shift_limit = 5
	sat_shift_limit = 30		b_shift_limit = 5
	val_shift_limit = 20		$p = 0.2$
	$p = 0.5$		

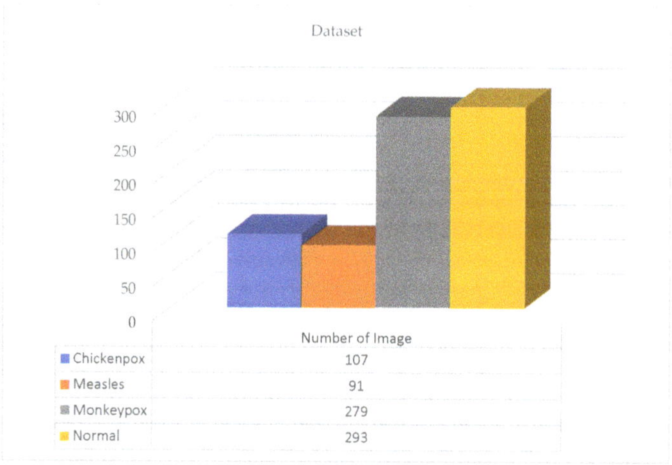

Figure 1. Original first version of the dataset.

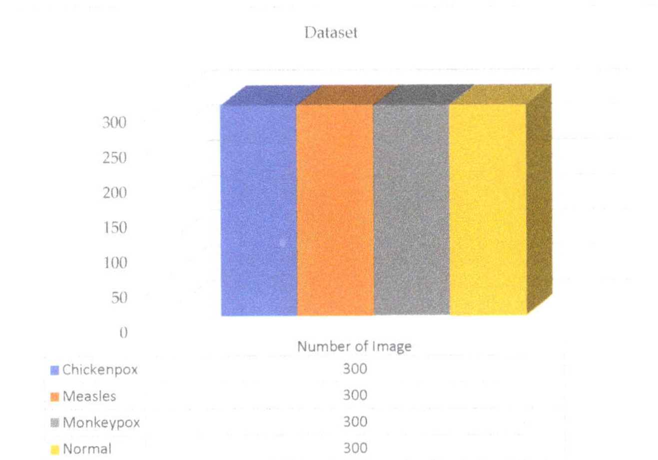

Figure 2. The new version of the dataset after data augmentation.

A total of 770 skin image datasets are available in the initial version of the dataset, including 293 normal, 279 monkeypox, 91 measles and 197 chickenpox images. Before data augmentation, a total of 240 images, 60 from each class in the original dataset, were selected for use in the test and validation dataset. Data augmentation was applied to the remaining images from the original dataset and a train dataset containing 960 images was obtained. Thanks to this method, the images in the test and validation set are not included in the train dataset. In this way, the success of this study and the designed models were handled in a more realistic way. After the data augmentation operations obtained by performing data preprocessing, a new dataset with a total of 1200 skin images, 300 in each class, was obtained. A sample image of both the original images and the images after data processing for each class of the dataset used in this study are given in Figures 3 and 4, respectively.

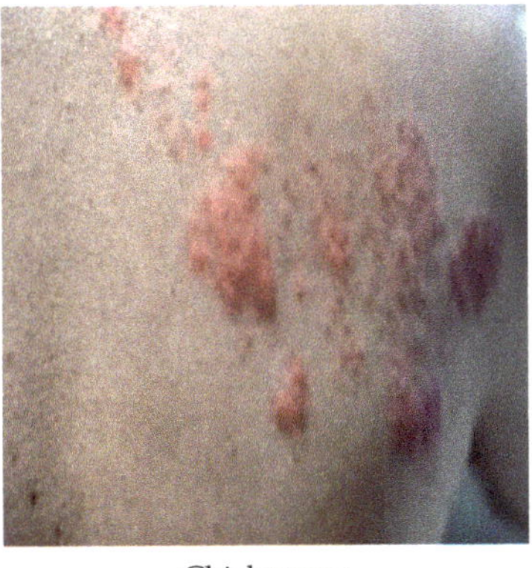

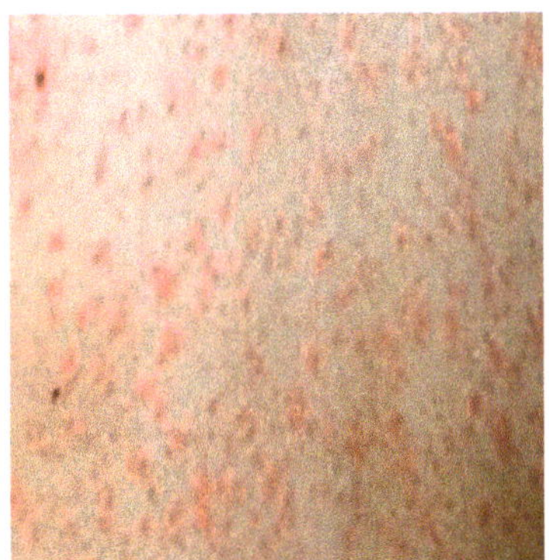

Chickenpox Measles

Figure 3. *Cont.*

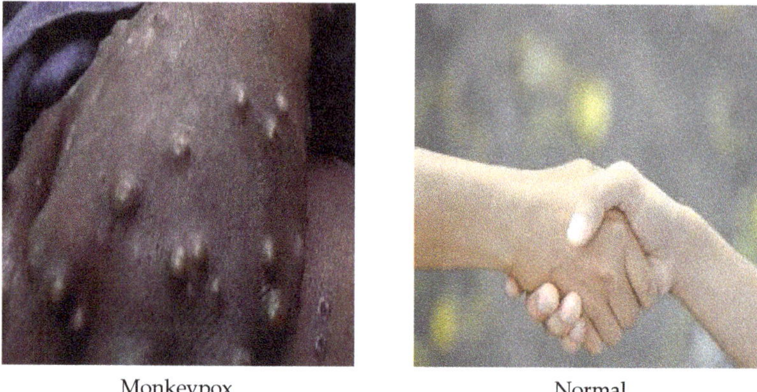

Monkeypox   Normal

Figure 3. A sample of the classes of the dataset in its original state.

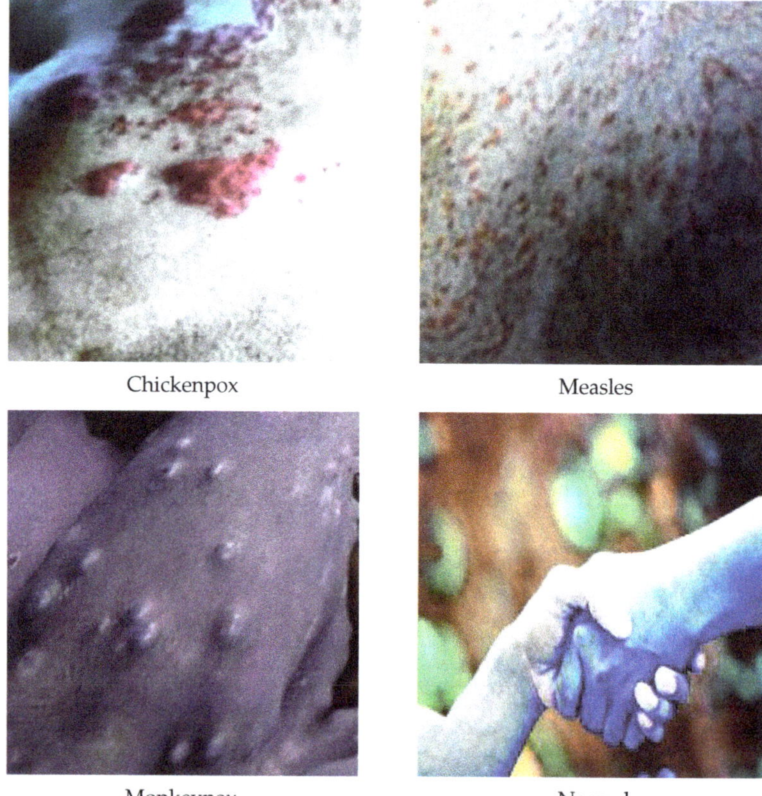

Chickenpox   Measles

Monkeypox   Normal

Figure 4. A sample of classes after dataset preprocessing.

In the new version of the dataset with data augmentation and data preprocessing, the training, validation and test distributions required for network training and classification in deep learning models are 80%, 10% and 10%, respectively. The images in each class were randomly determined in this data percentage distribution. No splitting occurred in the augmented dataset. A total of 30 test and 30 validation images were randomly selected for each class from the original dataset. The purpose of the random selection is that the researcher does not have the images in the test and validation dataset relatively easily. After this step, a test and validation dataset containing 120 images in total was obtained. A training dataset containing 960 images is required to ensure 80% training, 10% validation and 10% testing. Therefore, these 960 training datasets were obtained by augmenting the remaining 530 images in the original dataset. There is no imbalance as the test dataset contains 30 images from each class. Information on the amount and distribution of the data for each class is also shown in Figure 5 below.

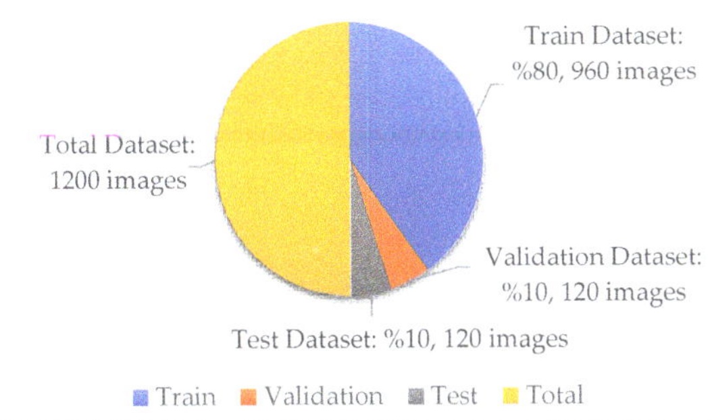

Figure 5. Dataset quantity and distribution.

In the open source dataset used in this study, there are 300 images for each class—240 in the training data, 30 in the test data, and 30 in the validation data. A total of 960 images were used for training in the dataset. No augmentation was made to analyze the classification results performed with the test dataset more realistically and accurately. Since the dataset distribution was determined as 80%, 10%, and 10%, the size of the training dataset was determined in this way.

First of all, a total of 7 different state-of-the-art deep learning models were used: CSPDarkNet with 53 layers, MnasNet with 100 layers, SE-ResNet with 50 layers, Xcep-tion with 71 layers, and InceptionV4, MobileNetV3, and RepVGG with different layer values. In addition to using these deep learning models for classification, a unique hybrid model was created by combining the best two CNN models with the LSTM model. All deep learning models that were customized and used in the classification process are given below as subheadings.

3.1. CSPDarkNet

DarkNet is a convolutional neural network used as a backbone in the YOLO object detection model. This backbone, which contains 3×3 and 1×1 convolutional layers, has different types depending on the number of layers [26]. Cross Stage Partial Network (CSPNet) is a backbone that can be applied in many different deep learning models and makes the model lightweighted [27]. In the YOLOv4 object detection model, CSPDarkNet with 53 layers was used as the backbone [28]. In addition to being used as a backbone in object detection models, it is also used in classification problems since it is a convolution neural network. In this study, CSPDarkNet-53 model is used for monkeypox detection by modifying the last layer.

3.2. InceptionV4

InceptionV4 is a convolutional neural network with more inception modules compared to its predecessor InceptionV3. InceptionV4 is an inception variant of the hybrid inception version Inception-ResNetV2 which does not include residual connections [29]. InceptionV4 model architecture used in this study was used for monkeypox detection.

3.3. MnasNet

MnasNet is a convolutional neural network whose main building block is the in-verted residual block in MobileNetV2 and proposes an automated mobile neural architecture search approach [30]. The MnasNet model used in this study has 100 layers and the number of features in the last layer is adapted for multi-class classification in accordance with the monkeypox dataset classes.

3.4. MobileNetV3

MobileNetV3 is a convolutional neural network that boasts an efficient design incorporating squeeze-and-excitation modules, making it suitable for various tasks such as classification, segmentation, and detection. This network has two variants, MobileNetV3-Large and MobileNetV3-Small, which cater to different levels of resource usage. On the ImageNet dataset for classification and the COCO dataset for detection, MobileNetV3 demonstrates improved performance compared to its predecessor, MobileNetV2 [31]. In this study, the MobileNetV3-Large model with 100 layers was adapted and used for monkeypox detection.

3.5. RepVGG

RepVGG is fundamentally a deep learning model that employs 3×3 convolution layers and ReLU non-linear activation functions. It features two primary types, RepVGG-A and RepVGG-B, each with distinct subtypes corresponding to the layers within each stage [32]. The RepVGG-B0 model, with its varying number of layers among the subtypes, was adapted to accommodate the specific task of monkeypox detection in this current study.

3.6. SE ResNet

SE ResNet is a variant of the ResNet model and is a deep learning model that in-cludes squeeze-and-excitation blocks. This model, which uses the SE ResNet module instead of the original ResNet module, gives better classification performance than many models on the ImageNet dataset [33]. A modified SE ResNet model architecture was used for monkeypox detection. In monkeypox detection using the 50-layer SE ResNet model, the number of features was reduced to 4 in the last layer in accordance with the multi-class classification and the number of classes was equalized.

3.7. Xception

The Xception model is a convolutional neural network that includes depthwise separable convolution layers instead of the inception module and uses model parameters more efficiently compared to the InceptionV3 model. The Xception deep learning model, which stands out with its better performance than the InceptionV3 model, especially on the ImageNet database, can be used for many image classification problems [34]. In this study, Xception is used by modifying the last layer to generate an output with 4 classes suitable for monkeypox detection.

3.8. LSTM

The LSTM model is a deep learning model, which is a type of recurrent neural networks. Its basic architecture consists of input, recurrent LSTM and output layers, respectively. LSTMs actually address the vanishing gradient problem. The recurrent connections in the LSTM layer are cyclic [35,36]. In this study, the LSTM model is used as an encoder network immediately after the CNN structure in the developed hybrid model. The architectural details of the LSTM used are described in detail in the experiments section.

4. Experiments

In the classification studies for monkeypox detection, seven different deep learning models with different layers and structures were used alone. The training process was carried out in this study by adapting pretrained deep learning models that utilized transfer learning from the ImageNet dataset. The initial 1000-class structure in the final layers was transformed to a four-class configuration, tailored to the dataset employed in the current research. After data augmentation and preprocessing, the results of these classification processes were analyzed and the best two CNN models were determined. These models were combined with a LSTM encoder network and a hybrid artificial intelligence system for monkeypox detection was developed. The proposed approach for monkeypox detection is presented in Figure 6 below.

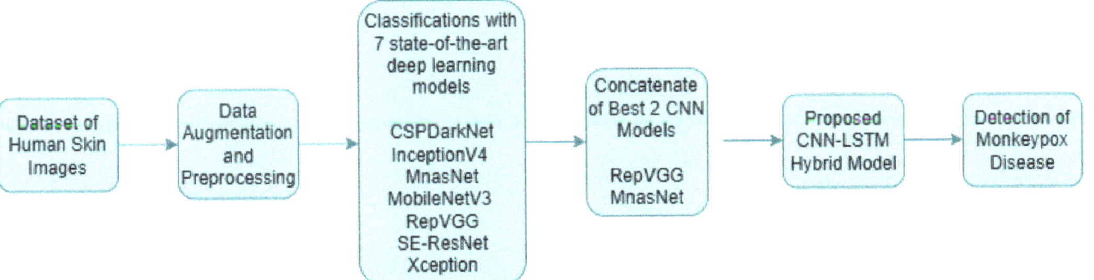

Figure 6. Proposed approach for monkeypox disease detection.

The block diagram of the hybrid artificial intelligence system proposed within the scope of this study is given in Figure 7 below. The "Image" section refers to human skin images utilized in this research. Following the necessary augmentation and preprocessing of the dataset images, they are fed into two distinct encoders. "Encoder 0" corresponds to the RepVGG-B0 deep learning model, whereas "Encoder 1" denotes the MnasNet-100 deep learning model. Upon entering the artificial intelligence system, the two encoders yield "Features 0" and "Features 1", comprising 1280 features for RepVGG and MnasNet, respectively. Subsequently, a concatenation operation is performed on both models' features, resulting in 2560 combined features, as indicated in the "Total Features" section. This novel CNN encoder structure is then integrated with an LSTM model. Following the LSTM outputs, referred to as "LSTM Features", a "Dropout FCs" layer with a ratio of 0.1 is connected to the "FC Layer". Finally, the monkeypox detection process is executed through the "Prediction" output. The structure of the proposed hybrid model is further detailed in Algorithm 1 below.

Algorithm 1 Proposed CNN–LSTM Hybrid Model

Input: test_dataset
Process:
    for image in test_dataset:
      features_0 = Model_I (image)
      features_1 = Model_II (image)
      total_features = concat (features_0, features_1)
      features_lstm = LSTM (total_features)
      out = nn.Linear (1024, 256) (features_lstm)
      Dropout (0.1)
      out = nn.Linear (256, 128) (features_lstm)
      Dropout (0.1)
    prediction = nn.Linear (128, num_classes) (features_lstm)
Output: prediction

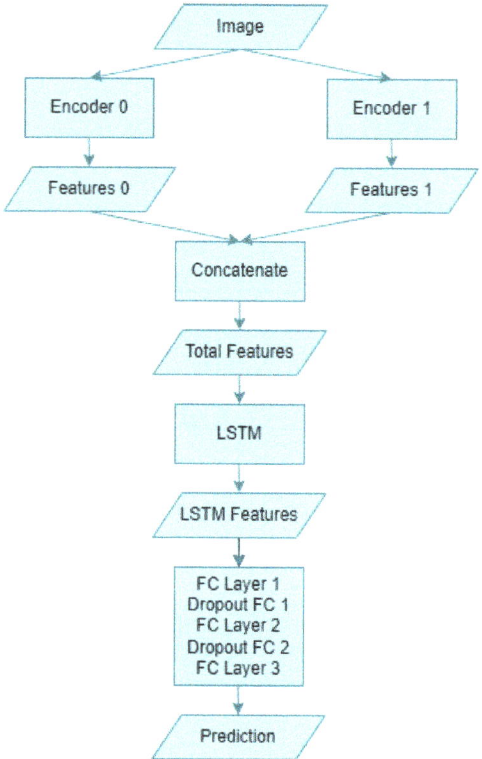

Figure 7. Block diagram of the proposed hybrid artificial intelligence system.

The operation of the above proposed algorithm is as follows: Images from any dataset are sent to both CNN architectures, respectively, and two different feature maps are obtained. Then, a single vector is obtained by combining both feature maps. This feature vector obtained is given as an input to an LSTM network and the LSTM network is provided to perform a feature extraction. The final feature vector obtained is passed through two layers and the classification process is performed.

Classification was performed using the Google Colab environment. All classifications in Colab are based on PyTorch, an open source machine learning framework. In addition, torch was used for the LSTM model, timm [37] for CNN encoder, albumentations [38] for data augmentation, and splitfolders for dataset generation. The parameters used in all artificial intelligence models for monkeypox detection are learning rate 0.001, epoch 100, batch size 8, optimizer Adam, loss function cross entropy loss.

4.1. Evaluation Metrics

There are many evaluation metrics in the literature to clearly evaluate the results obtained in binary and/or multi-class classification problems. In order to accurately analyze the results of multi-class classification for monkeypox detection, many possible evaluation metrics have been obtained in this study. These metrics are confusion matrices consisting of true-positive (TP), false-positive (FP), true-negative (TN) and false-negative (TN) values for each class; precision, recall, f1 score, ROC curve, AUC score obtained for each class; and accuracy, Cohen's kappa score and Matthews correlation coefficient score obtained using training, validation and test data. Equations (1)–(9) were taken into account in the calculation of all metrics.

$$\text{Accuracy} = P_0 = \frac{TP + TN}{TP + TN + FP + FN} \tag{1}$$

$$P_{\text{positive}} = \frac{(TP + FP)(TP + FN)}{(TP + TN + FP + FN)^2} \tag{2}$$

$$P_{\text{negative}} = \frac{(FN + TN)(FP + TN)}{(TP + TN + FP + FN)^2} \tag{3}$$

$$P_e = P_{\text{positive}} + P_{\text{negative}} \tag{4}$$

$$\text{Cohen's kappa} = \frac{P_0 - P_e}{1 - P_e} \tag{5}$$

$$\text{Matthews correlation} = \frac{(TN * TP) - (FP * FN)}{\sqrt{(TN + FN)(FP + TP)(TN + FP)(FN + TP)}} \tag{6}$$

$$\text{Precision} = \frac{TP}{TP + FP} \tag{7}$$

$$\text{Recall} = \frac{TP}{TP + FN} \tag{8}$$

$$\text{F1 score} = \frac{2TP}{2TP + FP + FN} \tag{9}$$

4.2. Monkeypox Detection Results of Deep Learning Models

The mean accuracy with standard deviation ($\pm$SD), highest accuracy, Cohen's kappa, and Matthews correlation coefficient (MCC) scores obtained in the training phase for seven different state-of-the-art deep learning models used in monkeypox detection and the precision, recall, f1 score and AUC score values in the monkeypox class are given in Table 2 below. Epoch change graphs of accuracy for training are included in Figure A1 in Appendix A.

Table 2. Training results of deep learning models (SD: Standard Deviation, AUC: Area Under the ROC (Receiver Operator Characteristic) Curve, MCC: Matthews Correlation Coefficient).

Model Name	Mean Accuracy ($\pm$SD)	Highest Accuracy	Precision	Recall	F1 Score	AUC Score	Cohen's Kappa	MCC
CSPDarkNet	0.9417 ($\pm$0.0525)	0.9906	0.96	0.90	0.92	0.911	0.9263	0.9268
InceptionV4	0.8808 ($\pm$0.0821)	0.9875	0.98	0.80	0.88	0.865	0.8888	0.8905
MnasNet	0.9645 ($\pm$0.0410)	1.0000	0.94	0.84	0.89	0.917	0.875	0.8761
MobileNetV3	0.9615 ($\pm$0.0435)	0.9979	0.99	0.89	0.94	0.877	0.95	0.9506
RepVGG	0.9176 ($\pm$0.0828)	0.9938	0.96	0.88	0.92	0.984	0.9166	0.9181
SE-ResNet	0.9624 ($\pm$0.0439)	0.9990	0.99	1.00	0.99	0.997	0.9847	0.9847
Xception	0.9330 ($\pm$0.0705)	0.9948	1.00	0.95	0.98	0.984	0.9805	0.9806

The training results in the table above show that network training was performed in the best way in the MnasNet model with the highest accuracy value. The precision, recall, f1 score and AUC score values and mean accuracy with standard deviation ($\pm$SD), highest accuracy, Cohen's kappa, Matthews correlation coefficient scores in the monkeypox class obtained for the validation phase in deep learning models used for monkeypox detection are given in Table 3 below. Epoch change graphs of accuracy for validation are included in Figure A2 in Appendix B.

Table 3. Validation results of deep learning models (SD: Standard Deviation, AUC: Area Under the ROC (Receiver Operator Characteristic) Curve, MCC: Matthews Correlation Coefficient).

Model Name	Mean Accuracy (±SD)	Highest Accuracy	Precision	Recall	F1 Score	AUC Score	Cohen's Kappa	MCC
CSPDarkNet	0.8023 (±0.0534)	0.9083	0.89	0.83	0.86	0.843	0.8777	0.8789
InceptionV4	0.7670 (±0.0572)	0.8417	0.82	0.93	0.87	0.860	0.7888	0.7957
MnasNet	0.8027 (±0.0483)	0.9083	0.90	0.90	0.90	0.929	0.8777	0.8778
MobileNetV3	0.8010 (±0.0502)	0.9000	0.88	0.97	0.92	0.930	0.8666	0.8701
RepVGG	0.7714 (±0.0711)	0.8833	0.88	0.93	0.90	0.980	0.8444	0.8470
SE-ResNet	0.8043 (±0.0422)	0.8750	0.79	0.90	0.84	0.940	0.8333	0.8373
Xception	0.7782 (±0.0520)	0.8667	0.86	1.00	0.92	0.979	0.8222	0.8270

Table 3 shows that the best-performing models are CSPDarkNet and MnasNet for accuracy, Cohen's kappa and Matthews correlation coefficient scores. Best epoch of accuracy for training and validation is included in Table A1 in Appendix C. Table 4 shows the accuracy with standard deviation (±SD), Cohen's kappa, Matthews correlation coefficient scores for the classifications performed on the test data after training and validation, as well as the precision, recall, f1 score and AUC score values in the monkeypox class.

Table 4. Test results of deep learning models (SD: Standard Deviation, AUC: Area Under the ROC (Receiver Operator Characteristic) Curve, MCC: Matthews Correlation Coefficient).

Model Name	Accuracy (±SD)	Precision	Recall	F1 Score	AUC Score	Cohen's Kappa	MCC
CSPDarkNet	0.80 (±0.0408)	0.76	0.83	0.79	0.813	0.7333	0.7364
InceptionV4	0.74 (±0.0528)	0.88	0.70	0.78	0.809	0.6555	0.6712
MnasNet	0.84 (±0.0348)	0.84	0.87	0.85	0.873	0.7888	0.7901
MobileNetV3	0.79 (±0.0499)	0.79	0.87	0.83	0.873	0.7222	0.7277
RepVGG	0.85 (±0.0290)	0.84	0.87	0.85	0.961	0.8	0.8025
SE-ResNet	0.73 (±0.0295)	0.67	0.87	0.75	0.892	0.6444	0.6508
Xception	0.73 (±0.0396)	0.73	0.80	0.76	0.939	0.6444	0.6552

In the multi-class classification process for monkeypox detection, the highest accuracy values among seven different deep learning models were obtained as 0.85 in RepVGG and 0.84 in MnasNet. The ROC curves obtained for each class with deep learning models on the test dataset are given in Figure 8 below.

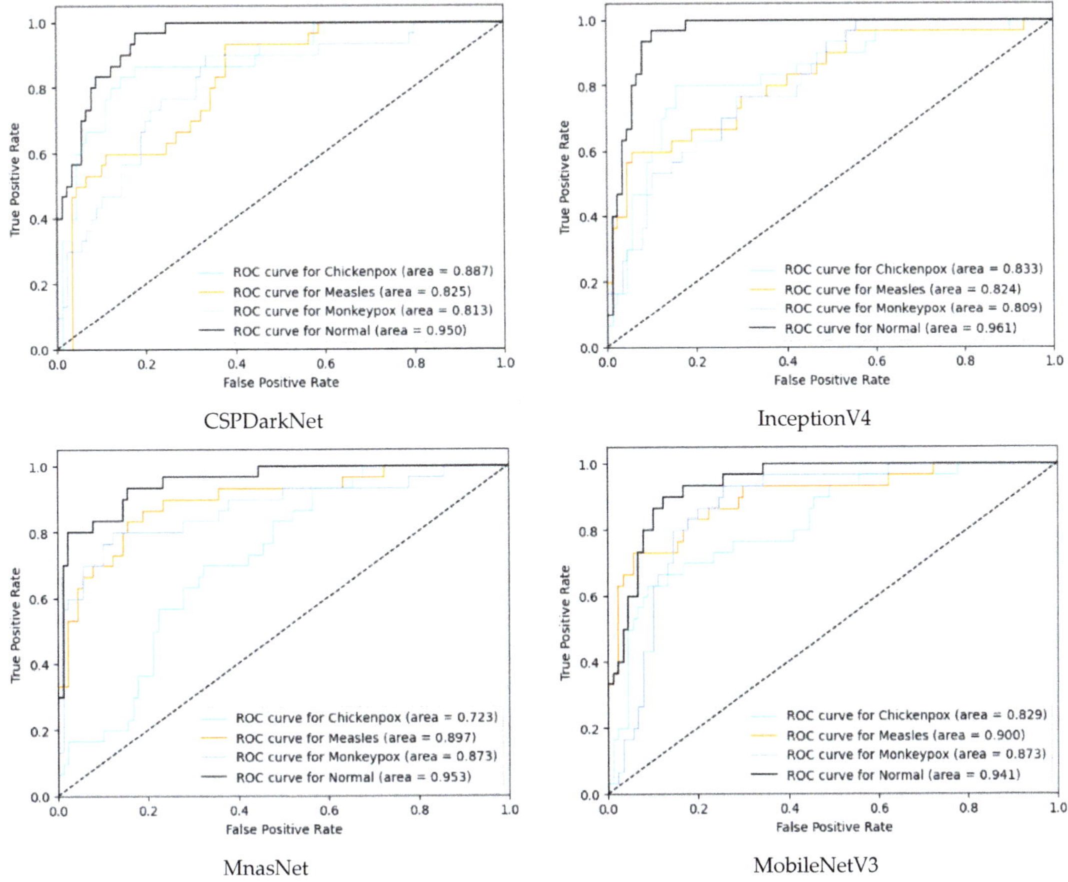

Figure 8. *Cont.*

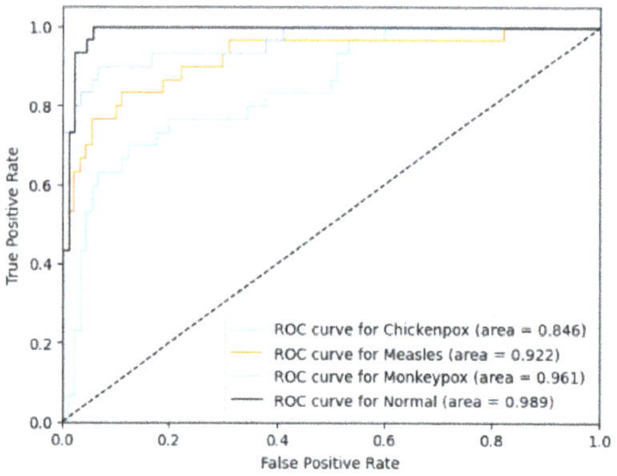

RepVGG

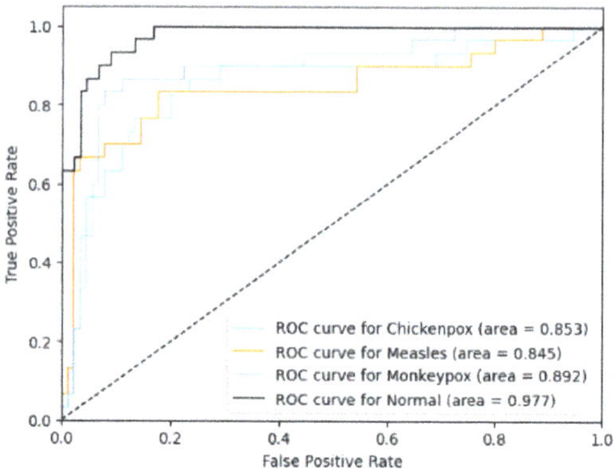

SE-ResNet

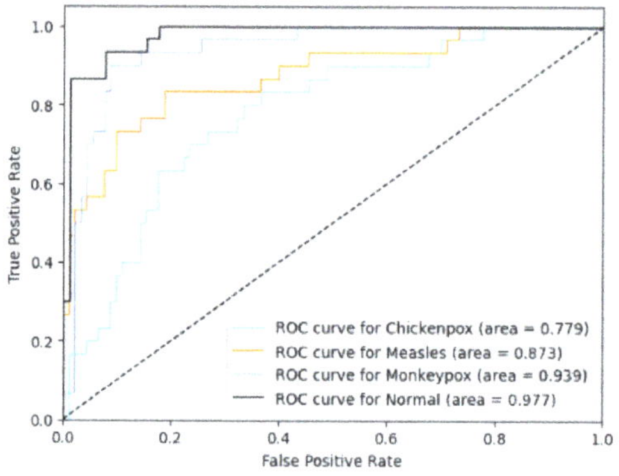

Xception

Figure 8. ROC curves in test (ROC: Receiver Operator Characteristic).

Among the deep learning models used in classification, the ROC curves in the monkeypox class show that the two highest AUC scores are in the RepVGG and Xception models. The confusion matrices obtained for the test dataset are given in Figure 9 below.

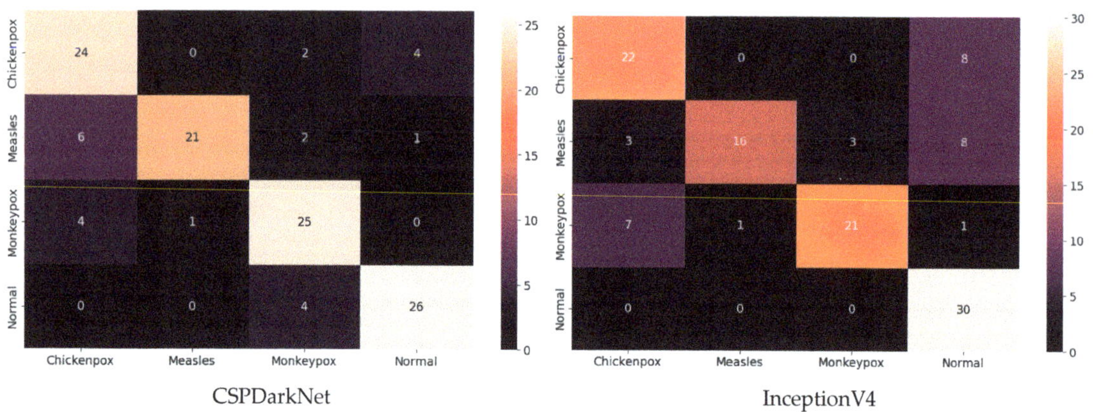

CSPDarkNet

InceptionV4

Figure 9. *Cont.*

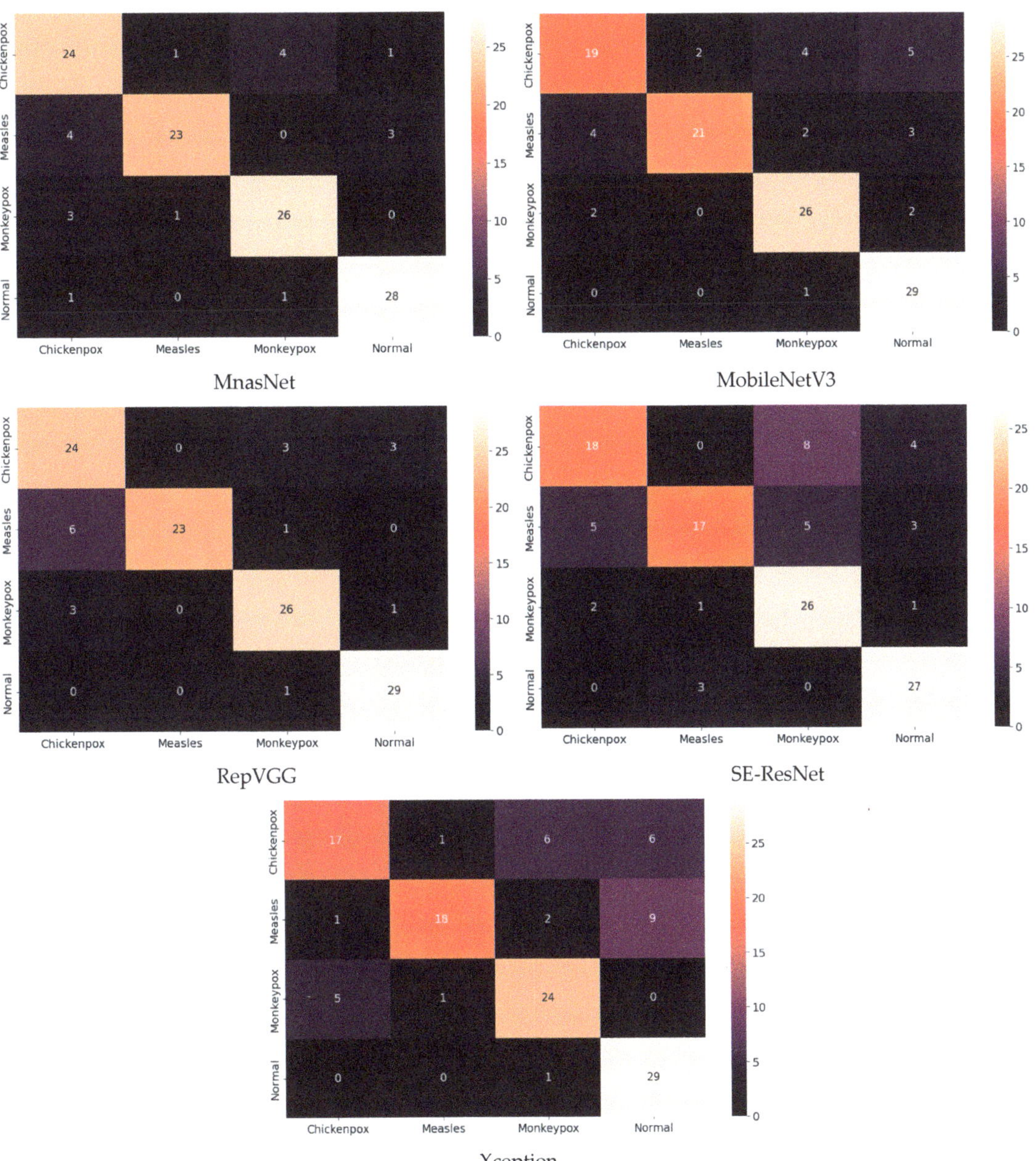

Figure 9. Confusion matrices in test.

The classification results obtained using the test dataset show that the models to be used in the CNN part of the hybrid model should be RepVGG and MnasNet to further improve classification accuracy.

4.3. Monkeypox Detection Results of the Proposed Hibrid Deep Learning Model

The proposed CNN–LSTM hybrid deep learning model for monkeypox detection achieved the following scores on the test dataset: 0.87 accuracy, 0.8222 Cohen's kappa, and

0.8240 Matthews correlation coefficient score. Furthermore, for the monkeypox class, the model attained 0.93 precision, 0.87 recall, 0.90 f1 score, and 0.9344 AUC score values. Below, Figure 10 shows the ROC curve for the proposed hybrid deep learning model and Figure 11 shows the confusion matrix.

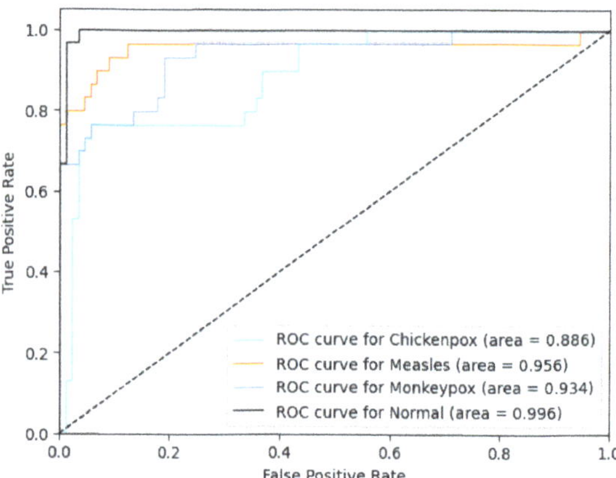

Figure 10. ROC curve in the proposed hybrid model (ROC: Receiver Operator Characteristic).

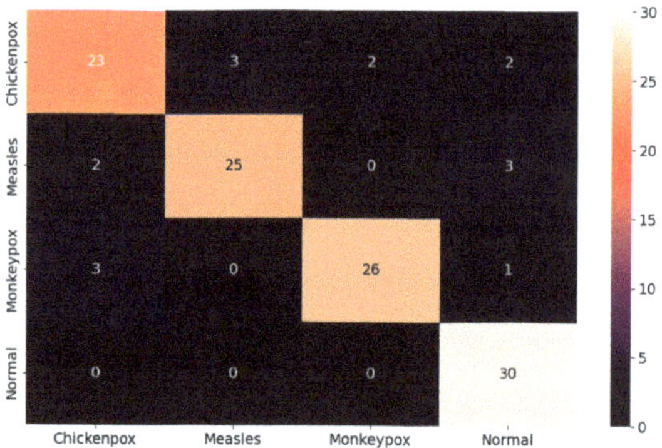

Figure 11. Confusion matrix in the proposed hybrid model.

Two deep learning models, RepVGG and MnasNet, which produced the highest results among the seven different models employed for monkeypox detection, were utilized in the proposed hybrid deep learning model within the scope of this study. The evaluation metric results for the test dataset can be found in Figure 12 and Table 5 below. The results show an increase in accuracy, Cohen's kappa and Matthews correlation coefficient scores with the hybrid model.

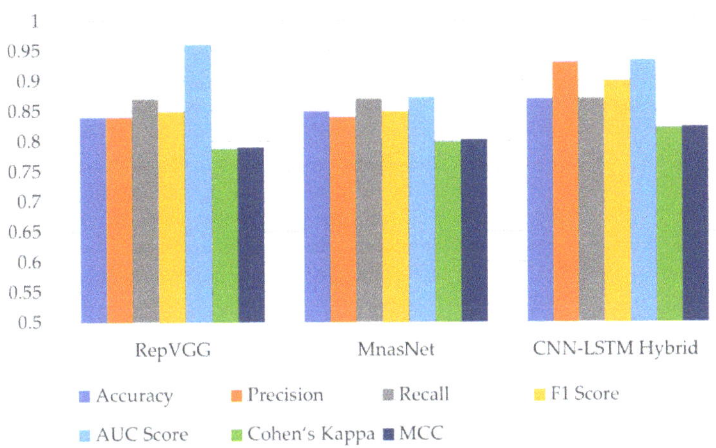

Figure 12. Test results of RepVGG, MnasNet and the proposed hybrid deep learning model (AUC: Area Under the ROC (Receiver Operator Characteristic) Curve, MCC: Matthews Correlation Coefficient).

Table 5. Test results of RepVGG, MnasNet and the proposed hybrid deep learning model (SD: Standard Deviation, AUC: Area Under the ROC (Receiver Operator Characteristic) Curve, MCC: Matthews Correlation Coefficient).

Model Name	Accuracy (±SD)	Precision	Recall	F1 Score	AUC Score	Cohen's Kappa	MCC
RepVGG	0.84 (±0.0290)	0.84	0.87	0.85	0.961	0.7888	0.7901
MnasNet	0.85 (±0.0348)	0.84	0.87	0.85	0.873	0.8	0.8025
Proposed CNN–LSTM hybrid model	0.87 (±0.0352)	0.93	0.87	0.90	0.934	0.8222	0.8240

There are many independent variables such as the dataset used in studies on similar subjects, batch sizes and image sizes that change depending on the performance of the devices used during model training, and hyper parameters (optimizer, learning rate, mini batch size) preferred during model training. In two different studies using the same model, different results can be achieved by using different batch sizes. However, this does not mean that one of the models is worse. In this context, since the classification results obtained depend on the dataset, it is more appropriate to evaluate it in itself. In this study, it was found that hybrid models achieve higher performance than conventional models.

5. Conclusions and Future Works

In this study, firstly, data augmentation and preprocessing operations were per-formed on open source and 4-class human skin images in order to make the dataset balanced. In the created balanced dataset, classification was performed with seven different pretrained deep learning models. Each of these various deep learning models used in this study was used pretrained in ImageNet. The structure, which has 1000 classes in ImageNet, has been made into 4 classes to be suitable for operation. While machine learning algorithms and traditional neural networks process the image as a single input, convolutional neural networks use a moving filter to allow the model to learn local features such as edges and corners. Convolutional neural network architectures can have a very deep structure, containing tens or even hundreds of layers, making it easier to learn complex features in the data compared to other methods. Therefore, better results were obtained using convolutional neural networks in this study. The results obtained were analyzed and a hybrid deep learning model was created by using the best two CNN models and LSTM encoder together in order to further improve monkeypox detection. A 140-layer RepVGG-B0 and a 100-layer MnasNet100 were used in the CNN part of the CNN–LSTM hybrid model proposed in this study. In the LSTM part, there are four layers. The final classifier network of the hybrid model consists of two layers. With this hybrid artificial intelligence system created for monkeypox disease detection, the highest classification results were obtained, 0.87, 0.8222 and 0.8240 in test accuracy, Cohen's kappa and Matthews correlation coefficient scores, respectively. Since hybrid systems are designed by combining different types of models, they can learn more generalizable features and thus overfitting is prevented. Likewise, combining different architectures gives reliable results with higher accuracy for the problem being dealt with. For this reason, a hybrid artificial intelligence system was used in this study. The contributions of this study to the literature are listed below.

- In order to analyze the classification results correctly, the imbalance in the dataset was eliminated with various data augmentation methods and the dataset was balanced.
- The augmentation procedures for the new balanced dataset were applied only to the training dataset. Thus, since the validation and test datasets were in its original state, the evaluation metrics obtained in the classification could be analyzed in a more realistic way.
- In order to detect monkeypox, many different state-of-the-art deep learning models were used, adapted to multi-class classification.
- The classification results of deep learning models with different layers and structures were analyzed with many different evaluation metrics and the two most appropriate CNN models were determined.
- A study-specific hybrid deep learning model was developed with CNN models and LSTM encoder models.
- With the proposed CNN–LSTM hybrid artificial intelligence system, the highest test accuracy, Cohen's kappa and Matthews correlation coefficient scores in monkeypox detection were obtained.

In future studies, machine learning models can be utilized for monkeypox disease detection alongside the deep learning models used in this study and the hybrid model developed in this study. In addition to the multi-class classification, which is a more comprehensive classification problem, binary classifications can be performed for different human skin diseases. In the future, an online web interface, an offline graphical user interface and/or a mobile application for monkeypox detection can be developed for real-time use by physicians.

Funding: This research received no external funding.

Institutional Review Board Statement: Not applicable.

Informed Consent Statement: Not applicable.

Data Availability Statement: Data used in this study are available at: https://www.kaggle.com/datasets/dipuiucse/monkeypoxskinimagedataset (accessed on 1 September 2022).

Conflicts of Interest: The author declares no conflict of interest.

Appendix A. Epoch Change Graphs of Accuracy for Training

Epoch change graphs of accuracy for training are given in Figure A1 below. The training accuracies given in Table 2 are the highest values in the training accuracy and epoch change graphs given in Figure A1.

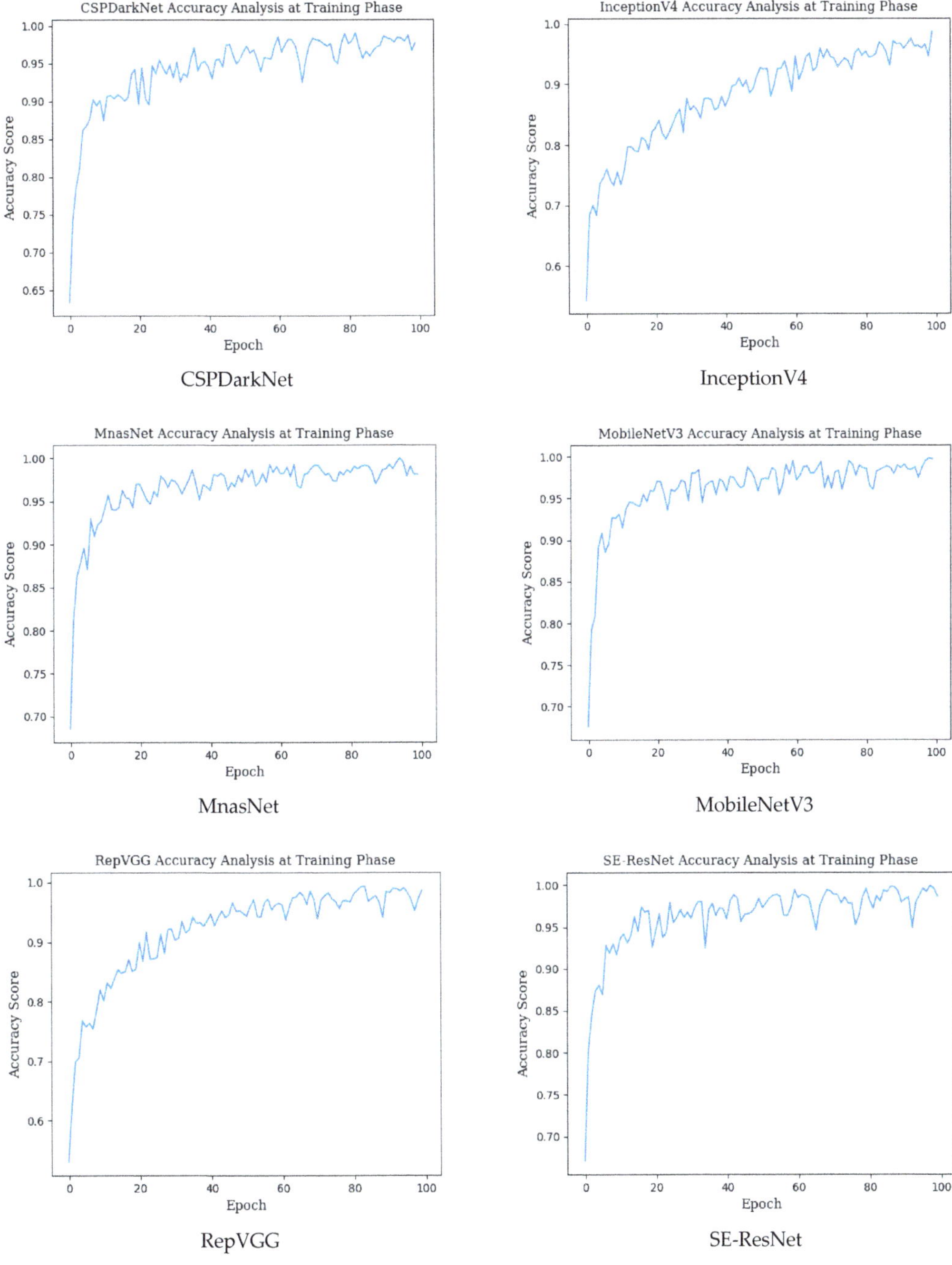

CSPDarkNet

InceptionV4

MnasNet

MobileNetV3

RepVGG

SE-ResNet

Figure A1. *Cont.*

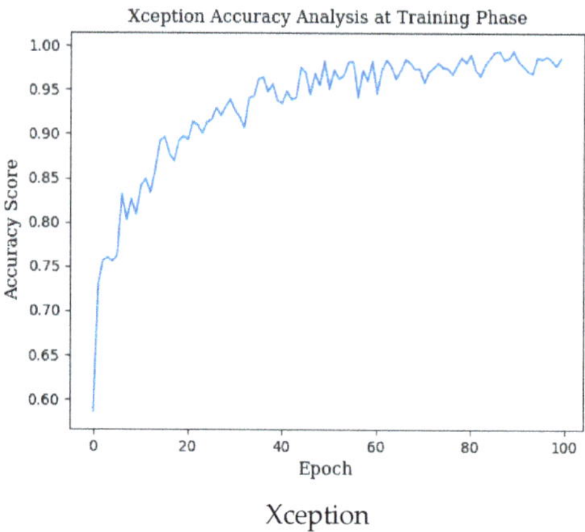

Xception

Figure A1. Epoch change graphs of accuracy for training.

Appendix B. Epoch Change Graphs of Accuracy for Validation

Epoch change graphs of accuracy for validation are given in Figure A2 below. The validation accuracies given in Table 3 are the highest values in the validation accuracy and epoch change graphs given in Figure A2.

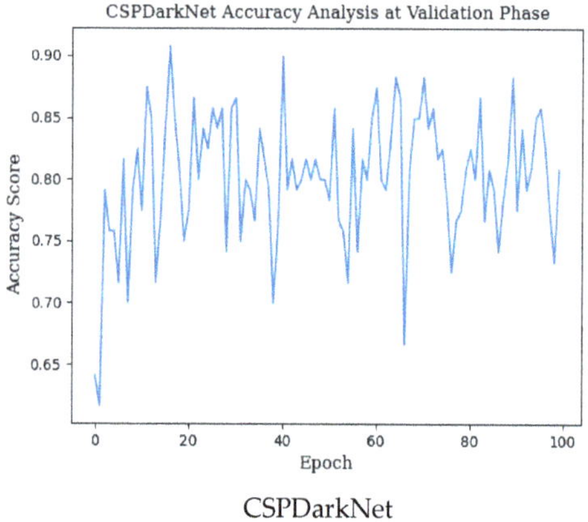

CSPDarkNet

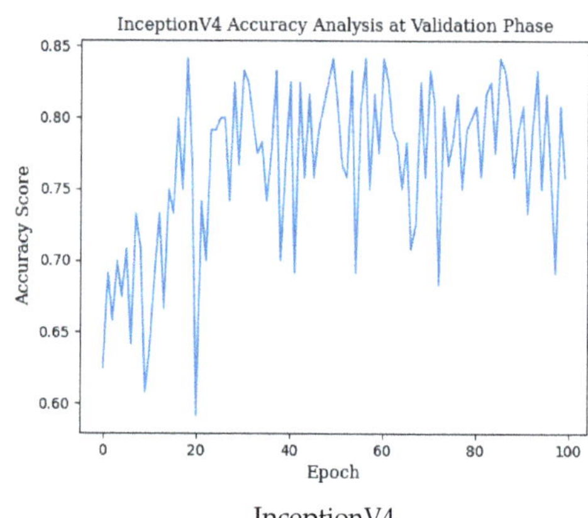

InceptionV4

Figure A2. *Cont.*

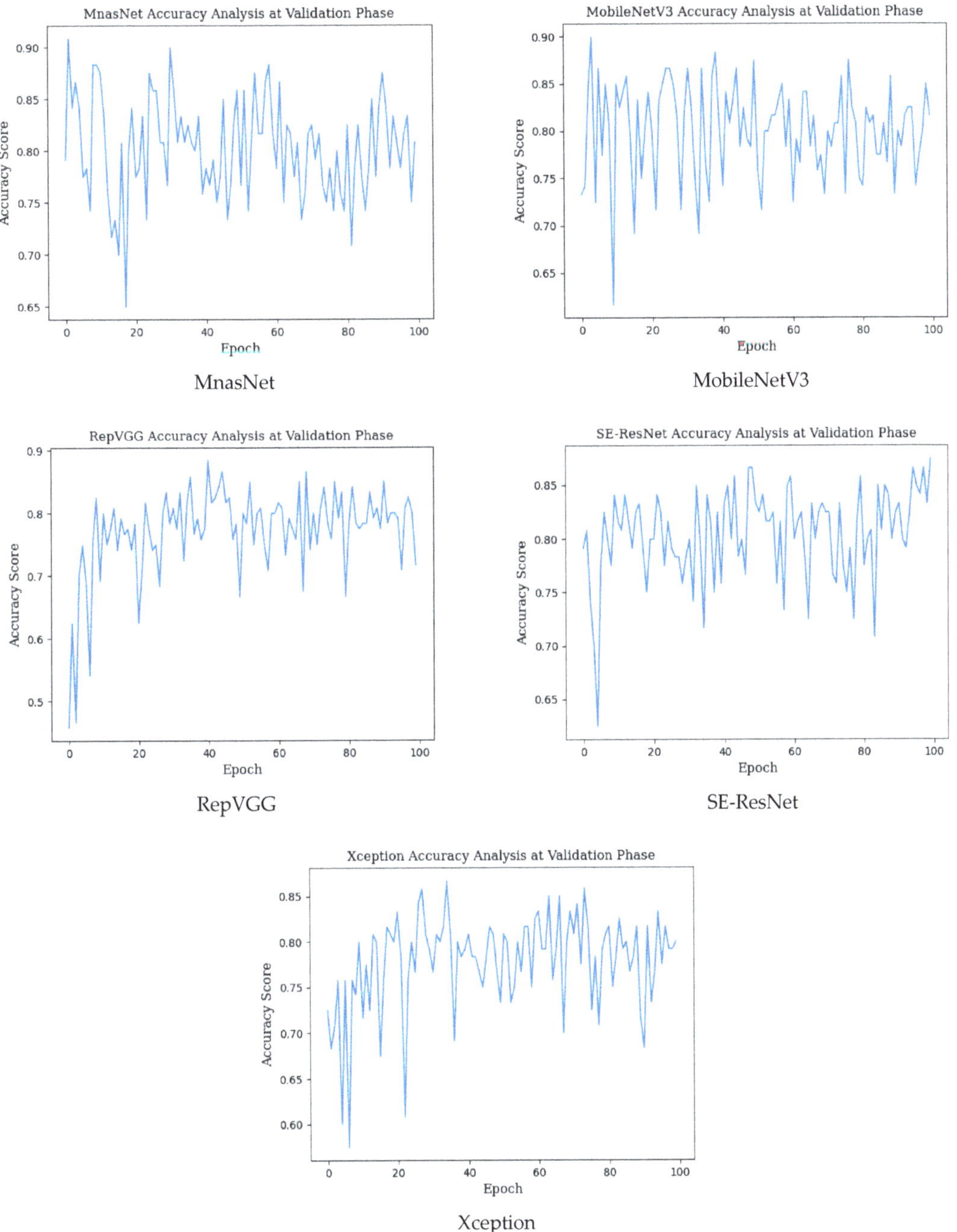

Figure A2. Epoch change graphs of accuracy for validation.

Appendix C. Best Epoch of Accuracy for Training and Validation

Best epoch of accuracy for training and validation is given in Table A1 below. The training accuracy given in Table 2 and the validation accuracy given in Table 3 are the values corresponding to the best epochs given in Table A1. When the epoch values with the highest accuracy given in this table are examined, it is observed that network training in the training part is completed with a rate close to 100%. When the validation part is examined, it is seen that the highest scores are mostly obtained within the first 50 epochs. Classification processes in the test dataset were carried out using weights at the highest epoch values obtained in validation, and no network training was performed in the test dataset. The independence of the test dataset from the validation dataset made the results more realistic and reliable.

Table A1. Best epoch of accuracy for training and validation.

Model Name	Best Epoch of Accuracy for Training	Best Epoch of Accuracy for Validation
CSPDarkNet	83	17
InceptionV4	100	19
MnasNet	95	3
MobileNetV3	99	4
RepVGG	84	41
SE-ResNet	98	100
Xception	87	35

References

1. Monkeypox. Available online: https://www.who.int/news-room/fact-sheets/detail/monkeypox (accessed on 1 January 2023).
2. Abdelhamid, A.A.; El-Kenawy, E.-S.M.; Khodadadi, N.; Mirjalili, S.; Khafaga, D.S.; Alharbi, A.H.; Ibrahim, A.; Eid, M.M.; Saber, M. Classification of Monkeypox Images Based on Transfer Learning and the Al-Biruni Earth Radius Optimization Algorithm. *Mathematics* **2022**, *10*, 3614. [CrossRef]
3. Almutairi, S.A. DL-MDF-OH^2: Optimized Deep Learning-Based Monkeypox Diagnostic Framework Using the Metaheuristic Harris Hawks Optimizer Algorithm. *Electronics* **2022**, *11*, 4077. [CrossRef]
4. Dwivedi, M.; Tiwari, R.G.; Ujjwal, N. Deep Learning Methods for Early Detection of Monkeypox Skin Lesion. In Proceedings of the 8th International Conference on Signal Processing and Communication (ICSC), Noida, India, 1–3 December 2022.
5. Gairola, A.K.; Kumar, V. Monkeypox Disease Diagnosis using Machine Learning Approach. In Proceedings of the 8th International Conference on Signal Processing and Communication (ICSC), Noida, India, 1–3 December 2022.
6. Irmak, M.C.; Aydin, T.; Yağanoğlu, M. Monkeypox skin lesion detection with MobileNetV2 and VGGNet models. In Proceedings of the 2022 Medical Technologies Congress (TIPTEKNO), Antalya, Turkey, 31 October–2 November 2022.
7. Khafaga, D.S.; Ibrahim, A.; El-Kenawy, E.-S.M.; Abdelhamid, A.A.; Karim, F.K.; Mirjalili, S.; Khodadadi, N.; Lim, W.H.; Eid, M.M.; Ghoneim, M.E. An Al-Biruni Earth Radius Optimization-Based Deep Convolutional Neural Network for Classifying Monkeypox Disease. *Diagnostics* **2022**, *12*, 2892. [CrossRef] [PubMed]
8. Singh, U.; Songare, L.S. Analysis and Detection of Monkeypox using the GoogLeNet Model. In Proceedings of the 2022 International Conference on Automation, Computing and Renewable Systems (ICACRS), Pudukkottai, India, 13–15 December 2022.
9. Sitaula, C.; Shahi, T.B. Monkeypox virus detection using pre-trained deep learning-based approaches. *J. Med. Syst.* **2022**, *46*, 78. [CrossRef] [PubMed]
10. Sahin, V.H.; Oztel, I.; Yolcu Oztel, G. Human monkeypox classification from skin lesion images with deep pre-trained network using mobile application. *J. Med. Syst.* **2022**, *46*, 79. [CrossRef]
11. Ahsan, M.M.; Uddin, M.R.; Ali, M.S.; Islam, M.K.; Farjana, M.; Sakib, A.N.; Momin, K.A.; Luna, S.A. Deep transfer learning approaches for Monkeypox disease diagnosis. *Expert Syst. Appl.* **2023**, *216*, 119483. [CrossRef]
12. Altun, M.; Gürüler, H.; Özkaraca, O.; Khan, F.; Khan, J.; Lee, Y. Monkeypox Detection Using CNN with Transfer Learning. *Sensors* **2023**, *23*, 1783. [CrossRef]
13. Uzun Ozsahin, D.; Mustapha, M.T.; Uzun, B.; Duwa, B.; Ozsahin, I. Computer-Aided Detection and Classification of Monkeypox and Chickenpox Lesion in Human Subjects Using Deep Learning Framework. *Diagnostics* **2023**, *13*, 292. [CrossRef]
14. Saleh, A.I.; Rabie, A.H. Human monkeypox diagnose (HMD) strategy based on data mining and artificial intelligence techniques. *Comput. Biol. Med.* **2023**, *152*, 106383. [CrossRef]
15. Almufareh, M.F.; Tehsin, S.; Humayun, M.; Kausar, S. A Transfer Learning Approach for Clinical Detection Support of Monkeypox Skin Lesions. *Diagnostics* **2023**, *13*, 1503. [CrossRef]
16. Alrusaini, O.A. Deep Learning Models for the Detection of Monkeypox Skin Lesion on Digital Skin Images. *Int. J. Adv. Comput. Sci. Appl.* **2023**, *14*, 637–644. [CrossRef]

17. Ariansyah, M.H.; Winarno, S.; Sani, R.R. Monkeypox and Measles Detection using CNN with VGG-16 Transfer Learning. *J. Comput. Res. Innov.* **2023**, *8*, 32–44. [CrossRef]
18. Bala, D.; Hossain, M.S.; Hossain, M.A.; Abdullah, M.I.; Rahman, M.M.; Manavalan, B.; Gu, N.; Islam, M.S.; Huang, Z. MonkeyNet: A robust deep convolutional neural network for monkeypox disease detection and classification. *Neural Netw.* **2023**, *161*, 757–775. [CrossRef]
19. Çelik, M.; Özkan, İ. Detection of Monkeypox Among Different Pox Diseases with Different Pre-Trained Deep Learning Models. *J. Inst. Sci. Technol.* **2023**, *13*, 10–21. [CrossRef]
20. Gupta, A.; Bhagat, M.; Jain, V. Blockchain-enabled healthcare monitoring system for early Monkeypox detection. *J. Supercomput.* **2023**, 1–25. [CrossRef]
21. Pramanik, R.; Banerjee, B.; Efimenko, G.; Kaplun, D.; Sarkar, R. Monkeypox detection from skin lesion images using an amalgamation of CNN models aided with Beta function-based normalization scheme. *PLoS ONE* **2023**, *18*, e0281815. [CrossRef]
22. Thieme, A.H.; Zheng, Y.; Machiraju, G.; Sadee, C.; Mittermaier, M.; Gertler, M.; Salinas, J.L.; Srinivasan, K.; Gyawali, P.; Carrillo-Perez, F.; et al. A deep-learning algorithm to classify skin lesions from mpox virus infection. *Nat. Med.* **2023**, *29*, 738–747. [CrossRef]
23. Velu, M.; Dhanaraj, R.K.; Balusamy, B.; Kadry, S.; Yu, Y.; Nadeem, A.; Rauf, H.T. Human Pathogenic Monkeypox Disease Recognition Using Q-Learning Approach. *Diagnostics* **2023**, *13*, 1491. [CrossRef]
24. Yasmin, F.; Hassan, M.M.; Hasan, M.; Zaman, S.; Kaushal, C.; El-Shafai, W.; Soliman, N.F. PoxNet22: A Fine-Tuned Model for the Classification of Monkeypox Disease Using Transfer Learning. *IEEE Access* **2023**, *11*, 24053–24076. [CrossRef]
25. Monkeypox Skin Images Dataset (MSID). Available online: https://www.kaggle.com/datasets/dipuiucse/monkeypoxskinimagedataset (accessed on 1 September 2022).
26. Redmon, J.; Farhadi, A.; Redmon, J.; Farhadi, A. Yolov3: An incremental improvement. *arXiv* **2018**, arXiv:1804.02767.
27. Wang, C.Y.; Liao, H.Y.M.; Wu, Y.H.; Chen, P.Y.; Hsieh, J.W.; Yeh, I.H. CSPNet: A new backbone that can enhance learning capability of CNN. In Proceedings of the IEEE/CVF Conference on Computer Vision and Pattern Recognition Workshops (CVPRW), Virtual, 14–19 June 2020.
28. Bochkovskiy, A.; Wang, C.Y.; Liao, H.Y.M. Yolov4: Optimal speed and accuracy of object detection. *arXiv* **2020**, arXiv:2004.10934.
29. Szegedy, C.; Ioffe, S.; Vanhoucke, V.; Alemi, A. Inception-v4, inception-resnet and the impact of residual connections on learning. In Proceedings of the AAAI Conference on Artificial Intelligence, San Francisco, CA, USA, 4–9 February 2017.
30. Tan, M.; Chen, B.; Pang, R.; Vasudevan, V.; Sandler, M.; Howard, A.; Le, Q.V. Mnasnet: Platform-aware neural architecture search for mobile. In Proceedings of the IEEE/CVF Conference on Computer Vision and Pattern Recognition (CVPR), Long Beach, CA, USA, 15–20 June 2019.
31. Howard, A.; Sandler, M.; Chu, G.; Chen, L.C.; Chen, B.; Tan, M.; Wang, W.; Zhu, Y.; Pang, R.; Vasudevan, V.; et al. Searching for mobilenetv3. In Proceedings of the IEEE/CVF International Conference on Computer Vision (ICCV), Seoul, Republic of Korea, 27 October–2 November 2019.
32. Ding, X.; Zhang, X.; Ma, N.; Han, J.; Ding, G.; Sun, J. Repvgg: Making vgg-style convnets great again. In Proceedings of the IEEE/CVF Conference on Computer Vision and Pattern Recognition (CVPR), Virtual, 20–25 June 2021.
33. Hu, J.; Shen, L.; Sun, G. Squeeze-and-excitation networks. In Proceedings of the IEEE Conference on Computer Vision and Pattern Recognition, Salt Lake City, UT, USA, 18–23 June 2018.
34. Chollet, F. Xception: Deep learning with depthwise separable convolutions. In Proceedings of the IEEE Conference on Computer Vision and Pattern Recognition (CVPR), Honolulu, HI, USA, 21–26 July 2017.
35. Hochreiter, S.; Schmidhuber, J. Long short-term memory. *Neural Comput.* **1997**, *9*, 1735. [CrossRef] [PubMed]
36. Sak, H.; Senior, A.; Beaufays, F. Long short-term memory based recurrent neural network architectures for large vocabulary speech recognition. *arXiv* **2014**, arXiv:1402.1128.
37. Pytorch Image Models. Available online: https://github.com/rwightman/pytorch-image-models (accessed on 1 September 2022).
38. Buslaev, A.; Iglovikov, V.I.; Khvedchenya, E.; Parinov, A.; Druzhinin, M.; Kalinin, A.A. Albumentations: Fast and Flexible Image Augmentations. *Information* **2020**, *11*, 125. [CrossRef]

Review

Computational Intelligence-Based Disease Severity Identification: A Review of Multidisciplinary Domains

Suman Bhakar [1], Deepak Sinwar [1], Nitesh Pradhan [2], Vijaypal Singh Dhaka [1], Ivan Cherrez-Ojeda [3], Amna Parveen [4,*] and Muhammad Umair Hassan [5,*]

[1] Department of Computer and Communication Engineering, Manipal University Jaipur, Dehmi Kalan, Jaipur 303007, Rajasthan, India; sumanbhakar2016@gmail.com (S.B.); deepak.sinwar@gmail.com (D.S.); prof.dhaka@gmail.com (V.S.D.)

[2] Department of Computer Science and Engineering, Manipal University Jaipur, Dehmi Kalan, Jaipur 303007, Rajasthan, India; nitesh.pradhan943@gmail.com

[3] Allergy and Pulmonology, Espíritu Santo University, Samborondón 0901-952, Ecuador; ivancherrez@gmail.com

[4] College of Pharmacy, Gachon University, Medical Campus, No. 191, Hambakmoero, Yeonsu-gu, Incheon 21936, Republic of Korea

[5] Department of ICT and Natural Sciences, Norwegian University of Science and Technology (NTNU), 6009 Ålesund, Norway

* Correspondence: amnaparvin@gmail.com (A.P.); muhammad.u.hassan@ntnu.no (M.U.H.)

Citation: Bhakar, S.; Sinwar, D.; Pradhan, N.; Dhaka, V.S.; Cherrez-Ojeda, I.; Parveen, A.; Hassan, M.U. Computational Intelligence-Based Disease Severity Identification: A Review of Multidisciplinary Domains. *Diagnostics* **2023**, *13*, 1212. https://doi.org/10.3390/diagnostics13071212

Academic Editor: Mugahed A. Al-antari

Received: 29 January 2023
Revised: 6 March 2023
Accepted: 8 March 2023
Published: 23 March 2023

Abstract: Disease severity identification using computational intelligence-based approaches is gaining popularity nowadays. Artificial intelligence and deep-learning-assisted approaches are proving to be significant in the rapid and accurate diagnosis of several diseases. In addition to disease identification, these approaches have the potential to identify the severity of a disease. The problem of disease severity identification can be considered multi-class classification, where the class labels are the severity levels of the disease. Plenty of computational intelligence-based solutions have been presented by researchers for severity identification. This paper presents a comprehensive review of recent approaches for identifying disease severity levels using computational intelligence-based approaches. We followed the PRISMA guidelines and compiled several works related to the severity identification of multidisciplinary diseases of the last decade from well-known publishers, such as MDPI, Springer, IEEE, Elsevier, etc. This article is devoted toward the severity identification of two main diseases, viz. Parkinson's Disease and Diabetic Retinopathy. However, severity identification of a few other diseases, such as COVID-19, autonomic nervous system dysfunction, tuberculosis, sepsis, sleep apnea, psychosis, traumatic brain injury, breast cancer, knee osteoarthritis, and Alzheimer's disease, was also briefly covered. Each work has been carefully examined against its methodology, dataset used, and the type of disease on several performance metrics, accuracy, specificity, etc. In addition to this, we also presented a few public repositories that can be utilized to conduct research on disease severity identification. We hope that this review not only acts as a compendium but also provides insights to the researchers working on disease severity identification using computational intelligence-based approaches.

Keywords: disease severity; deep learning; machine learning; Parkinson's disease; diabetic retinopathy; Alzheimer's disease; CNN

PACS: J0101

---

1. Introduction

Early and accurate diagnosis of diseases is essential for the right treatment. In addition to accurate and rapid diagnosis, the severity identification using computational intelligence-based approaches is becoming popular and challenging nowadays. Traditional computational approaches (i.e., classification) are mainly focused on solving two-class classification problems, i.e., positivity or negativity of disease, or presence or absence of certain values. However, nowadays, with the advancements in deep learning technologies, one can easily diagnose the disease and its severity. Most of the work on severity identification is based on recent deep-learning-based models. The training of these models depends on the labeling of disease severity levels by expert personnel. However, the process of multi-class manual labeling is quite tedious, time-consuming, and non-quantitative [1].

In this paper, the problem of severity identification is addressed with the help of multi-class classification. A comprehensive review of various research articles concentrating on disease severity identification using computational intelligence-based approaches is presented. Research articles focused on the severity identification of Parkinson's Disease (PD) and Diabetic Retinopathy (DR) are mainly considered for this study. We followed the PRISMA statement to prepare this review on the severity identification of diseases using computational intelligence-based approaches. The search terms/combinations to search sources for this study followed search phrases such as "(disease AND severity AND deep learning)", "(severity identification AND computational intelligence)", "(Diabetic Retinopathy AND severity AND artificial intelligence)", "(Parkinson's Disease AND severity AND artificial intelligence)", etc. The search strategy followed by the identification and analysis of sources for this study is also depicted in Figure 1. In addition to this, we briefly surveyed a few articles on the severity identification of some other diseases, i.e., COVID-19, Knee Osteoarthritis (KOA) [2], Autonomic Nervous System Dysfunction (ANSD) [3], Tuberculosis [4], and Sepsis [5], etc. It is evident that radiology is widely used for the diagnosis of various critical diseases. Some computational approaches also consider radiological images for disease identification. Radiology is one discipline of medicine that uses imaging technologies to diagnose diseases [6]. Radiology is divided into two main classes, viz. Diagnostic Radiology and Interventional Radiology [7]. Diagnostic radiology provides structures inside the body, whereas interventional radiology is associated with minimally invasive procedures.

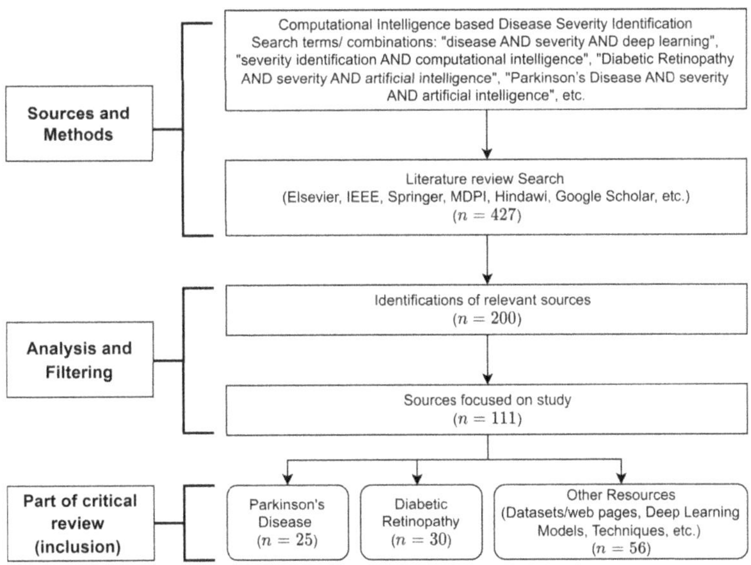

Figure 1. Strategy for inclusion of sources for this study.

Due to the recent advances in deep learning and machine learning, the potential of computational approaches regarding the recognition of complex patterns from radiological images has increased to a great extent. Nowadays, the integration of computational approaches and radiological imaging technologies is gaining tremendous popularity and becoming an active research area. Undoubtedly, future clinical decision support systems and monitoring systems will be equipped with state-of-the-art artificial intelligence. It is observed that plenty of deep-learning- and machine-learning-based research work has been carried out on radiological imaging. Deep-learning-based disease identification follows several steps, viz. data collection, labeling, classification, and model evaluation. These models can be optimized by fine-tuning the parameters, as depicted in Figure 2.

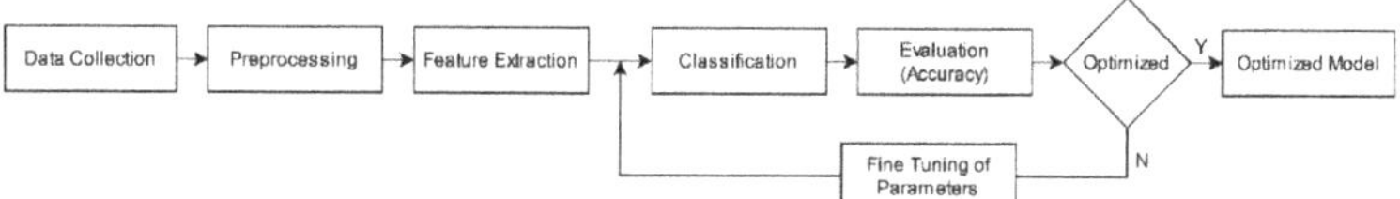

Figure 2. Process of optimizing a classification model for disease identification.

In the case of machine-learning-based disease identification, the non-imaging data values and efficient algorithms play an important role in decision support systems. Machine-learning and deep-learning techniques have numerous applications in the medical domain. The work embodied in this paper mainly focuses on diagnosing Parkinson's disease, Diabetic Retinopathy (DR), and some other diseases (infectious diseases, tuberculosis, COVID-19, sepsis, etc.) using computational approaches. The subsequent sections will highlight some of the work conducted by researchers to diagnose these diseases. In short, the major contributions of this paper are highlighted as follows:

- In-depth analysis of several recent pieces of work for disease severity identification using computational intelligence-based approaches.
- A comprehensive discussion on the challenges and issues of each approach for severity identification.
- Classification of several works according to major disease types such as Parkinson's Disease and Diabetic Retinopathy.
- Presentation of several public repositories for conducting disease severity identification research.

The remainder of this paper is organized as follows. Sections 2 and 3 discuss some of the work related to detecting the severity of Parkinson's Disease and Diabetic Retinopathy, respectively. Works based on severity identification of a few other diseases, i.e., COVID-19, autonomic nervous system dysfunction, tuberculosis, sepsis, sleep apnea, psychosis, traumatic brain injury, breast cancer, knee osteoarthritis, and Alzheimer's disease, are briefly presented in Section 4. A few public repositories are depicted in Section 5. Finally, Section 6 presents concluding remarks along with future directions for severity identification using computational intelligence-based approaches.

2. Severity Identification of Parkinson's Disease

Movement disorders caused by PD may not remain the same in different patients. Thus, it is essential to develop an automated tool to evaluate a patient's gait. Xia et al. [8] presented a novel gait evaluation approach (known as "dual-modal attention-enhanced deep learning network"), which not only distinguishes between normal gaits and PD gaits but also computes the severity of PD by quantification of gaits. The system is capable of modeling both left and right gaits separately. Multiple 1D vertical ground reaction force (VGRF) signals achieve the segmentation of left and right samples. A CNN-LSTM-based dual-modal attention-enhanced network was utilized to analyze the gait movements on the gait dataset [9] with two severity levels, viz. Hoehn and Yahr (H&Y) and the Unified Parkinson's Disease Rating Scale (UPDRS). Their architecture utilizes an input with the dimensions B × 150 × 9 × 1, where B indicates the batch size of samples, 150 indicates the period of a sample, and 9 indicates the number of VGRF signals. Their CNN consists of three layers in which every convolution operation is followed by the ReLU activation function for feature extraction. However, pooling is not incorporated due to the limited data samples. After the last convolution, the output of the feature map comprises dimensions of B × 150 × 9 × C3. Using flattening, the feature map 9 × C3 is converted into a tensor, i.e., C4, which was fed to an attention-enhanced LSTM (AE-LSTM). The AE-LSTM concatenates the branches and passes them to the fully connected (FC) layer. Finally, the severity of PD is achieved using probability distribution by mapping the output of FC using a SoftMax classifier. Experimental results claim 99.01% accuracy in classifying PD patients into different severity levels.

Pereira et al. [10] have reviewed several papers to predict PD at the earliest stage. After reviewing the papers, the authors have concluded that there are still many problems that need to be addressed, so they proposed image processing techniques to address these existing problems. For this experiment, handed datasets are utilized, collected from Brazil University. It contains the meander and spiral images gathered through the handwritten exam and 92 handwritten exams conducted on healthy people (control group) and PD patients. Handwritten Trace (HT) and Exam Template (ET) features are extracted through the blurring method. The feature extraction technique is applied to compare and evaluate both the HT and ET features. The Support Vector Machine (SVM) with some modifications, Naïve Bayes (NV) technique, and Optimum path forest (OPF) pattern recognition methods are used for the severity classification. The experimental results show 67% accuracy in identifying the precise class to predict the stage of the severity. As per the amount of information concerned for PD identification, meander images represent more information than spiral images. Although they presented an automated system that diagnosed the PR at an early stage, the performance can be improved by considering large as well as consistent datasets.

Prashanth et al. [11] addressed the fact that if PD disease is detected at an early stage, it can be cured by the proper therapies and medicines. In this regard, they utilized Single-Photon Emission Computed Tomography (SPECT) along with 123I-Ioflupane to diagnose the PD disease at an earlier stage on the PPMI database. The dataset contains the Striatum Binding Ratio (SBR) value of 179 normal people and 369 PD patients in the initial stage. The logical regression is applied for the calculation of the significant numerical features. The visualization of each SBR feature is calculated through histograms. The notched plots mark the patients separately in normal, PD, and early-stage categorization. The classifications and prediction have been acquired through the Support Vector Machine (SVM) and Logistic Regression (LR). The SVM uses a linear kernel to classify the decision boundary through by input features. The binomial logistic regression model uses the logit transformation method to develop the prediction model to predict the risk factor in PD patients. The experimental results report that the SVM classification method has achieved 96.14% accuracy and 95.03% specificity for the classification of PD patients. Although this system provided high performance and distinguished early PD patients from normal patients, the system can be enhanced through the Scans Without Evidence of Dopaminergic Deficit (SWEDD) and other validation approaches.

Parkinson's Disease can be identified on various input signals, as depicted in Figure 3. In this regard, Cernak et al. [12] proposed a model to identify voice characteristics to predict the PD patient's information. They utilized the read Voice Quality (VQ) datasets by Kane (2012) and Laver (1980). They covered the five non-model vocalizations, viz. creaky, breathiness, falsetto, harsh, and tense. To study the vocalization features, the Spanish database contains the speech recording detail of PD patients and a healthy control group. With the help of statistical measures, the authors differentiated the model and non-model vocalization. They computed the probability of the vocalization features through a machine-learning-based approach. The Euclidean distance calculates the similarity of the model in PD, and the alignment of the non-model is calculated through the inverse distance. The vocalization analysis section is computed through the Deep Neural Network (DNN). Further, the binary classification method was utilized to identify the probability of a specific vocalization class. They also applied the acoustic model for the phonic configurations. The experimental results reported the characteristics of PD patients: the composition of a maximum of 30% of breathy voice and a minimum of 12% of harsh voice. The system provided the accuracy of the vocalization speech based on the voice quality, but analysis of the speech was limited due to available datasets.

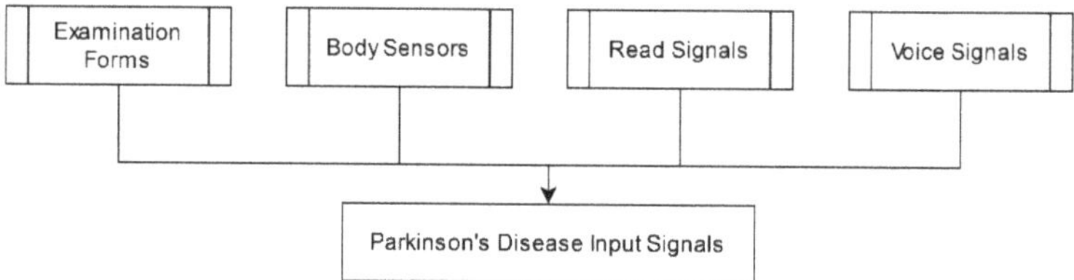

Figure 3. Various inputs to Parkinson's Disease diagnosis.

Lahmiri et al. [13] also proposed a method to detect PD through voice patterns. They utilized the 195 vowels and voices data set comprising 147 PD-affected and 48 healthy patients. The Wilcoxon and ROC techniques were used to identify eight different patterns. The well-established SVM classification technique was applied to classify the PD patient and the healthy one. The system reported a 92.21% accuracy, 82.79% specificity, and 99.63% sensitivity. Although this automated system provided a good performance through voice patterns only, the researchers may combine some other parameters for the identification of PD patients at an early stage because voice is not the only symptom that characterizes PD.

Ertuğrul et al. [14] presented a machine-learning model to detect PD disease at an earlier stage. Initially, the data are collected from the gaitpdb datasets that contain information about healthy people and PD patients. Eight sensors are placed under the foot for 2 min, and the recorded sensor information is converted into the LBP domain and processed through shifted 1D-LBP. The LBP signal value lies between 0 and 255, matched with a special and distinct pattern formed through the shifted 1D-LBP signal. Then, the histogram technique illustrated the 256 different signal patterns according to their corresponding signal. The statistical features such as correlation, entropy, and skewness are computed through the 1D-LBP histogram sensor. The classification and design features were processed through the machine-learning approach. The experiment evaluation on 10-fold cross-validation reported an accuracy of 88.89% and a sensitivity of 0.89. The authors implemented the proposed system on biomedical information, and in addition to this, some other symptoms such as speech may be considered in the future.

Marek et al. [15] stated that PD detection at the earliest age is crucial because there is no accurate method to detect PD. Either motor symptoms or non-motor symptoms can be detected through PD diseases. They proposed an automated multi-modal feature and machine-learning techniques based on non-motor symptoms for detecting PD. Based on biomarkers, the feature description is processed through the REM sleep Behavior Disorder Screening Questionnaire (RBDSQ) and CerebroSpinal Fluid (CSF). The Wilcoxon sum test is applied for the feature analysis. The PD classification is achieved through SVM, random forest, and logistic regression. The experimental result reported a 96.0% accuracy for the tested dataset.

Acharya et al. [16] differentiated PD patients from normal persons by drawing movements. They investigated handwriting markers for muscular movements and interpretation of other activities of the patients. To experiment with this model, the dataset was categorized into two parts, i.e., 20 healthy and 57 PD patients. The data pre-processing was achieved through five different score vectors. The Normalized Velocity Variability (NVV) is applied to identify the speed of the pen of the subject. They applied the NVVALL score to focus on healthy and PD patients. The receiver operating characteristic (ROC) was observed to be 0.9354. The UPDRS score represented the writing behavior of PD patients on the Hoehn (H) and Yahr (Y) scale. Naïve Bayes, Adaboost, and logistic regression methods were applied for the PD classification. The experimental results reported the highest accuracy of 90.90% through Naïve bays and the lowest accuracy of 86.36% through the SVM classifier.

Nilashi et al. [17] presented a new automated method to predict and monitor PD disease patients with characteristic motor and total UPDRS. Clustering was applied to form a cluster with similar characteristics and merge similar features into one cluster. Thus, in the output, different clusters were created of different sizes. A self-organizing map (SOM)-based cluster approach effectively handled the large datasets and provided similar clusters. The R^2 method was utilized to evaluate the value of the SOM. In addition, the PCA method was applied for the feature analysis of the cluster approach. Further, the deep belief network was also applied to identify PD patients better. The RMSE method was applied to find the exact and accurate information about PD patients. They also included the SVR [18] and ANFIS [19] learning techniques and presented an accuracy of 89.4%.

Sztaho et al. [20] proposed a method to detect the severity level of Parkinson's disease through speech signals. To implement this method, the authors used the Hungarian speech database that consists of the speech signals of 51 patients. The severity of patients was classified according to the Hoehn (H) and Yahr (Y) scales. The sound card was utilized to record the speech of patients. The feature extraction technique was utilized to categorize speech, such as pause ratio and speech speed. The authors implemented this method using two types of detection methods, viz. binary classification and regression. The classification method was processed by the K-Nearest Neighbor (K-NN) method and SVM. They utilized two types of regression methods, viz. linear regression and support vector regression. The Root Mean Square Error (RMSE) was used to evaluate the performance of the regression method. The binary classification method reported an overall accuracy of 83.56% for the read text, 85.11% for the speech signal, and 84.62% for both.

Xia et al. [8] proposed a dual model based on the deep-learning method to detect the characteristics of Parkinson's disease from the gait signals. The left and right gaits were recorded by the VGRF tool. The severity level is identified with the help of the Hoehn (H) and Yahr (Y) scales. They applied an N-size vector for feature extraction and selection through this vector gait cycle detection, which is processed by fixing the N = 150. The dual-mode consists of two-channel levels for processing separate signals. The VGF gait signals are first passed through the two-layer CNN model to understand the features of gait signals, followed by LSTM for temporal features. Further, they utilized the attention method, which provided meaningful information on the subject that can be accessed with the help of a score. A Fully Connected layer (FC) was incorporated to combine both left and right gait signals, followed by final classification through the SoftMax layer. The efficacy of the model was measured using a five-fold cross-validation approach. The model experimentally reported an accuracy of 99.31% and a sensitivity of 99.23%.

Park et al. [21] compared the performance of the PD diagnosis system through SVM with the two methods, viz. Multiple Layer Perceptron (MLP) and Radial Basis Function Network (RBN). Seventy-four-year-old data are utilized to implement this method, and the signal Electromyograph (EMG) is recorded through the AgCI conductor. In the pre-processing stage, signals are firstly filtered into 3 to 10 Hz by a type-2 filter followed by Fast Fourier Transformation (FFT) to identify the same frequency band of the tremor. After these steps, EMG signals are classified into two stages, viz. experienced and visual signal to detect the exact tremor status. The MLP network consists of the input layer, hidden layer, and output layer, and it is used to reduce the overfitting issue in the datasets. The status of tremors is detected through −1 and 1. On the other hand, the radial basis function utilized the fuzzy c-mean clustering method to identify the initial stages of the cluster. Overall, 81.14% accuracy was reported using the SVM classification of tremor status.

Hariharan et al. [22] presented an intelligent system based on a hybrid model. They initially incorporated the Gaussian mixture method as a pre-processing step to remove the unwanted noise present in the dataset. They also utilized two types of feature reduction methods, viz. PCA (Principal Component Analysis) to identify the hidden features presented in the datasets and LDA for mapping 22 features into a one-dimensional space. General Regression Neural Network (GRNN), Probabilistic Neural Network (PNN), and SVM were utilized for the severity classification of PD. The promising classification was reported based on the cross-validation method.

On the other hand, Balaji E. et al. [23] proposed a machine-learning model that can assist clinicians in detecting the stages of PD through gait information. Gait information provides all mobility information about healthy people and PD-affected people. This model is trained and tested with the public datasets based on the gait pattern provided by Physionet. VGRF is placed under the foot to provide gait information through different sensors. The feature extraction process is achieved using statistical and kinematic feature extraction approaches. The statistical feature extraction process is used to identify the four levels of PD through H and Y scales. It created a 16×166 matrix based on the sensor and subject-level PD severity. In contrast, the kinematic features were used to identify PD patients' steps, swing time, and speed. A 10-fold cross-validation is adopted in which 90% of data are used for training purposes and the remaining for testing purposes. Decision Tree (DT), SVM, Bayes, and Ensemble classifier were utilized for the classification. Experimental evaluation reported that the Decision Tree (DT) classifier has the highest accuracy of 99.04%, the sensitivity of 99.06%, and the specificity of 99.08%.

Kim et al. [24] presented a novel approach based on CNN to detect the severity rate of Parkinson's disease by performing tremor quantification from raw datasets. For experimental evaluation, 92 PD patients' tremor sensor datasets were collected using a wrist sensor device as wearable equipment. A neurologist was provided with the information on PD on four-level severity, i.e., normal to severe, based on the unified Parkinson's disease rating scale (UPDRS). In addition, they designed a neural network to assess the severity in PD patients. In this network, 2D images are used as input for the convolution layer, and a 3×50 convolution filter combines both local and sensor information. They processed the input signals computed by the wrist sensor in the form of gyroscope signals and accelerometer signals. Experimental evaluation depicted a classification accuracy of 85%.

Oung et al. [25] addressed that the existing system does not differentiate between people infected with Parkinson's Disease (PD) and healthy people. Therefore, to handle this issue, they proposed a multi-class classification system to classify PD severity levels (low, mid, high) and a healthy control group. For experimental evaluation, datasets of 65 persons of different ages were collected from the Neurology hospitals and the severity level in Hoehn (H) and Yahr (Y) was rated through the UPDRS measure. The dataset signal is assorted through two stages, i.e., motion and speech-based signals. The speech signals were recorded through the Motion Node Bus (MNB) from the IMU wearable device, and the speech signals were recorded through the audio sensor, i.e., a headset placed at 5 cm away from the mouth. The authors acquired the Empirical Wavelet Transform (EWT) to decompose the motion signals to find the approximate information from the detailed information, and the Empirical Wavelet Packet Transform (EWPT) was developed to decompose the speech signals. The EWPT method uses Fast Fourier transform (FFT) to obtain the exact frequency, i.e., lies between 0 and π. Feature extraction was processed through the Hilbert transform based on amplitude and frequency. Extracted features are categorized into three groups: speech signals, motion signals, and a mix of motion and speech. They employed Probabilistic Neural Network (PNN), Extreme Learning Machine (ELM), and K-Nearest Neighbor (kNN) for the classification. Experimental evaluation reported an accuracy of 90% on classification using an Extreme Learning Machine (ELM) for both motion and audio signals.

Recent studies analyzed that it is hard to diagnose PD at an earlier stage. Many remote detecting tests were utilized to detect the PD severity and realized that variables in gait signals could easily distinguish PD patients from healthy ones. In this regard, Cantürk et al. [26] proposed a system to detect PD patients' severity using gait signals. Their system was trained and tested with 306 publicly available signals with 93 PD patients and 73 healthy subjects based on different categories. The gait system was measured through Ultraflex Computer Dyno Graphy (UCDG) with eight sensors placed under the foot. The Fuzzy Recurrence Plots (FRP) convert the signals into texture representations for both PD and healthy patients. Further, AlexNet was applied to extract the deep features, followed by implementing SVM and k-Nearest Neighbor (kNN) for binary and multi-class classification. The experimental result of the kNN method reported an accuracy of 99%, whereas the SVM reported 98%.

Zhao et al. [27] presented a machine-learning method to detect the severity level of PD from the gait data. This is the hybrid technique consisting of both Long Short-Term Memory (LSTM) and a Convolutional Neural Network (CNN) to recognize the spatial time-based pattern through the gait data. The hybrid model has five convolution layers and two layers of LSTM to detect the severity rate in PD patients. The authors acquired two convolution layers of 5×5, in which the first layer is mapped with 32 features and the second one is mapped with 64 features. LSTM and CNN are trained and tested on the PhysioNet [28] dataset. The pre-processing and L2 normalization were applied to reshape the datasets into $100 \times 19 \times N$ (N −"Ga:13592, Si:7744, Ju:11734"). Further core parameters of LSTM were transformed to achieve better classification results into four levels, viz. normal (severity 0), severity 2, severity 2.5, and severity 3. Final classification was achieved using the SoftMax layer. The model reported 98.70% accuracy for the first dataset, 98.41% for the second dataset, and 98.88% for the third dataset. However, this method provided better accuracies in PD detection, and this model is the baseline for detecting the PR disease.

An automated machine-learning-based method is proposed to detect and identify the level of severity of Parkinson's disease from the gait data by Maachi et al. [29]. They employed a Deep Neural Network with the help of a 1D convolution Neural Network. This algorithm has divided the information into two parts, viz. Parkinson's and a control group. For the experiment, publicly available datasets are used and cited from the PhysioNet. The datasets contain 93 patients with Parkinson's disease and 73 patients in the control groups. The Vertical Ground Reaction Force (VGRF) based on 18-1D signals provides the information of a recorded walk with the foot sensors positioned below the foot. The VGRF signal is divided into datasets into m-parts that are based on subject categorization. Further, these parts are the input of the proposed method of DNN. The DNN method is processed with two parts, viz. 18 parallel 1D and a fully connected network. The feature extraction is processed through the 18 1D-CNN. The Parallel 1D network has taken input from the VGRF signal and processed it through the four convolution layers, which are fully connected. Further, this layer has extracted the features used to help categorize the PD and control groups. The output layer generates one neuron to detect the disease and five neurons to classify the level of severity that were categorized into five classes based on some criteria. This method reports an accuracy of 98.7% in detecting the severity and 85.8% accuracy in the classification of the severity level.

Prashanth et al. [30] addressed different stages of PD as a very important factor in a medical decision. The subject's disordering features were measured by UPDRS, but it does not give information about the PD stage. In this paper, they proposed a new model based on machine-learning to detect the PD and different stages of PD (early, normal, and moderate). This hybrid model supports SVM, AdaBoost, and RUSBoost-based and ordinal logistic regression (OLR) classifiers. It utilized the Parkinson's Progression Markers Initiative (PPMI) datasets with 197 healthy and 434 PD subjects. The statistical analyzer is used to classify the features into three categories based on a filter. They used classification algorithms such as random forests, SVM, and logistic regression to classify the PD stages. The validation of the performance was measured by the 10-fold cross method. The experimental results indicated that AdaBoost reports the highest detection accuracy of 97.46% for the normal PD subject, and SVM reports 98.04% for the early stage of PD detection. Although automated detection improves the stage of PD, there is a need to address more stages for PD patients.

Prashanth et al. [31] also presented a prediction model based on machine-learning to distinguish healthy and early PD patients. The dataset utilized for the experiment is from the Parkinson's Progression Markers Initiative (PPMI). They further applied the Patient

Questionnaire (PQ) to analyze the dataset. In PPMI, data are arranged in the longitudinal format, so they performed the record and subject-wise cross-validations. The dataset is divided into 90% training sets, and the remaining are test sets. To remove the redundancy and select the appropriate features, they have used three different selection methods, viz. Wilcoxon rank, Least Absolute Shrinkage and Selection Operator (LASSO), and Principal Component Analysis (PCA). The Wilcoxon rank method is acquired for the significant features through the sum test. The LASSO method is also applied to shrink the datasets, and the PCA method is the reduction approach used for decomposing the multivariate datasets into one manner format. The authors have processed the logistic regression, SVM, random forests, and boosted trees for the classifications. The experimental results indicated 96.50% accuracy using SVM through the subject-wise validation.

Aydın et al. [32] presented the Hilbert–Huang Transform (HHT) method to detect the severity of Parkinson's Disease (PD) from the gait pattern. The datasets are utilized from the PhysioNET [28], and the signals, such as step swing time, are measured through the VGRF sensor. The authors applied three types of feature selection techniques, i.e., the filter approach, the wrapper approach, and the embedded approach. The filter approach is used to identify the common characteristic of the training datasets. The wrapper feature selection approach is applied for mapping with relevance and extracting the optimal features, and the last approach is applied to check the performance of the features. They also applied the feature creation method, and a 10-fold cross-validation approach checks the performance of this method. The regression tree classification approach is processed to distinguish PD patients from healthy ones. The experimental results showed that the accuracy of the proposed system is 98.79%, sensitivity is 98.92%, and specificity is 98.61%. The performance analysis of some PD identification approaches is depicted in Table 1. On the other hand, a systematic review of AI-based approaches for the diagnosis of PD is presented by Saravanan et al. [33].

Table 1. Performance analysis of various Parkinson's Disease (PD) identification approaches.

References	Input	Features Extraction Approach	Classifier	Performance Accuracy (%)
Pereira et al. (2016) [10]	Spiral, Meander images	Zhang–Suen-based thinning algorithm	NB, OPF, SVM	67.00
Cantürk (2021) [26]	Gait Signals	Alexnet	SVM, kNN	99.00
Xia et al. (2019) [8]	Gait information	CNN 2D	CNN & LSTM	99.31
Zhao et al. (2018) [27]	Gait information	CNN model	CNN & LSTM	97.86
Hariharan et al. (2014) [22]	Speech samples	PCA, LDA, SFS	LS-SVM, PNN, and GRNN	100.00
Prashanth et al. (2014) [11]	SPECT images	LR	SVM, LR	96.14
Sztaho et al. (2017) [20]	Speech Rhythm	Feature Vector	SVM, Deep learning	94.87
Maachi et al. (2020) [29]	Gait signals	Manual method	Deep 1D-convent	98.70
Lahmiri and Shmuel (2019) [13]	Voice pattern	Wilcoxon-based	SVM	92.21
Ertuğrul et al. (2016) [14]	Gait signals	1D-LBP	LR, MLP, NB, BAyesNT	88.90
Yurdakul et al. (2020) [34]	Gait Signals	Local Binary Patterns	Generalized Linear Regression Analysis (GLRA) and SVM	98.30
Oung et al. (2018) [25]	Speech and Motion signal	Wavelet Energy and Entropy	kNN, PNN, ELM	95.93
Prashanth and Roy (2018) [30]	Motor signals	Wilcoxon rank-sum test	SVM, Random Forest, probabilistic ADAboost-based ensemble	97.46
Aydın and Aslan (2021) [32]	Gait Pattern	One R Attribute Evaluation and vibes algorithm	Hilbert-Huang transform	98.79
Kim et al. (2018) [24]	Wrist sensor pattern	Convolutional filters of CNN	CNN	85.00
Balaji E. et al. (2020) [23]	Gait signals	Statistical analysis	DT, BC, EC and SVM	99.50

Discussion

As stated earlier, the movement disorders caused by Parkinson's Disease are not uniform in all patients. Deep-learning models play crucial roles in developing automated tools for evaluating a patient's gait. It is obvious that to cure any disease, its detection must take place at the early stages. To detect PD at an early stage, both artificial intelligence and machine-learning-based techniques are contributing to a great extent, e.g., feature extractions and pattern recognition from motor symptoms, voice pattern recognitions, etc. Plenty of work has been carried out to identify PD at an early stage, but this field is still in its infancy stage. It is observed that very few works are available on PD identification using non-motor symptoms, and the availability of PD datasets is not adequate to develop automated models. The researchers may consider these issues while developing an automated model for the detection of PD at an early stage with excellent efficiency.

3. Severity Identification of Diabetic Retinopathy

Excessive glucose growth in the blood causes diabetes that subsequently harms other components of the human body, i.e., eyesight loss, kidney malfunctioning, nerve failure, damage to blood vessels, etc. This excessive amount of glucose leads to damage to the retina's blood vessel, which is the main cause of Diabetic Retinopathy (DR) disease. Blurriness, color difficulty, floaters, and dark vision are early symptoms of DR disease. It has become one of the major reasons globally for visual losses. Timely diagnosis and subsequent treatment of its several stages/severities can save visual loss to some extent. Several computational models are presented by plenty of researchers for the detection of DR from fundus images. Shankar et al. [35] presented a novel automated model called HPTI-v4 (Hyperparameter Tuning Inception-v4) DR detection from color fundus images. Initially, the contrast of fundus images is enhanced using Contrast Limited Adaptive Histogram Equalization (CLAHE) [36] followed by histogram-based segmentation. HPTI-v4 then processes the segmented images for feature extraction followed by a Multi-Layer Perceptron (MLP) classifier. Experimental results on the MESSIDOR (Methods to Evaluate Segmentation and Indexing Techniques in the field of Retinal Ophthalmology) DR dataset exhibited that HPTI-v4 outperforms other state-of-the-art deep-learning models (i.e., ResNet [37], GoogleNet [38], VGGNet-16 [39], VGGNetCOVID-19 [39], VGGNet-s, AlexNet [40], Modified AlexNet, and DNN-MSO). The dataset consists of 1200 posterior pole eye fundus images that were mainly classified into four classes, viz. normal, stage-1 (images with some microaneurysms), stage-2 (image with both microaneurysms and hemorrhages), and stage-3 (images with high microaneurysms and hemorrhages). In addition to HPTI-v4, 10-fold cross-validation was used to subdivide the dataset into training and testing sets; and Bayesian optimization was employed for selecting an optimal set of hyperparameters. The proposed HPTI-v4 obtained the highest accuracy of 99.49% as compared to other models under consideration.

Wang et al. [41] presented a hierarchical multi-task framework based on deep learning for simultaneously detecting DR features and severity levels. Severity levels in DR are characterized by the presence of various signs in the fundus images. DR severity identification becomes easier if the DR disease-related signs are present in the fundus images. Earlier, Wang et al. [42] investigated the feasibility of diagnosing DR severity levels and the presence of DR-related features. Their hierarchical multi-task framework consists of two main tasks, viz. severity diagnosis of DR and identification of DR-related features. Their architecture consists of one backbone squeeze-and-excitation (SE) network [43] for feature extraction and two neural networks (one for DR-related feature extraction and the other for severity detection). To validate their framework, an experimental evaluation was conducted on two independent test sets, followed by a grader study to compare the performance of the proposed framework with experienced ophthalmologists. Results depicted that the proposed model was able to improve the performance of traditional machine-learning-based approaches.

Torre et al. [44] developed a new method to detect diabetic retinopathy using a deep-learning classifier. For the system's good performance, they categorized the retinal images according to the level of severity. To experiment with this model, the EyePacs dataset [45] from Kaggle was utilized. The ophthalmologists classified the images into different criteria based on the grading scale. The authors also applied the deep-learning model modifications to classify the retinal images. They used the optimal retina images whose diameter size is 640 pixels. The Rectified Linear Unit (ReLU) is applied for the activation function with an epoch size of 30. The multi-class classification model is used to obtain better identification of the disease as well as severity levels. The experimental results reported a specificity of 91.1% and a sensitivity of 90.8%. The main advantage of this model is classifying the five-severity levels of DR disease and identifying the score levels of each class.

Shankar et al. [46] proposed a Synergic Deep-Learning Model (SDL) to classify the severity level of diabetic retinopathy fundus images. The model utilized the MESSI-DOR [47] dataset, which contains 1200 color fundus images. The first step was to apply the pre-processing in which each image was converted into RGB format. Then the segmentation was performed through a histogram to fetch the green color of the image for further information. The SDL model has processed the classification to classify the DR image into different stages. Different performance matrices, such as accuracy, sensitivity, and specificity, are used to evaluate the system. The model experimentally proved 99.28% accuracy for the classification, 98.54% sensitivity, and 99.38% specificity.

Nowadays, few smartphone-based systems help in performing the retinal screening of diabetic patients. Still, the accuracy of DR identification is based on the quality of the image and the region of the view. Therefore, the smartphone system must consider a highly compact design to provide accurate information. In this regard, Hacisoftaoglu et al. [48] presented a new system to detect DR based on a smartphone-based system through a deep learning approach. The system uses transfer learning approaches such as GoogLeNet [38] and ResNet [37]. The validation of the experiment has been processed through different datasets, i.e., MESSIDOR and EyePacs. The experimental result reported an AUC of 0.99, a sensitivity of 98.2%, a classification accuracy of 98.6%, and a specificity of 99.1%. On the other hand, Son et al. [49] developed a new model that helps to identify the abnormality in DR patients based on retinal images. They utilized three datasets for the validation of the approach; 103262 images from 309786 were used to develop the model, and for the testing, other external datasets were used. Finally, the MESSIDOR dataset [47] was used for comparison purposes. The deep-learning model has been applied to classify the abnormality in retinal images. The classification output has oscillated from 0 to 1, which shows the probability of finding the existence of abnormalities. The experimental evaluation reported an ROC value of 96.2% to 99.2%. The proposed deep-learning model not only categorizes the finding by accuracy but also calculates the salient features of the images.

A new automated method based on deep CNN for detection of DR is proposed in [50]. They utilized two datasets to validate their study: the EyePACS and MESSIDOR-1 & 2. The pre-processing has been performed in both online and offline stages. In the online stage, the image is cropped in the desired shape, followed by the removal of the black border of the image, whereas in offline mode, the pre-processing has been processed by the augmentation method. The results of the model show better accuracy on the same public datasets compared to other existing algorithms. Moreover, the suitable preliminary process for screening larger numbers of patients for an automated system is batch processing and minimum assumption time. The efficient screening process helps to obtain the model's best results. The model was found to enhance the 0.92 AUC for the MESSIDOR-2 dataset [51] with a sensitivity of 81.02% and a specificity of 86.09%.

The automated NAS (Neural Architecture Search) machine-learning model [52] predicts the DR patients with no and severe stages of DR disease. To train and validate the model, a Kaggle dataset comprising the information of 3662 images was used. Out of 3662 images, 3113 images were used for the training data sets, and the remaining were used for the testing datasets. Harikrishnan et al. [52] first applied the pre-processing steps to remove the unwanted noise and other information. They resized the image in a particular format, then applied the Gaussian filter to improve the image quality. The NAS acquired the RNN (Recurrent Neural Network) to add more functions with different combinations to obtain the optimal solution. The accuracy of the model was reported to be 75%. To develop this model, the learning rate was set as 0.0001, and the initial weight was chosen as the net image weight. The authors observed that the model obtained the minimum accuracy when including the dense layer without a pre-processing stage. The proposed model was validated through the existing database based on E-Ophtha Exude. This model also reported a sensitivity of 76.6% and a specificity of 77.1%.

Washburn et al. [53] proposed a new system design to detect the retinal image at the earliest phase. The model utilized the public retinal image datasets. They applied the image acquisition for the screening method with an existing database. The next step was pre-processing, which helped in improving the images and removing the unwanted noise, which consisted of three steps, viz. converting color space, filtering, and enhancement of image for the quality of the retinal image. The region-based segmentation was processed to identify the boundary of the backside images. The Gabor wavelets were applied for the feature extraction approach to extract useful information from large datasets. Further, the adaptive boost classification was applied to obtain a better prediction result for the retinal images. The system experimentally reported an accuracy of 98.4%, a specificity of 98.8%, and a sensitivity of 98.4%.

Li et al. [54] developed a new optical coherence tomography system based on deep learning to diagnose diabetic retinopathy at an earlier stage. The system was validated with OCT images collected from the Wenzhou Medical University (WMU). The dataset consists of 4168 OCT images collected from 155 patients. A total of 1112 images out of the 4168 images belong to DR grade 1 and 1856 to DR grade 0. The pre-processing was performed by resizing the OCT images to 224 × 224. The OrgNet and segmentation calculated the deep characteristics to obtain an extra feature for better classifications. In this work, the feature merging was processed through the summation method in place of concatenation. The augmentation technique was processed to enhance the neural network environment. The system is provided with the DR multi-classification, such as grades 0 and 1. An accuracy of 92%, specificity of 90%, and sensitivity of 0.95 were recorded for grade 0 DR classification.

The fundus image is the perquisition stage to calculate the accurate severity rate of DR. The manual scoring procedure is considered challenging because of the dissimilarity in morphology, number, and image size. In this regard, Sambyal et al. [55] presented an automated method based on segmentation that helps detect the boundaries and helps ophthalmologists quickly detect the DR with severity grades. They developed an improved U-Net architecture inspired by U-Net [56] that is pretrained on ResNet34 [37]. It contains the encoder and decoder at their left and right parts, respectively, resulting in better system performance. This method is also useful for improving the result compared to the existing method. The system is validated on two public datasets, viz. e-ophtha [57] and IDRiD [58]. The experimental result reported 99.88% accuracy, 99.85% sensitivity, and 99.95% specificity for the IDRiD Dataset. For the e-ophtha datasets, the accuracy was 99.98%, with a sensitivity of 99.88%.

Quellec et al. [59] proposed a machine-learning-based solution for diabetic retinopathy detection at an early stage. The authors utilized heat map concepts to identify the importance of a particular pixel in an image. To produce a good quality heat map, they trained the ConvNets network with the help of the backpropagation method. Three different categories of the dataset were used in this study (i.e., Kaggle Diabetic Retinopathy, DiaretDB1 [60], and 'e-ophtha'). The proposed method is validated on approximately 90,000 fundus images. They followed data augmentation and pre-processing processes to transform images (i.e., 448 × 448 pixels). To train the dataset, the three ConvNets were trained to detect diabetic retinopathy. The performance of the proposed model on different datasets was found to be 0.954, 0.955, and 0.949, respectively.

Liu et al. [61] proposed a weighted path CNN (WPCNN) model to detect the diabetic retinopathy with severity levels. The system was validated through the raw database comprising 60,000 images categorized into 0 and 1 on severity scales. The authors divided the datasets into 80% and 20% training and testing sets. They scaled and resized the images to 299 × 299 through the pre-processing steps. The data augmentation method was applied to fit the image at standard formation such as right, up, left side, etc. The convolution layer processed the feature extraction through CNN and extracted the noteworthy features from the retinal fundus images. During the experimental setup, the authors suggested an over-fitting issue if the size of the network expands. The coefficient of the WPCNN was enhanced by using the backpropagation method. The system experimentally reported 94.02% accuracy in comparison to the existing models. It also achieved an AUC of 0.9823 and an F1-score of 0.9087, the highest compared to the existing methods. Although this proposed system achieved a good performance, the authors pointed out that adding more features to the automated system can improve the overall performance of the system.

Hua et al. [62] introduced a trilogy of skip-connection deep networks (Tri-SDN) to analyze the DR images. The new attribute based on EMR was introduced to identify the risk probability to increase the system's performance. In the first phase, the feature extraction was performed from the ImageNet database. The ResNet [37] is pre-trained by the multiple convolution layers. Further, the corresponding vector mapped the feature map to identify the risk factor in the DR images. The deep learning network was built with the two skip connection blocks to identify the characteristics of the retinal images. The authors also applied the EMR-based value to identify the risk factor of the severity because it provides the numerical value, and in this work, 22 risk factors are involved. The EMR-based value is used for the DR orientation characteristic to improve the performance. The system was validated with the historical information of the 96 patients collected from the medical university in South Korea. The system experimentally reported an accuracy of 90.6%, a sensitivity of 96.5%, and an of 88.8% of AUROC, which is higher than the existing models such as random forest and 11-layer CNN. Although the system provides a good performance compared to the existing algorithm, it needs to add more retinal images to make the system more efficient so that the ophthalmologists can make easy decisions.

Reddy et al. [63] claimed that DR could be easily detected through different machine-learning algorithms. For this experiment, they used the DIARETDb1 [60] data set containing 89 images, out of which 5 are of the normal stage, and the rest of the 84 images are Mild Non-proliferative DR (NPDR) cases. The pre-processing was achieved through the grey scaling method, image copper, and image resizing to remove the noise and improve detection accuracy. They applied the segmentation technique to visualize the blood vessels in the retina. Further, the region growing technique was utilized to identify whether the pixels belong to the same region or different regions. The clustering method was applied for the data analysis. Feature extraction was applied to generalize and extract different features from the data sets. For classifications, the authors employed the SVM, k-NN, and probabilistic neural network (PNN) techniques. Different matrices such as accuracy, TPR, and FPR were employed to evaluate different classifiers. They experimentally determined the best accuracy (96.57%) through cross-validation using SVM.

Wu et al. [64] proposed an automated hierarchically Coarse-to-Fine network (CF-DRNet) tool to detect the DR, as depicted in Figure 4. They applied a convolution neural network to classify severity, viz. no DR, mild DR, moderate DR, severe DR, and proliferative DR. The experiment was performed on 88,400 fundus image datasets taken from Kaggle. This technique integrates three steps in which the first step performed the pre-processing, the second phase performed the CF-DRNet module, and the last stage performed the aggregation concept. The pre-processing was performed through image enhancement, image normalization, and data augmentation. Image enhancement is applied to remove unwanted noise with varying luminous factors. Image normalization is used to reduce the complexity and normalize the pixels of images in the coarse network. Data augmentation is performed to reduce over-fitting and imbalance issues in the datasets. Further, the CF-DRNet is applied to check the presence of DR. For better detection, it is classified into two different networks, such as the coarse and fine networks. Then, the aggregation method is applied to determine the level of DR and No DR. The authors experimentally claimed that CF-DRNet reported the highest accuracy of 83.10%, sensitivity of 53.99%, and specificity of 91.22%.

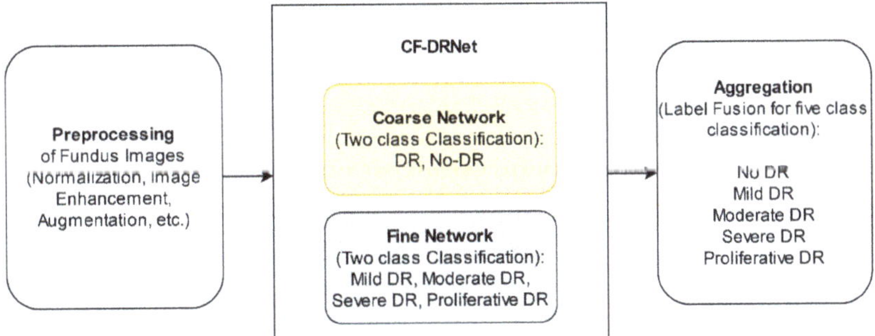

Figure 4. Coarse–Fine Diabetic Retinopathy Network [64].

The detection of diabetes with different severity levels is a complicated system; hence, it is very difficult and time-consuming. In this regard, Pratt et al. [65] proposed a machine-learning-based approach with 75% accuracy for the diagnosis of diabetes in five levels of severity classifications. To train the network, the KAGGLE dataset with 80,000 retinal images was utilized. The color normalization has been processed through OpenCV for the categorization of the data into a different age, group, and authenticity. Further, they resized the images into 512×512 pixels for the identification of complex features. Stochastic gradient descent was utilized for training the datasets with a 0.0001 learning rate for five epochs. To concede, the first 10,290 images were pre-trained through the CNN network to classify the severity levels. Further, 5000 images took 188 s for the validation process. This technique achieved a specificity and sensitivity of 95% and 30%, respectively.

On the other hand, Yun et al. [66] proposed a backpropagation method to classify the DR into four categories, viz. normal, severe, moderate, and proliferative DR. The authors used 124 retinal images from Singapore University to process the work. This method was trained with 27 samples as training sets and the remaining as testing data samples. A feed-forward neural network was utilized to classify images into different classes. The pre-processing of images has been carried through the histogram and binarization process. Further, the ANOVA process extracted the features of retinal images into different areas and categories. The authors evaluated the model's performance on three matrices, i.e., accuracy, specificity, and sensitivity. The method reported 80% accuracy, 90% sensitivity and 100% specificity.

Akram et al. [67] developed a multi-model for categorizing severity levels of DR into normal, mild, moderate, and severe Non-Proliferative Diabetic Retinopathy (NPDR). This model is the hybrid of medoids and the Gaussian Mixture Model (GMM) for the best classification and solving the overfitting issue. The mean-based approach was utilized to remove the noise and background. The segmentation process has been processed through Gabor and the multi-layer thresholding processes. For processing, the authors utilized datasets such as DRIVE and STARE, which are easily available in the public domain. They divided the datasets into two parts: an image and a lesion. Further, the feature vector was used to classify the severity of NPDR through color and intensity factors. The performance was evaluated on accuracy, sensitivity, specificity, and AUC metrics. The model reported 97.56% accuracy and 97.39% sensitivity, with 98.02% specificity.

Mookiah et al. [68] proposed a system for automated classification of normal, Non-Proliferate Diabetes Retinopathy (NPDR) and Proliferated Diabetes Retinopathy (PDR) using retinal images. They applied pre-processing techniques such as Wiener filtering, gray level shading correction using low pass filtering, and contrast enhancement to remove noise and uneven illumination. They also removed the optical disk to reduce the number of false positives reported while detecting the lesions. The authors applied A-IFS Histon and the 2D Gabor-matched filter approach for segmentation. Further, they extracted the features such as blood vessel area, exudate area, bifurcation point count, Local Binary Pattern (LBP) energy, LBP entropy, Laws mask energy, and entropies from the fundus images. The authors employed Probabilistic Neural Network (PNN), Decision Tree (DT), and SVM for the classification. They applied genetic algorithm and Particle Swarm Optimization (PSO) algorithms to optimize the efficacy of the classifiers. The authors experimentally determined the threshold value as 0.0104 and claimed that PNN reports the highest accuracy of 96.15%, sensitivity of 96.27%, and specificity of 96.08%.

Chowdhury et al. [69] developed a method to detect DR through four levels categorized as normal, PDR, average PDR, and acute PDR. The categorization into four levels was completed through a random forest classifier. The pre-processing was achieved through a contrast enhancement technique which helped in extracting the RGB value from 120 retinal images. The contrast augmentation was completed through adaptive thresholding to remove the unwanted noise. For conversion to a binary image, the global threshold technique was adopted. The authors utilized a feed-forward neural network based on three-layer architecture. This technique reported 90% accuracy in normal cases, 87.5% accuracy in the case of acute NPDR cases, and 90% sensitivity and 100% specificity classification. Table 2 depicts the comparison of a few works based on the severity identification for Diabetic Retinopathy. On the other hand, Kaur et al. [70] presented a systematic survey of computational methods for DR diagnosis based on fundus image analysis.

Table 2. Performance analysis of some diabetic retinopathy identification approaches.

References	Input Image	Features Extraction Approach	Classifier	Performance Accuracy (%)
Hua et al. (2019) [62]	Fundus Images	ResNet 50	Tri-SDN	90.60
Wu et al. (2020) [64]	Fundus Images	Resnet	CF-DRNet	83.10
Pratt et al. (2016) [65]	Fundus Images	PCA	CNN	75.00
Chowdhury et al. (2019) [69]	Fundus Images	Feature Vector	RF, NB	93.58
Li et al. (2019) [54]	OCT images	Org_Net and Seg_Net	OCTD_Net	92.00
Hacisoftaoglu et al. (2020) [48]	Fundus images	Not mentioned	SVM, NB, RF	98.60
Akram et al. (2014) [67]	Fundus Images	Gabor filter	GMM and m-Mediods	97.56
Mookiah et al. (2013) [68]	Fundus images	LBP, LTE	PNN, DT, SVM	96.15
Sambyal et al. (2020) [55]	DR Images	ResNet 34	Modified U-net with ResNet	99.88
Liu et al. (2019) [61]	Fundus images	WP-CNN	WP-CNN	94.23
Washburn et al. (2020) [53]	Color retinal images	Gabor wavelets	AdaBoost	98.40
Yun et al. (2008) [66]	Retinal optical images	Imaging technique	Neural Network	84.00

Discussion

As mentioned earlier, early detection of the DR can help patients to recover quicker. In this regard, most of the work toward DR detection involves extracting features and classification using SVM and machine-learning-based models. The level of severity has defined various stages of the disease. Based on the literature analysis, it can be stated that although various techniques are still used to detect the disease, there is a need to improve the system in terms of complexity, detection time, and severity stages. Many existing techniques had worked on small datasets, and most of the algorithms did not elaborate on the method of feature extraction approaches. Therefore, there is a need to develop a hybrid as well as an efficient computed model to identify the severity of the DR disease at an earlier stage.

4. Severity Identification of Some Other Diseases

Infectious diseases, e.g., Tetanus and Hand Foot and Mouth Disease (HFMD), have a significant influence on the low- and middle-income countries [3]. Mortalities due to infectious diseases are associated with Autonomic Nervous System Dysfunction (ANSD). In addition to clinical examinations, the development of some automated computerized system is essential for the severity analysis of ANSD. In this regard, Tadesse et al. [3] presented an automated system to diagnose the severity of HFMD based on the fusion of multi-modal physiological data collected via low-cost wearable devices. For rapid diagnosis of severity levels of HFMD, their multi-layer decision system comprises an on-site triage process followed by a longitudinal model and the fusion of a multi-modal framework. Finally, deep-learning-assisted mapping of time-series physiological signals with images was obtained using spectrogram representations.

Mithra and Emmanuel [4] proposed a Gaussian Decision Tree-based Deep Belief Network (GDT-DBN) for the detection of the degree of infection in the patients of Tuberculosis (TB), as depicted in Figure 5. This network is the hybrid of a Deep Belief Network (DBN), Decision Tree (DT) and Gaussian model. Initially, the sputum smear image was used as an input to the system, followed by color space transformation. For segmentation, thresholding-based mechanism was adopted. Once the segmentation is achieved, the important features (e.g., length density, local direction pattern, histogram, etc.) were extracted. The authors used the ZNSM-iDB [71] dataset comprising microscopic digital images for training and testing of the model. A two-level classification was achieved using the proposed GDT-DBN classifier. However, it is ineffective in distinguishing abnormal mycobacteria from mycobacteria TB substances due to a similarity in their geometrical structure. As mentioned earlier about the immense popularity of deep-learning-based approaches in severity identification, Alebiosu et al. [72] presented a novel DAvoU-Net segmentation framework for improving the severity assessment of tuberculosis. Experimental evaluations on the ImageCLEF 2019 TB dataset showed promising results as compared to seven other models under consideration.

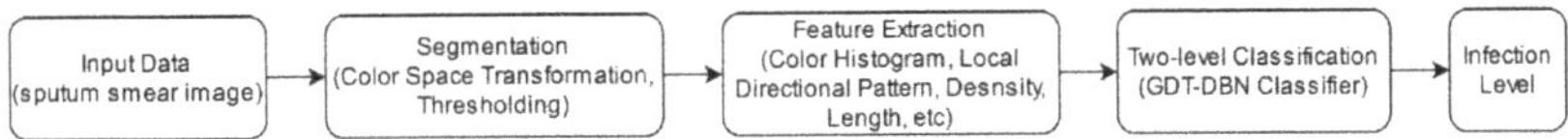

Figure 5. Block diagram of GDT-DBN classification for TB infection level identification [4].

Sepsis is a fatal disease if not detected at an early phase. Sequential Organ Failure Assessment (SOFA) is used to determine the level of Sepsis, but this method is totally dependent upon the laboratory measurements. In this regard, Aşuroğlu [5] presented a regression-based method to detect the level of sepsis. They used the Mart In Intensive Care (MIMIC)-III dataset [73] for experimental evaluations and binary classification for the prediction of sepsis. This model consumed less time and provided an AUC of 0.98, which is higher than other existing models. However, due to the large number of samples in the dataset, it seems difficult to balance the sepsis and un-sepsis samples, thereby causing a delay.

COVID-19 is a contagious disease that has spread all over the world, affecting the human body and health, and as such, it is very necessary to identify the level of severity at an early stage. Deep-learning-based approaches proved to be significant in the diagnosis of COVID-19 at earlier stages. CT-Scans are helpful in providing information about the severity of COVID-19 patients in medical reports. Cai et al. [74] presented a deep-learning-based approach for recognition of the infection region. Initially, patient data (RT-PCR, CT Samples) were collected and examined at different levels of severity, i.e., moderate, severe, and critical. In addition, the clinical data, including routine blood tests, clinical symptoms, demographic data, and treatments, etc., were also considered for the same reason. The

3DQI tool [75] was utilized for lesion quantifications, followed by data analysis with respect to disease severity and clinical outcomes. Chi-squared test, Student's t-test and other ML models are applied for the analysis of clinical data. Two U-Net models were employed for performance analysis on 99 chest CT scans. The mean Dice Similarity Coefficient (DSC) is found to be 0.981 for lung segmentation and 0.778 for lesion segmentation. On the other hand, Yao et al. [76] proposed a machine-learning-based model to detect the severity of COVID-19. The level of severity of COVID-19 in a person is recognized by SVM with 32 features. The algorithm was used on 137 COVID-19 patients, which were confirmed by Huazhong University. Among this dataset, only 17 patients were diagnosed with mild cases, 45 cases were diagnosed with moderate cases, and the remaining 75 patients were severely infected by this disease. The samples were categorized into 80% of testing and 20% of training sets. Feature extraction has been processed through the conservative recursive features (cREF) technique to enhance the performance of the model by eliminating redundant features. The model exhibited 81% accuracy and 0.699 specificity. Roy et al. [77] presented a novel deep-learning model called Reg-STN (Regularized-Spatial Transformer Network) based on Spatial Transformer Networks (STNs) [78] for analyzing Lung Ultrasonography (LUS) images. Disease severity was predicted for each input frame of LUS images. Each frame of the LUS image was classified into four different severity levels. In addition to this, they implemented a fully annotated database called "Italian COVID-19 Lung Ultrasound DataBase (ICLUS-DB)" [79] that consists of four-level scale labels. STNs are composed of three components: (i) a localization network that is responsible for the prediction of affine transformations, (ii) a grid generator for selecting grid coordinates from images, and (iii) a sampler for wrapping the input image. The evaluation of their method was conducted for accurate prediction and localization of COVID-19 at both the frame level and video level. On the other hand, Lai et al. [80] presented a combination of ML- and DL-based approaches for detecting novel coronavirus-infected pneumonia (NCIP) from CT images. Their model is based on a few-shot learning approach. For the segmentation of lung regions from CT images, a pre-trained network is utilized. Segmentation not only reduces the lesion detection but also the computation time, thereby avoiding false positives. For lesion detection and prediction, a multitask DCNN based on U-Net was utilized. Experimental results on a real patient's data revealed Area Under the Curve (AUC) of 0.91. Fouzia Altaf et al. [81] introduced a transfer learning concept by implementing augmented ensemble transfer learning that gives better results as compared to conventional transfer learning. To implement an efficient deep transfer learning model, they also modified the architecture of the existing network by adding an extra layer to change the dimensionality between the input image and the target image. They tested their model on the pre-trained ImageNet model. The authors used two different publicly available datasets for their execution purpose, namely Chest-Xray 14 radiographs and COVID-19 radiographs. Results on the Chest-Xray 14 dataset indicated a 50% reduction in the error rate compared to the baseline transfer learning technique. Another dataset was used for a binary problem as well as a multi-class classification problem. The modified trained model secured a 99.49% accuracy for the binary classification and 99.24% accuracy for multi-class classification. Zekuan Yu [82] identified 19 severity levels in CT scans through the classification of deep features. A total of 729 2D axial plan slices with 246 severe cases and 483 non-severe cases were employed in this study. By taking advantage of the pre-trained deep neural network, four pre-trained off-the-shelf deep models (Inception-V3, ResNet-50, ResNet-101, DenseNet-201) were exploited to extract the features from these CT scans. To identify the severe and non-severe COVID-19 cases, the features were then fed to multiple classifiers. Three validation strategies (holdout validation, tenfold cross-validation, and leave-one-out) were employed to validate the feasibility of the proposed pipelines. Experimental evaluations represented promising results as the DenseNet-201 with cubic SVM model achieved the best performance. Specifically, it achieved the highest severity classification accuracy of 95.20% and 95.34% for 10-fold cross-validation and leave-one-out, respectively. The established pipeline was able to achieve a rapid and accurate identification of the severity of COVID-19. This may assist physicians in making more efficient and reliable decisions. Many other works on COVID-19 diagnosis using AI-based approaches [83–85] were published by researchers. Chahar et al. [86] and Sinwar et al. [87] presented a survey of such learning models.

Taehoon Kim et al. [88] implemented a machine-learning model to identify the severity of sleep disorder breathing (sleep apnea). As a dataset, they considered patients that were presented at a sleep center with snoring while breathing during sleep. The authors developed four categories (i.e., normal, mild, moderate, and severe) of severity based on their Apnea Hypopnea Index (AHI) value among 120 patients. To capture the breathing sound, they used polysomnography, which records the sound using four different methods, as mentioned in Figure 6.

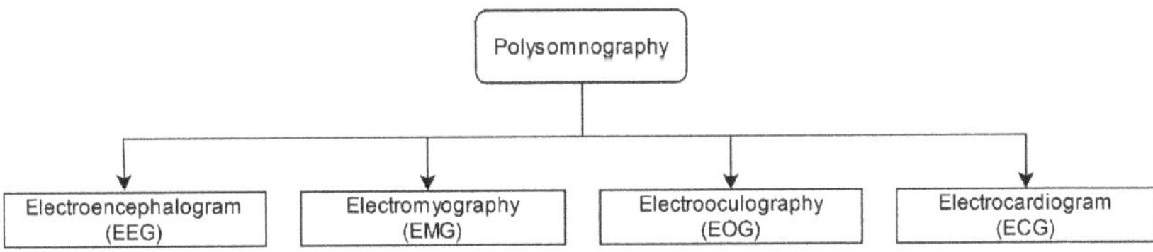

Figure 6. Different categories to measure the sound using polysomnography [88].

The recorded sound also had some noise components (i.e., machine noise, conversion noise); thus, the authors utilized two different filtering processes (i.e., spectral subtraction filtering and sleep stage filtering) to capture the useful information from the recorded sound. After filtering the sound, various audio features were extracted, and then four group and binary classification algorithms were applied to it. As a result, they scored 88.3% accuracy for the four-stage classifier and 92.5% accuracy for binary classification.

Linda A. Antonucci et al. [89] illustrated a machine-learning model to identify psychosis at an early stage. To implement the model, the authors used a support vector classifier and cross-validation section. They trained the model on approximately 105 samples composed of 71 samples of healthy controls, and 34 were psychosis samples. A total of three tests were evaluated on the samples, namely the discovery sample (healthy controls vs. psychosis), clinical validation sample (healthy controls vs. early stage of disease), and validation of familial risk (healthy controls vs. familial high risk). The resultant accuracy achieved for all three above-mentioned tests was found to be 72.2%, 63.5%, and 44.2%, respectively. The performance of the system may be improved with the help of a large dataset because a small dataset may lead to an overfitting issue.

Ahmad Abujaber et al. [90] implemented two different machine-learning models (i.e., linear regression and Artificial Neural Network) to predict the severity level of traumatic brain injury. The authors included 785 patients' (581 survived and 204 deceased) data as a dataset in their research. Pre-processing steps were also applied to the gathered dataset in the form of cleaning and transformation. The trained model achieved an accuracy of 87% using LR and 80.9% using ANN. In the end, they concluded that the LR model provided good results compared to an ANN.

Zeng Z et al. [91] identified local recurrences in breast cancer using Electronic Health Records (EHRs). They reviewed the development corpus of 50 progress notes and extracted partial sentences that indicated breast cancer local recurrence. MetaMaps were used to process these partial sentences to obtain a set of Unified Medical Language Systems (UMLS). After using MetaMaps on patients' progress notes, the sets that came under positive concept sets were retained. An SVM was trained to identify the local recurrences using these features with the pathology records of each patient. The model was compared with three baseline classifiers using either full MetaMap concepts, filtered MetaMap concepts, or bag of words. The model achieved the best AUC of 0.93 in cross-validation and 0.87 in held-out testing. This model provides an automated way to identify local breast cancer recurrences as compared to a labor-intensive chart review. By minimally adapting the positive concept set, the study can be replicated at other institutions with a moderately sized training dataset.

Kwon et al. [92] presented an automated classification of Knee Osteoarthritis (KOA) by combining both deep-learning and machine-learning approaches. Their automated system is based on the Kellgren–Lawrence (KL) grading system, gait analysis data, and radiographical images. Inception-ResNet-v2 was utilized for extracting relevant features from radiographical images followed by KOA multi-classification using SVM. Experimental results on both radiographical images and gait data indicated that both radiographical images and gait data are complementary for KOA classification.

Alzheimer's disease (AD) is one of the most common neurodegenerative diseases in the world. Currently, the diagnosis of AD is carried out using Mini-Mental State Exam (MMSE), which is quite a complex and time-consuming process. Martinez-Murcia et al. [93] presented an autoencoder-based deep-learning methodology to find out the relationship between neurodegeneration and cognitive symptoms. For the analysis and visualization of distortion of extracted features, regression and SVM-based classification techniques were employed. Experimental results on the ADNI dataset revealed the classification accuracy to be 84%. On the other hand, Sethuraman et al. [94] evaluated the severity of Alzheimer's Disease using Biomarkers. They utilized an ADNI Dataset [95] that comprises neuroimages of persons affected by AD. Their deep-learning-based model showed a performance accuracy of 96.61%.

Discussion

Computational intelligence-based methods are used in a variety of ways to strengthen the medical field. It is hard to imagine the existence of the medical field and the subsequent treatment of several critical diseases without CI-based methods. In this section, a critical review of various CI-based methods for identifying the severity of diseases is presented. A variety of diseases (e.g., COVID-19, sleep disorder, psychosis, brain diseases, breast cancer, knee osteoarthritis, sepsis, tuberculosis, etc.) are covered for severity identification by different researchers using various techniques (deep belief networks, decision tree, Chi-squared test, Student's t-test, regression, deep learning, etc.). The performance of these systems on a single type of data (e.g., imaging data, sensor data, etc.) is found to be satisfactory. However, in the future, hybrid systems (comprising several types of data as well as an ensemble of several techniques) need to be deployed to strengthen the medical field.

5. Some Public Repositories for Disease Severity Identification

The dataset plays a very crucial role in analyzing the performance of disease identification methods. Table 3 presents some public repositories that can be utilized to conduct disease severity identification tasks, mainly on DR, PD, and COVID-19.

Table 3. Public datasets available for conducting disease severity identification tasks.

Contributor	Name of Database	Modality	Disease
EyePACS [45]	EyePACS	Fundus Images	DR
Decencière et al. [51]	MESSIDOR	Fundus Images	DR
Porwal et al. [58]	IDRiD fundus	Fundus Images	DR
Kauppi et al. [96]	DIARETDB0	Fundus Images	DR
Kauppi 2007 [96]	DIARETDB1	Fundus Images	DR
· S. R. Rath [97]	Diabetic Retinopathy	Fundus Images	DR
Chalakkal et al. [98]	UoA-DR	Fundus Images	DR
J. Staal et al. [99]	DRIVE	Fundus Images	DR
M. Goldbaum [100]	STARE	Fundus Images	DR
Decencière [101]	E-ophtha	Fundus Images	DR
Clayton et al. [102]	HandPD	Handwriting images	PD
Goldberger et al. [9]	PhysioNET	Spiral & meander image	PD
Alam et al. [103]	VGRF	Gait information	PD

Table 3. *Cont.*

Contributor	Name of Database	Modality	Disease
Acharya et al. [16], University of Bonn [104]	EEG time series data	EEG Signals	Epilepsy
ICLUS [79]	ICLUS—Italian COVID-19 Lung Ultrasound project	Ultrasound	
Cohen et al. [105]	COVID-19 image data collection	X-ray, CT	COVID-19 and other associated diseases
Xuehai He et al. [106]	CT-Dataset: a CT scan dataset about COVID-19	CT	COVID-19
Wang et al. [107]	COVIDx	X-ray	COVID-19, Pneumonia, Normal
RSNA [108]	COVID-19 Imaging Data Sets	X-ray, CT	COVID-19, Pneumonia
Chowdhury et al. [109,110]	COVID-19 Radiography Database	X-ray	COVID-19, Pneumonia, Normal
Eduardo Soares et al. [111]	SARS-CoV-2 CT-scan dataset	CT	COVID-19, Normal

The EyePACS dataset [51] is found to be one of the famous datasets for performing DR identification. It consists of approximately five million retinal images captured on different degrees of DR. In addition to retinal images, fundus images are also playing vital roles in the identification of diabetic retinopathy. MESSIDOR [58], IDRiD fundus [96], DIARETDB0 [96], DIARETDB1 [97], and E-ophtha [16] are a few famous repositories that contain fundus images to accomplish DR identification tasks. To perform COVID-19 identification from chest X-ray and CT-scan images at an early stage, several COVID-19 datasets [105–110] were made available to the public.

6. Conclusions

There is no doubt that on-time disease severity identification can save the lives of human beings. Many researchers have used artificial intelligence and machine-learning-based techniques to identify the severity level of different categories of diseases based on their symptoms. The study embodied in this paper was focused mainly on two diseases, viz. Parkinson's Disease and Diabetic Retinopathy. However, severity identification of a few other diseases, such as COVID-19, autonomic nervous system dysfunction, tuberculosis, sepsis, sleep apnea, psychosis, traumatic brain injury, breast cancer, knee osteoarthritis, and Alzheimer's disease, was also briefly covered. For severity identification, the task of multi-level classification was adopted. Based on patterns in the input data, the multiple output classes indicated different severity levels of the disease. Hoehn (H) and Yahr (Y), through the UPDRS measure, was found to be utilized mainly for severity identification. It was observed from the literature on Parkinson's Disease (PD) that there is a huge scope to improve the accuracy using non-motor symptoms. On the other hand, for severity identification of Diabetic Retinopathy (DR), a scope to reduce the algorithmic complexity and detection rate was observed. For rapid diagnosis of COVID-19, researchers applied various models (e.g., Inception-V3, ResNet, DenseNet, etc.) to a patient's X-ray and CT scan images. This article also provided the information of some public repositories for conducting disease severity identification tasks on DR, PD, and COVID-19. It is evident that that deep-learning models provide several advantages, viz. rapid diagnosis of diseases, automatic feature extraction, learning from examples, etc. In addition to these, they also suffer from several drawbacks, viz. lack of transparency, inefficiency in processing low-quality images, a massive amount of data required for better accuracy, etc. It can be stated that not only the development of automated disease severity identification is in its infancy stage, but also the development of massive as well as hybrid datasets enriched with epidemic characteristics. There is no doubt that deep-learning approaches have the capability of rapid diagnosis of disease, but imaging data alone do not serve this purpose. Thus, the integration of clinical and statistical observations with computational intelligence-based approaches is essential not only for an enhancement in the accuracy of computations, severity identification, and subsequent validation of results but also for minimizing outbreaks.

Author Contributions: For preparing this manuscript, S.B. and D.S. were involved in the conceptualization, identification of sources for this study, and the investigation of research gaps. They also prepared the original draft of the manuscript. V.S.D., N.P., and I.C.-O. were involved in the validation of the concepts and methodology adopted. D.S. and M.U.H. were involved in the review and editing of the original manuscript. A.P. and M.U.H. were involved in the project administration and funding acquisition. All authors have read and agreed to the published version of the manuscript.

Funding: This research received no external funding.

Institutional Review Board Statement: Not applicable.

Informed Consent Statement: This article does not contain any studies with human participants or animals performed by any of the authors.

Data Availability Statement: All data are included in the main manuscript.

Conflicts of Interest: The authors declare no conflict of interest.

References

1. Wu, D.; Gong, K.; Arru, C.D.; Homayounieh, F.; Bizzo, B.; Buch, V.; Ren, H.; Kim, K.; Neumark, N.; Xu, P.; et al. Severity and consolidation quantification of COVID-19 from CT images using deep learning based on hybrid weak labels. *IEEE J. Biomed. Health Inform.* **2020**, *24*, 3529–3538. [CrossRef] [PubMed]

2. Nguyen, H.H.; Saarakkala, S.; Blaschko, M.B.; Tiulpin, A. Semixup: In-and out-of-manifold regularization for deep semi-supervised knee osteoarthritis severity grading from plain radiographs. *IEEE Trans. Med. Imaging* **2020**, *39*, 4346–4356. [CrossRef] [PubMed]

3. Tadesse, G.A.; Javed, H.; Thanh, N.L.N.; Thi, H.D.H.; Thwaites, L.; Clifton, D.A.; Zhu, T. Multi-modal diagnosis of infectious diseases in the developing world. *IEEE J. Biomed. Health Inform.* **2020**, *24*, 2131–2141. [CrossRef] [PubMed]

4. Mithra, K.; Emmanuel, W.S. Gaussian model based hybrid technique for infection level identification in TB diagnosis. *J. King Saud Univ.-Comput. Inf. Sci.* **2021**, *33*, 988–998. [CrossRef]

5. Aşuroğlu, T.; Oğul, H. A deep learning approach for sepsis monitoring via severity score estimation. *Comput. Methods Programs Biomed.* **2021**, *198*, 105816. [CrossRef]

6. Zhang, Z.; Sejdić, E. Radiological images and machine learning: Trends, perspectives, and prospects. *Comput. Biol. Med.* **2019**, *108*, 354–370. [CrossRef]

7. MedlinePlus. Imaging and Radiology. Available online: https://medlineplus.gov/ency/article/007451.htm (accessed on 12 September 2021).

8. Xia, Y.; Yao, Z.; Ye, Q.; Cheng, N. A dual-modal attention-enhanced deep learning network for quantification of Parkinson's disease characteristics. *IEEE Trans. Neural Syst. Rehabil. Eng.* **2019**, *28*, 42–51. [CrossRef]

9. Goldberger, A.L.; Amaral, L.A.; Glass, L.; Hausdorff, J.M.; Ivanov, P.C.; Mark, R.G.; Mietus, J.E.; Moody, G.B.; Peng, C.K.; Stanley, H.E. PhysioBank, PhysioToolkit, and PhysioNet: Components of a new research resource for complex physiologic signals. *Circulation* **2000**, *101*, e215–e220. [CrossRef]

10. Pereira, C.R.; Pereira, D.R.; Silva, F.A.; Masieiro, J.P.; Weber, S.A.; Hook, C.; Papa, J.P. A new computer vision-based approach to aid the diagnosis of Parkinson's disease. *Comput. Methods Programs Biomed.* **2016**, *136*, 79–88. [CrossRef]

11. Prashanth, R.; Roy, S.D.; Mandal, P.K.; Ghosh, S. Automatic classification and prediction models for early Parkinson's disease diagnosis from SPECT imaging. *Expert Syst. Appl.* **2014**, *41*, 3333–3342. [CrossRef]

12. Cernak, M.; Orozco-Arroyave, J.R.; Rudzicz, F.; Christensen, H.; Vásquez-Correa, J.C.; Nöth, E. Characterisation of voice quality of Parkinson's disease using differential phonological posterior features. *Comput. Speech Lang.* **2017**, *46*, 196–208. [CrossRef]

13. Lahmiri, S.; Shmuel, A. Detection of Parkinson's disease based on voice patterns ranking and optimized support vector machine. *Biomed. Signal Process. Control* **2019**, *49*, 427–433. [CrossRef]

14. Ertuğrul, Ö.F.; Kaya, Y.; Tekin, R.; Almalı, M.N. Detection of Parkinson's disease by shifted one dimensional local binary patterns from gait. *Expert Syst. Appl.* **2016**, *56*, 156–163. [CrossRef]

15. Marek, K.; Jennings, D.; Lasch, S.; Siderowf, A.; Tanner, C.; Simuni, T.; Coffey, C.; Kieburtz, K.; Flagg, E.; Chowdhury, S.; et al. The Parkinson progression marker initiative (PPMI). *Prog. Neurobiol.* **2011**, *95*, 629–635. [CrossRef] [PubMed]

16. Acharya, U.R.; Molinari, F.; Sree, S.V.; Chattopadhyay, S.; Ng, K.H.; Suri, J.S. Automated diagnosis of epileptic EEG using entropies. *Biomed. Signal Process. Control* **2012**, *7*, 401–408. [CrossRef]

17. Nilashi, M.; Ahmadi, H.; Sheikhtaheri, A.; Naemi, R.; Alotaibi, R.; Alarood, A.A.; Munshi, A.; Rashid, T.A.; Zhao, J. Remote tracking of Parkinson's disease progression using ensembles of deep belief network and self-organizing map. *Expert Syst. Appl.* **2020**, *159*, 113562. [CrossRef]

18. Awad, M.; Khanna, R. Support vector regression. In *Efficient Learning Machines*; Apress: Berkeley, CA, USA, 2015; pp. 67–80.

19. Jang, J.S. ANFIS: Adaptive-network-based fuzzy inference system. *IEEE Trans. Syst. Man Cybern.* **1993**, *23*, 665–685. [CrossRef]
20. Sztahó, D.; Tulics, M.G.; Vicsi, K.; Valálik, I. Automatic estimation of severity of parkinson's disease based on speech rhythm related features. In Proceedings of the 8th IEEE International Conference on Cognitive Infocommunications (CogInfoCom 2017), Debrecen, Hungary, 11–14 September 2017; pp. 11–16.
21. Park, D.H.; Kim, H.K.; Choi, I.Y.; Kim, J.K. A literature review and classification of recommender systems research. *Expert Syst. Appl.* **2012**, *39*, 10059–10072. [CrossRef]
22. Hariharan, M.; Polat, K.; Sindhu, R. A new hybrid intelligent system for accurate detection of Parkinson's disease. *Comput. Methods Programs Biomed.* **2014**, *113*, 904–913. [CrossRef]
23. Balaji, E.; Brindha, D.; Balakrishnan, R. Supervised machine learning based gait classification system for early detection and stage classification of Parkinson's disease. *Appl. Soft Comput.* **2020**, *94*, 106494.
24. Kim, H.B.; Lee, W.W.; Kim, A.; Lee, H.J.; Park, H.Y.; Jeon, H.S.; Kim, S.K.; Jeon, B.; Park, K.S. Wrist sensor-based tremor severity quantification in Parkinson's disease using convolutional neural network. *Comput. Biol. Med.* **2018**, *95*, 140–146. [CrossRef] [PubMed]
25. Oung, Q.W.; Muthusamy, H.; Basah, S.N.; Lee, H.; Vijean, V. Empirical wavelet transform based features for classification of Parkinson's disease severity. *J. Med. Syst.* **2018**, *42*, 1–17. [CrossRef] [PubMed]
26. Cantürk, İ. A computerized method to assess Parkinson's disease severity from gait variability based on gender. *Biomed. Signal Process. Control* **2021**, *66*, 102497. [CrossRef]
27. Zhao, A.; Qi, L.; Li, J.; Dong, J.; Yu, H. A hybrid spatio-temporal model for detection and severity rating of Parkinson's disease from gait data. *Neurocomputing* **2018**, *315*, 1–8. [CrossRef]
28. PhysioNet: The Research Resource for Complex Physiologic Signals. Available online: https://physionet.org/ (accessed on 12 September 2021).
29. El Maachi, I.; Bilodeau, G.A.; Bouachir, W. Deep 1D-Convnet for accurate Parkinson disease detection and severity prediction from gait. *Expert Syst. Appl.* **2020**, *143*, 113075. [CrossRef]
30. Prashanth, R.; Roy, S.D. Novel and improved stage estimation in Parkinson's disease using clinical scales and machine learning. *Neurocomputing* **2018**, *305*, 78–103. [CrossRef]
31. Prashanth, R.; Roy, S.D. Early detection of Parkinson's disease through patient questionnaire and predictive modelling. *Int. J. Med. Inform.* **2018**, *119*, 75–87. [CrossRef]
32. Aydın, F.; Aslan, Z. Recognizing Parkinson's disease gait patterns by vibes algorithm and Hilbert-Huang transform. *Eng. Sci. Technol. Int. J.* **2021**, *24*, 112–125. [CrossRef]
33. Saravanan, S.; Ramkumar, K.; Adalarasu, K.; Sivanandam, V.; Kumar, S.R.; Stalin, S.; Amirtharajan, R. A Systematic Review of Artificial Intelligence (AI) Based Approaches for the Diagnosis of Parkinson's Disease. *Arch. Comput. Methods Eng.* **2022**, *29*, 3639–3653. [CrossRef]
34. Yurdakul, O.C.; Subathra, M.; George, S.T. Detection of parkinson's disease from gait using neighborhood representation local binary patterns. *Biomed. Signal Process. Control* **2020**, *62*, 102070. [CrossRef]
35. Shankar, K.; Zhang, Y.; Liu, Y.; Wu, L.; Chen, C.H. Hyperparameter tuning deep learning for diabetic retinopathy fundus image classification. *IEEE Access* **2020**, *8*, 118164–118173. [CrossRef]
36. Welikala, R.; Fraz, M.; Williamson, T.; Barman, S. The automated detection of proliferative diabetic retinopathy using dual ensemble classification. *Int. J. Diagn. Imaging* **2015**, *2*, 64–71. [CrossRef]
37. He, K.; Zhang, X.; Ren, S.; Sun, J. Deep residual learning for image recognition. In Proceedings of the IEEE Conference on Computer Vision and Pattern Recognition, Las Vegas, NV, USA, 27–30 June 2016; pp. 770–778.
38. Szegedy, C.; Liu, W.; Jia, Y.; Sermanet, P.; Reed, S.; Anguelov, D.; Erhan, D.; Vanhoucke, V.; Rabinovich, A. Going deeper with convolutions. In Proceedings of the IEEE Conference on Computer Vision and Pattern Recognition, Boston, MA, USA, 7–12 June 2015; pp. 1–9.
39. Simonyan, K.; Zisserman, A. Very deep convolutional networks for large-scale image recognition. *arXiv* **2014**, arXiv:1409.1556.
40. Krizhevsky, A.; Sutskever, I.; Hinton, G.E. Imagenet classification with deep convolutional neural networks. *Adv. Neural Inf. Process. Syst.* **2012**, *25*, 84–90. [CrossRef]
41. Wang, J.; Bai, Y.; Xia, B. Simultaneous diagnosis of severity and features of diabetic retinopathy in fundus photography using deep learning. *IEEE J. Biomed. Health Inform.* **2020**, *24*, 3397–3407. [CrossRef]
42. Wang, J.; Bai, Y.; Xia, B. Feasibility of diagnosing both severity and features of diabetic retinopathy in fundus photography. *IEEE Access* **2019**, *7*, 102589–102597. [CrossRef]
43. Hu, J.; Shen, L.; Sun, G. Squeeze-and-excitation networks. In Proceedings of the IEEE Conference on Computer Vision and Pattern Recognition, Salt Lake City, UT, USA, 18–23 June 2018; pp. 7132–7141.
44. De La Torre, J.; Valls, A.; Puig, D. A deep learning interpretable classifier for diabetic retinopathy disease grading. *Neurocomputing* **2020**, *396*, 465–476. [CrossRef]
45. Bhaskaranand, M.; Cuadros, J.; Ramachandra, C.; Bhat, S.; Nittala, M.G.; Sadda, S.; Solanki, K. EyeArt+ EyePACS: Automated retinal image analysis for diabetic retinopathy screening in a telemedicine system. In Proceedings of the Ophthalmic Medical Image Analysis International Workshop, OmIA, Munich, Germany, 9 October 2015; pp. 105–112.
46. Shankar, K.; Sait, A.R.W.; Gupta, D.; Lakshmanaprabu, S.; Khanna, A.; Pandey, H.M. Automated detection and classification of fundus diabetic retinopathy images using synergic deep learning model. *Pattern Recognit. Lett.* **2020**, *133*, 210–216. [CrossRef]

47. Messidor. ADCIS. Available online: https://www.adcis.net/en/third-party/messidor/ (accessed on 12 October 2021).

48. Hacisoftaoglu, R.E.; Karakaya, M.; Sallam, A.B. Deep learning frameworks for diabetic retinopathy detection with smartphone-based retinal imaging systems. *Pattern Recognit. Lett.* **2020**, *135*, 409–417. [CrossRef]

49. Son, J.; Shin, J.Y.; Chun, E.J.; Jung, K.H.; Park, K.H.; Park, S.J. Predicting high coronary artery calcium score from retinal fundus images with deep learning algorithms. *Transl. Vis. Sci. Technol.* **2020**, *9*, 28. [CrossRef]

50. Santa Cruz, J.F.H. An ensemble approach for multi-stage transfer learning models for COVID-19 detection from chest CT scans. *Intell.-Based Med.* **2021**, *5*, 100027. [CrossRef] [PubMed]

51. Decencière, E.; Zhang, X.; Cazuguel, G.; Lay, B.; Cochener, B.; Trone, C.; Gain, P.; Ordonez, R.; Massin, P.; Erginay, A.; et al. Feedback on a publicly distributed image database: The Messidor database. *Image Anal. Stereol.* **2014**, *33*, 231–234. [CrossRef]

52. Harikrishnan, V.; Vijarania, M.; Gambhir, A. Diabetic retinopathy identification using autoML. In *Computational Intelligence and Its Applications in Healthcare*; Elsevier: Amsterdam, The Netherlands, 2020; pp. 175–188.

53. Washburn, P.S. Investigation of severity level of diabetic retinopathy using adaboost classifier algorithm. *Mater. Today Proc.* **2020**, *33*, 3037–3042. [CrossRef]

54. Li, X.; Shen, L.; Shen, M.; Tan, F.; Qiu, C.S. Deep learning based early stage diabetic retinopathy detection using optical coherence tomography. *Neurocomputing* **2019**, *369*, 134–144. [CrossRef]

55. Sambyal, N.; Saini, P.; Syal, R.; Gupta, V. Modified U-Net architecture for semantic segmentation of diabetic retinopathy images. *Biocybern. Biomed. Eng.* **2020**, *40*, 1094–1109. [CrossRef]

56. Ronneberger, O.; Fischer, P.; Brox, T. U-net: Convolutional networks for biomedical image segmentation. In Proceedings of the International Conference on Medical image Computing and Computer-Assisted Intervention, Munich, Germany, 5–9 October 2015; Springer: Berlin/Heidelberg, Germany, 2015; pp. 234–241.

57. Nandy Pal, M.; Sarkar, A.; Gupta, A.; Banerjee, M. Deep CNN based microaneurysm-haemorrhage classification in retinal images considering local neighbourhoods. *Comput. Methods Biomech. Biomed. Eng. Imaging Vis.* **2022**, *10*, 157–171. [CrossRef]

58. Porwal, P.; Pachade, S.; Kamble, R.; Kokare, M.; Deshmukh, G.; Sahasrabuddhe, V.; Meriaudeau, F. Indian diabetic retinopathy image dataset (IDRiD): A database for diabetic retinopathy screening research. *Data* **2018**, *3*, 25. [CrossRef]

59. Quellec, G.; Charrière, K.; Boudi, Y.; Cochener, B.; Lamard, M. Deep image mining for diabetic retinopathy screening. *Med. Image Anal.* **2017**, *39*, 178–193. [CrossRef]

60. Kauppi, T.; Kalesnykiene, V.; Kamarainen, J.; Lensu, L.; Sorri, I.; Raninen, A.; Voutilainen, R.; Pietilä, J.; Kälviäinen, H.; Uusitalo, H. DIARETDB1 Standard Diabetic Retinopathy Database. *IMAGERET-Optimal Detect. Decis. Diagnosis Diabet. Retin.* **2007**, 15.1–15.10. Available online: https://www.it.lut.fi/project/imageret/diaretdb1/ (accessed on 12 October 2021).

61. Liu, Y.P.; Li, Z.; Xu, C.; Li, J.; Liang, R. Referable diabetic retinopathy identification from eye fundus images with weighted path for convolutional neural network. *Artif. Intell. Med.* **2019**, *99*, 101694. [CrossRef]

62. Hua, C.H.; Huynh-The, T.; Kim, K.; Yu, S.Y.; Le-Tien, T.; Park, G.H.; Bang, J.; Khan, W.A.; Bae, S.H.; Lee, S. Bimodal learning via trilogy of skip-connection deep networks for diabetic retinopathy risk progression identification. *Int. J. Med. Inform.* **2019**, *132*, 103926. [CrossRef] [PubMed]

63. Reddy, S.S.; Sethi, N.; Rajender, R.; Mahesh, G. Extensive analysis of machine learning algorithms to early detection of diabetic retinopathy. *Mater. Today Proc.* **2020.**. [CrossRef]

64. Wu, Z.; Shi, G.; Chen, Y.; Shi, F.; Chen, X.; Coatrieux, G.; Yang, J.; Luo, L.; Li, S. Coarse-to-fine classification for diabetic retinopathy grading using convolutional neural network. *Artif. Intell. Med.* **2020**, *108*, 101936. [CrossRef] [PubMed]

65. Pratt, H.; Coenen, F.; Broadbent, D.M. Convolutional Neural Networks For Diabetic Retinopathy. *Elsevier Procedia Comput. Sci.* **2016**, *90*, 200–205. [CrossRef]

66. Yun, W.L.; Rajendra Acharya, U.; Venkatesh, Y.; Chee, C.; Min, L.C.; Ng, E. Identification of different stages of diabetic retinopathy using retinal optical images. *Inf. Sci.* **2008**, *178*, 106–121. [CrossRef]

67. Akram, M.U.; Khalid, S.; Tariq, A.; Khan, S.A.; Azam, F. Detection and classification of retinal lesions for grading of diabetic retinopathy. *Comput. Biol. Med.* **2014**, *45*, 161–171. [CrossRef] [PubMed]

68. Mookiah, M.R.K.; Acharya, U.R.; Martis, R.J.; Chua, C.K.; Lim, C.M.; Ng, E.; Laude, A. Evolutionary algorithm based classifier parameter tuning for automatic diabetic retinopathy grading: A hybrid feature extraction approach. *Knowl.-Based Syst.* **2013**, *39*, 9–22. [CrossRef]

69. Chowdhury, A.R.; Chatterjee, T.; Banerjee, S. A random forest classifier-based approach in the detection of abnormalities in the retina. *Med. Biol. Eng. Comput.* **2019**, *57*, 193–203. [CrossRef]

70. Kaur, J.; Mittal, D.; Singla, R. Diabetic Retinopathy Diagnosis Through Computer-Aided Fundus Image Analysis: A Review. *Arch. Comput. Methods Eng.* **2021**, *29*, 1673–1711. [CrossRef]

71. Shah, M.I.; Mishra, S.; Yadav, V.K.; Chauhan, A.; Sarkar, M.; Sharma, S.K.; Rout, C. Ziehl–Neelsen sputum smear microscopy image database: A resource to facilitate automated bacilli detection for tuberculosis diagnosis. *J. Med. Imaging* **2017**, *4*, 027503. [CrossRef]

72. Olayemi Alebiosu, D.; Dharmaratne, A.; Hong Lim, C. Improving tuberculosis severity assessment in computed tomography images using novel DAvoU-Net segmentation and deep learning framework. *Expert Syst. Appl.* **2023**, *213*, 119287. [CrossRef]

73. MIMIC-III. Registry of Open Data on AWS. Available online: https://registry.opendata.aws/mimiciii/ (accessed on 4 April 2022).

74. Cai, W.; Liu, T.; Xue, X.; Luo, G.; Wang, X.; Shen, Y.; Fang, Q.; Sheng, J.; Chen, F.; Liang, T. CT quantification and machine-learning models for assessment of disease severity and prognosis of COVID-19 patients. *Acad. Radiol.* **2020**, *27*, 1665–1678. [CrossRef]

75. MGH; HMS. 3DQI: 3D Quantitative Imaging Laboratory. Available online: https://3dqi-lab.github.io/3dqi_website/ (accessed on 9 March 2023).

76. Yao, H.; Zhang, N.; Zhang, R.; Duan, M.; Xie, T.; Pan, J.; Peng, E.; Huang, J.; Zhang, Y.; Xu, X.; et al. Severity detection for the coronavirus disease 2019 (COVID-19) patients using a machine learning model based on the blood and urine tests. *Front. Cell Dev. Biol.* **2020**, *8*, 683. [CrossRef] [PubMed]

77. Roy, S.; Menapace, W.; Oei, S.; Luijten, B.; Fini, E.; Saltori, C.; Huijben, I.; Chennakeshava, N.; Mento, F.; Sentelli, A.; et al. Deep learning for classification and localization of COVID-19 markers in point-of-care lung ultrasound. *IEEE Trans. Med. Imaging* **2020**, *39*, 2676–2687. [CrossRef] [PubMed]

78. Jaderberg, M.; Simonyan, K.; Zisserman, A. Spatial transformer networks. *Adv. Neural Inf. Process. Syst.* **2015**, *28*, 1–9.

79. Trento, U. ICLUS—Italian Covid-19 Lung Ultrasound Project. Available online: https://www.disi.unitn.it/iclus (accessed on 20 August 2021).

80. Lai, Y.; Li, G.; Wu, D.; Lian, W.; Li, C.; Tian, J.; Ma, X.; Chen, H.; Xu, W.; Wei, J.; et al. 2019 Novel Coronavirus-Infected Pneumonia on CT: A Feasibility Study of Few-Shot Learning for Computerized Diagnosis of Emergency Diseases. *IEEE Access* **2020**, *8*, 194158–194165. [CrossRef]

81. Altaf, F.; Islam, S.; Janjua, N.K. A novel augmented deep transfer learning for classification of COVID-19 and other thoracic diseases from X-rays. *Neural Comput. Appl.* **2021**, *33*, 14037–14048. [CrossRef]

82. Yu, Z.; Li, X.; Sun, H.; Wang, J.; Zhao, T.; Chen, H.; Ma, Y.; Zhu, S.; Xie, Z. Rapid identification of COVID-19 severity in CT scans through classification of deep features. *BioMedical Eng. OnLine* **2020**, *19*, 1–13. [CrossRef]

83. Kumar, A.; Sinwar, D.; Saini, M. Study of several key parameters responsible for COVID-19 outbreak using multiple regression analysis and multi-layer feed forward neural network. *J. Interdiscip. Math.* **2021**, *24*, 53–75. [CrossRef]

84. Devi, M.; Maakar, S.K.; Sinwar, D.; Jangid, M.; Sangwan, P. Applications of flying ad-hoc network during COVID-19 pandemic. In *IOP Conference Series: Materials Science and Engineering*; IOP Publishing: Bristol, UK, 2021; Volume 1099, p. 012005.

85. Pandey, A.; Kedir, T.; Kumar, R.; Sinwar, D. Analyzing Effects of Temperature, Humidity, and Urban Population in the Initial Outbreak of COVID19 Pandemic in India. In *Data Engineering for Smart Systems*; Springer: Berlin/Heidelberg, Germany, 2022; pp. 469–478.

86. Chahar, S.; Roy, P.K. COVID-19: A Comprehensive Review of Learning Models. *Arch. Comput. Methods Eng.* **2022**, *29*, 1915–1940. [CrossRef]

87. Sinwar, D.; Dhaka, V.S.; Tesfaye, B.A.; Raghuwanshi, G.; Kumar, A.; Maakar, S.K.; Agrawal, S. Artificial Intelligence and Deep Learning Assisted Rapid Diagnosis of COVID-19 from Chest Radiographical Images: A Survey. *Contrast Media Mol. Imaging* **2022**, *2022*, 1306664. [CrossRef] [PubMed]

88. Kim, T.; Kim, J.W.; Lee, K. Detection of sleep disordered breathing severity using acoustic biomarker and machine learning techniques. *Biomed. Eng. Online* **2018**, *17*, 1–19. [CrossRef] [PubMed]

89. Antonucci, L.A.; Raio, A.; Pergola, G.; Gelao, B.; Papalino, M.; Rampino, A.; Andriola, I.; Blasi, G.; Bertolino, A. Machine learning-based ability to classify psychosis and early stages of disease through parenting and attachment-related variables is associated with social cognition. *BMC Psychol.* **2021**, *9*, 1–15. [CrossRef] [PubMed]

90. Abujaber, A.; Fadlalla, A.; Gammoh, D.; Abdelrahman, H.; Mollazehi, M.; El-Menyar, A. Prediction of in-hospital mortality in patients on mechanical ventilation post traumatic brain injury: Machine learning approach. *BMC Med. Inform. Decis. Mak.* **2020**, *20*, 1–11. [CrossRef]

91. Zeng, Z.; Espino, S.; Roy, A.; Li, X.; Khan, S.A.; Clare, S.E.; Jiang, X.; Neapolitan, R.; Luo, Y. Using natural language processing and machine learning to identify breast cancer local recurrence. *BMC Bioinform.* **2018**, *19*, 65–74. [CrossRef]

92. Kwon, S.B.; Han, H.S.; Lee, M.C.; Kim, H.C.; Ku, Y. Machine learning-based automatic classification of knee osteoarthritis severity using gait data and radiographic images. *IEEE Access* **2020**, *8*, 120597–120603. [CrossRef]

93. Martinez-Murcia, F.J.; Ortiz, A.; Gorriz, J.M.; Ramirez, J.; Castillo-Barnes, D. Studying the manifold structure of Alzheimer's disease: A deep learning approach using convolutional autoencoders. *IEEE J. Biomed. Health Inform.* **2019**, *24*, 17–26. [CrossRef]

94. Sethuraman, S.K.; Malaiyappan, N.; Ramalingam, R.; Basheer, S.; Rashid, M.; Ahmad, N. Predicting Alzheimer's Disease Using Deep Neuro-Functional Networks with Resting-State fMRI. *Electronics* **2023**, *12*, 1031. [CrossRef]

95. ADNI. Alzheimer's Disease Neuroimaging Initiative. Available online: https://adni.loni.usc.edu/ (accessed on 14 February 2022).

96. Kauppi, T.; Kalesnykiene, V.; Kamarainen, J.K.; Lensu, L.; Sorri, I.; Uusitalo, H.; Kälviäinen, H.; Pietilä, J. DIARETDB0: Evaluation database and methodology for diabetic retinopathy algorithms. *Mach. Vis. Pattern Recognit. Res. Group Lappeenranta Univ. Technol. Finl.* **2006**, *73*, 1–17.

97. Rath, S.R. Diabetic Retinopathy 224 × 224 (2019 Data). Available online: https://www.kaggle.com/sovitrath/diabetic-retinopathy-224x224-2019-data (accessed on 15 August 2021).

98. Chalakkal, R.J.; Abdulla, W.H.; Sinumol, S. Comparative analysis of university of Auckland diabetic retinopathy database. In Proceedings of the 9th International Conference on Signal Processing Systems, Auckland, New Zealand, 27–30 November 2017; pp. 235–239.

99. Staal, J.; Abràmoff, M.D.; Niemeijer, M.; Viergever, M.A.; Van Ginneken, B. Ridge-based vessel segmentation in color images of the retina. *IEEE Trans. Med. Imaging* **2004**, *23*, 501–509. [CrossRef]

100. Goldbaum, M. STructured Analysis of the Retina. Available online: https://cecas.clemson.edu/~ahoover/stare/ (accessed on 15 August 2021).

101. Decenciere, E.; Cazuguel, G.; Zhang, X.; Thibault, G.; Klein, J.C.; Meyer, F.; Marcotegui, B.; Quellec, G.; Lamard, M.; Danno, R.; et al. TeleOphta: Machine learning and image processing methods for teleophthalmology. *Irbm* **2013**, *34*, 196–203. [CrossRef]

102. Pereira, C.R.; Weber, S.A.; Hook, C.; Rosa, G.H.; Papa, J.P. Deep learning-aided Parkinson's disease diagnosis from handwritten dynamics. In Proceedings of the 2016 29th SIBGRAPI Conference on Graphics, Patterns and Images (SIBGRAPI), Sao Paulo, Brazil, 4–7 October 2016; pp. 340–346.

103. Alam, M.N.; Garg, A.; Munia, T.T.K.; Fazel-Rezai, R.; Tavakolian, K. Vertical ground reaction force marker for Parkinson's disease. *PLoS ONE* **2017**, *12*, e0175951. [CrossRef] [PubMed]

104. EEG Time Series Data, University of Bonn. Available online: http://www.meb.uni-bonn.de/epileptologie/science/physik/eegdata.html (accessed on 14 February 2022).

105. Cohen, J.P.; Morrison, P.; Dao, L. COVID-19 image data collection. *arXiv* **2020**, arXiv:2003.11597.

106. Zhao, J.; Zhang, Y.; He, X.; Xie, P. Covid-ct-dataset: A ct scan dataset about covid-19. *arXiv* **2020**, arXiv:2003.13865.

107. Wang, L.; Lin, Z.Q.; Wong, A. Covid-net: A tailored deep convolutional neural network design for detection of COVID-19 cases from chest X-ray images. *Sci. Rep.* **2020**, *10*, 1–12. [CrossRef] [PubMed]

108. RSNA. Radiological Society of North America COVID-19 Imaging Data Sets. Available online: https://www.rsna.org/covid-19 (accessed on 4 April 2022).

109. Chowdhury, M.E.; Rahman, T.; Khandakar, A.; Mazhar, R.; Kadir, M.A.; Mahbub, Z.B.; Islam, K.R.; Khan, M.S.; Iqbal, A.; Al Emadi, N.; et al. Can AI help in screening viral and COVID-19 pneumonia? *IEEE Access* **2020**, *8*, 132665–132676. [CrossRef]

110. Kaggle. COVID-19 Radiography Database. Available online: https://www.kaggle.com/tawsifurrahman/covid19-radiography-database (accessed on 6 April 2021).

111. Angelov, P.; Almeida Soares, E. SARS-CoV-2 CT-scan dataset: A large dataset of real patients CT scans for SARS-CoV-2 identification. *MedRxiv* **2020**. [CrossRef]

MDPI AG
Grosspeteranlage 5
4052 Basel
Switzerland
Tel.: +41 61 683 77 34

Diagnostics Editorial Office
E-mail: diagnostics@mdpi.com
www.mdpi.com/journal/diagnostics